CLASSICAL and FUZZY CONCEPTS in MATHEMATICAL LOGIC and APPLICATIONS

CLASSICAL and FUZZY
CONCEPTS in
MATHEMATICAL LOGIC
and APPLICATIONS

Mircea Reghiş
Eugene Roventa

CRC Press
Taylor & Francis Group
Boca Raton   London   New York

CRC Press is an imprint of the
Taylor & Francis Group, an **informa** business

Library of Congress Cataloging-in-Publication Data

Reghiş , Mircea.
    Classical and fuzzy concepts in mathematical logic and
applications / Mircea Reghiş , Eugene Roventa.
        p. cm.
    Includes bibliographical references and index.
    ISBN 0-8493-3197-8 (alk. paper)
    1. Logic, Symbolic and mathematical. 2. Fuzzy logic.
I. Roventa, Eugene. II. Title.
QA9.R357 1998
511.3—dc21
                                                                98-6348
                                                                CIP

This book contains information obtained from authentic and highly regarded sources. Reprinted material is quoted with permission, and sources are indicated. A wide variety of references are listed. Reasonable efforts have been made to publish reliable data and information, but the author and the publisher cannot assume responsibility for the validity of all materials or for the consequences of their use.

© 1998 by CRC Press LLC

No claim to original U.S. Government works
International Standard Book Number 0-8493-3197-8
Library of Congress Card Number 98-6348

Foreword

The role of logic in all scientific activities is indisputable. In one way or another, logic has become a cornerstone of all important achievements of science and technology. It has become a universal language with its beauty and appeal. This book, written by Professors Reghiş and Roventa, is about logic - not only two-valued logic but the fascinating world of many-valued and fuzzy logics as well. Especially, the latter have put logic in a new perspective that sheds light on new technological achievements. In this setting, I always recall the leitmotif of many-valued logic so succinctly put forward by Jan Lukasiewicz almost 70 year ago:

"...we might assume that a sentence, in the logical sense of the term, might have values other than falsehood or truth. A sentence, of which we do not know whether it is false or true, might have no value determined as truth or falsehood, but might have a third, undetermined, value. We might, for instance, consider that the sentence "in a year from now I shall be in Warsaw" is neither true nor false and has a third, undetermined value, which can be symbolized as 1/2. We might go still further and ascribe to sentences infinitely many values contained between falsehood and truth...In the logic of infinitely many values, it is assumed that sentences can take on values represented by rational numbers x that satisfy the condition $0 \leq x \leq 1$."

Multivalued logic has opened up new horizons and has changed the way we think about logic. This has been pushed even further when we recall what has happened in the setting of fuzzy sets, and fuzzy logic, in particular. The book is comprised of two main parts starting from propositional logic (Part I) and navigating the reader to predicate logic (Part II). When examining the contents of Professors Reghiş and Roventa's book I was struck by the vastness of the progress and impressive diversity of methodologies and ensuing techniques.

Viewed in this perspective, the book is unique. One cannot but be greatly impressed by the way in which all the methodological threads unveiled in the book lead to a coherent picture of today's status of the important role of logic - not only as an academic discipline but as an important and rich tool for knowledge-based systems.

The book by Professors Reghiş and Roventa is unique in many different ways. First, it presents a unified look at logic as an embodiment of scientific fundamentals. Second, the book is full of interesting and thought-provoking insights expressed in a formal yet lucid manner. The material is carefully organized and can be easily used as a text in many courses in science and engineering. It can also serve as useful reference material.

Professors Reghiş and Roventa deserve credit for producing this superb contribution; the text is definitely welcome by anyone who is interested in pursuing a comprehensive study in the area of logic both in its more theoretical endeavors as well as in the many diverse areas of applied research. The material herein presented will prove to be an invaluable reference for those interested in acquainting themselves with the field of contemporary logic.

Witold Pedrycz
Electrical and Computer Engineering
University of Manitoba
Winnipeg, Canada

Preface

"So far as the laws of mathematics refer to reality, they are not certain. And so far as they are certain, they do not refer to reality."

(ALBERT EINSTEIN)

One of most remarkable features of contemporary society is the unprecedented role of information in man's life. Sophisticated electronic machines and computing systems classify, store and process an enormous volume of information and launch it around the planet by interconnected global networks. Such a growing role of information is perfectly understandable if one thinks that information means knowledge and that the progress of civilization depends directly on the development of human knowledge. All activities in classifying, processing and storing information are performed by logical tools often using advanced results of mathematical logic.

Information is carried by language. More precisely each piece of information is expressed or codified in linguistic texts (written or spoken) belonging to some natural or artificial language. The simplest linguistic texts able to carry information are usually called sentences, propositions, phrases, etc. Roughly speaking, each sentence tells something about certain things. The "things" which a sentence is speaking about belong to a certain material or ideal "world" called the **universe of discourse**. It is important to notice that generally languages agree with many such universes. Moreover, in a certain sense, the language, once created, is independent of the universe of discourse. This means that the construction of sentences merely follows the syntactic rules of the language regardless of meanings the sentences may have; in more suggestive terms: a sentence can tell anything about any thing. It follows that sentences may have various truth values (true, false, very true, almost true, partially true, almost false, etc.).

The attribution of truth values to sentences sometimes constitutes a rather difficult problem. In principle, there are two ways to tackle this problem: the direct way, by experiment and the indirect one, by reasoning. The first way is usually adopted in everyday practice and in experimental sciences while the second characterizes theoretical thinking.

In the theoretical approach, truth values of sentences are deduced (computed) from other sentences by means of so-called inference rules. Inference rules represent generally accepted laws governing the transfer of truth values from some sentences to others, i.e., they are laws of correct reasoning. Since Greek antiquity the science of correct reasoning has been called Logic. There are several kinds of logics. The most important (and unanimously accepted) one is the classical (crisp) logic created in ancient times by Greek logicians, mainly by Aristotle (384 - 322 B.C.). In this basic logic sentences can have only two (extreme) truth values: true and false. That is why classical logic is called **bivalent logic** or also **crisp logic**. In the last century several new logics have arisen: many valued logic, intuitionistic logic, fuzzy logic, etc. Some basic elements of the last one will be discussed in two chapters of this textbook.

This textbook contains an introduction to mathematical logic in the classical sense (i.e., bivalent) and to some basic aspects of fuzzy logic. The first part of the textbook deals with propositional logic and the second one contains predicate logic. Each part presents the corresponding fuzzy aspects and some applications in computer science. The two parts of the textbook as well as the preliminary chapter are accompanied by exercises addressed to beginners. To make the textbook self-contained, the reader will find at the end two algebraic sections concerning Boolean algebras and MV-algebras and a section about the general ideas of fuzzy sets theory.

The purpose of the text is to help the reader grasp the notion of a formal language, to teach readers to reason correctly, to help them understand what a mathematical proof is and to make them aware of the applications of mathematical logic in various fields like logic networks, logic programming and approximate reasoning.

All the material in the textbook and the choice of topics reflect the personal opinion of the authors on the subject and the pedagogical approach to mathematical logic. The style of treatment is rather mathematical and based on precise definitions and rigorous proofs. Certain parallelism has been made between the crisp and fuzzy aspects of mathematical logic. The volume of the presented information is generally limited to the usual contents of logical books. Nevertheless, some topics are a little more advanced. However, the textbook is accessible to any person having mathematical knowledge at first level college; no other prerequisites are presupposed.

Taking into account that mathematical logic means a formal language $\mathcal{L}$ with a truth structure on $\mathcal{L}$, we have carefully defined both the formal language and the two versions of the truth structure: the semantic version and the syntactic one. In what concerns the semantical version we have focused our attention on the rigorous definition and construction of the notion of interpretation. The notion of interpretation is crucial to the definition of the semantic truth structure on $\mathcal{L}$: Boolean interpretations (for crisp logic) and interpretations valued in MV-algebras (for fuzzy logic).

We have chosen an approach to these notions which allows an algorithmic treatment and subsequently the possibility to use computers for finding and processing the truth values of sentences. In the literature there are many good books on mathematical (crisp) logic. In the elaboration of this textbook, the authors have profitably used the treatment of various topics contained in the books [3], [5], [14], [19], [21], [22], [25], [26], [28], [40], [47].

As far as fuzzy logic is concerned, we have adopted the point of view of the Czech school of logic. We agree with the initial ideas of J. A. Goguen and with J. Pavelka's approach to fuzzy propositional logic. This approach is well summarized in Novák's assertion [36] that "the term *fuzzy logic* very generally means many-valued logic with special properties". Actually, today fuzzy logic is a non-trivial crystallized formal theory with a formal language and well-defined semantic and syntactic truth structures. Thus, fuzzy logic is an interesting theory from a theoretical point of view providing at the same time good foundations for applications.

The list of references is not exhaustive and it contains only books and papers used in the elaboration of this textbook.

In the authors' opinion the textbook will be very useful for students at the advanced undergraduate level in mathematics, computer science and engineering; it could be helpful to professors, engineers and to any person interested in learning and applying logical concepts.

The authors would like to express their gratitude to Professor Witold Pedrycz from the University of Manitoba for his comments, suggestions and generous ideas concerning this work. The authors are also grateful to their colleagues (S. Guiasu, D. Tremaine, Y. Lespérance, M. Janta-Polczinsky from York University, R. Pateanu from IBM Toronto Laboratory, I. Caşu from University of Timişoara) for numerous useful suggestions and comments.

Finally, many thanks are due to the young mathematician I. Ugarcovici for his appreciable contribution to the technical editing of this textbook.

Timişoara - Toronto,   Mircea Reghiş
August, 1997   Eugene Roventa

Contents

Recommendations for readers

In order to facilitate the study of the book we propose the following "roadmap":

1. For undergraduate level: Sections 1.1, 2.1, 2.2, 3.1, 3.2, 4.1, 4.2, 7.1-7.5, 8, 9.1, 9.2, 10.1, 10.2, 11.1, 11.2, 11.3, 13.1, 13.2, 13.3.

2. For graduate level: all except Chapters 6 and 12.

3. For special graduate courses: all chapters.

Introduction to "Naive" Mathematical Logic

In daily life, the term "logic" is used in two ways: as a special behaviour of a human mind (human reasoning) and as science dealing with this behaviour. In the following we shall deal only with this second usage.

The beginnings of the scientific study of logic are lost in antiquity. Today everybody agrees that the creation of the science of logic and its development to maturity is due to the Greek logician and philosopher Aristotle (384-322 B.C.). Very soon after Aristotle, another group of logicians called *Stoics* discovered *the formalism,* i.e., the idea to study logic in its linguistic *form.* Consequently, Aristotle's "judgements" and "logical reasonings" became "propositions" or "sentences" respectively "logical discourses". Thus, began the millenary development of so-called *formal logic,* which finally led to the creation of mathematical logic. Mathematical logic, born after 1850, represents the extreme formalization of logic in mathematical spirit by means of mathematical tools.

(1) Sentences

The most elementary linguistic texts able to carry information are known in usual communication as propositions, sentences, phrases, assertions, statements; in mathematics they also appear under specific names as theorems, lemmas, corollaries, conjectures, and so on. In the following we shall call them **sentences**.

Examples of sentences:

A $\equiv$ The sheep is a carnivore;
B $\equiv$ The Cathedral of Köln is higher than the Dome of Milan;
C $\equiv$ Somebody is a liar;
D $\equiv$ (The number) x is prime;
E $\equiv$ x is less than 7;

F ≡ x belongs to y;

G ≡ (the line) d_1 is parallel to (the line) d_2.

Texts A, ..., G above are sentences regardless of their truth values.

REMARK 0.1   If we write A ≡ or B ≡ , etc., it means that we denote or we abbreviate by the letters A, B, etc. the linguistic texts which follow the sign ≡. ∎

REMARK 0.2   Some of the above sentences are mathematical ones and by using specific mathematical symbols we can write them in a condensed symbolic form such as :

E ≡ $x < 7$;

F ≡ $x \in y$;

G ≡ $d_1 \| d_2$.   ∎

(2) Logico-linguistic structure of a sentence

In the particular sentences A, B, ..., G considered in section (1) we can distinguish two kinds of subtexts (two kinds of parts) :

(a) one or more groups of linguistic signs, which designate an object in the universe of discourse. Such a subtext will be called a **subject** (of the given sentence);

(b) if we drop all subjects of a sentence we obtain a group of linguistic symbols which represent the pure assertion, the predicative part of the sentence. Such a subtext will be called **the predicator** (of the given sentence).

Notice that a sentence, from a logical point of view, may have one or several subjects; the predicator of the sentence is always unique. For instance, in the sentences A,..., G in section (1) we have:

Subjects	Predicator
A : The sheep	is carnivore
B : The Cathedral of Köln, the Dome of Milan	is higher than
C : somebody	is a liar
D : (the number) x	is prime
E : x, 7	is less than
F : x, y	belongs to
G : (the line)d_1, (the line)d_2	is parallel to

Schematically one can say that:

SENTENCE = SUBJECT(S) + PREDICATOR

If we denote by P the predicator of a given sentence U and by $s_1, s_2, \ldots, s_p$ the subjects of U ($p \geq 1$), then the sentence U can be denoted formally by $P(s_1, s_2, \ldots, s_p)$. In this case we say that the predicator P has p (free) places, or P has the arity p.

(3) Individual variables and constants, propositions and predicates

A subject in a given sentence may be specified or not. A specified subject will be called an **individual constant** (or a proper name); an unspecified subject, a generic one, will be called an **individual variable**. Usually (but not necessarily) individual variables are denoted by letters such as x, y, z, etc.; the individual constants can be denoted by a, b, c, etc. For example, in sentences A, B above we have the constant "the sheep" and the constants "The Cathedral of Köln", "The Dome of Milan". In sentences C, D, F, G we have individual variables: "somebody", "the number x", "x" and "y", "(the line) d_1" and "(the line) d_2". In E we have two subjects, one of them being the variable "x" and the other the constant "7".

(a) A sentence, all the subjects of which are constants (i.e., all subjects are specified) will be called a **proposition**.

(b) A sentence which contains at least one (individual) variable (one unspecified subject) will be called a **predicate**.

As a consequence, there are two kinds of sentences: propositions and predicates. For instance, in the list of sentences A,..., G above, A, B are propositions while C, D, E, F, G are predicates.

(4) Logico-linguistic operations with sentences

Sentences have a remarkable "algebraic" property: starting from one or several sentences we can construct new sentences by using a number of specific logical-linguistic operations, called **operators** and **quantifiers**. We list the most important of them:

OPERATORS	QUANTIFIERS
Negation : ¬ (not)	Existential quantifier ∃ (there is some, for some)
Disjunction :∨ (or)	Universal quantifier ∀ (for every, for all)
Conjunction : ∧ (and)	
Implication : ⇒ (if - then)	
Equivalence : ⇔ (iff ; if and only if)	

REMARK 0.3   Operators modify the predicator of the sentence, while quantifiers modify the subject.   ∎

The action of operators and quantifiers is illustrated by the following examples: let F, G be sentences and $H(x)$ be a predicate containing (at least) the individual variable x. By means of operators and quantifiers one can construct the following new sentences:

$\neg(F) \equiv$ not F (negation of F);

$(F) \vee (G) \equiv F$ or G (disjunction of F, G);

$(F) \wedge (G) \equiv F$ and G (conjunction of F, G);

$(F) \Rightarrow (G) \equiv F$ implies $G \equiv$ if F, then G (implication of F, G);

$(F) \Leftrightarrow (G) \equiv F$ is equivalent to $G \equiv F$ if and only if G (equivalence of F, G);

$\exists x[H(x)] \equiv$ there is x such that $H(x) \equiv$ for some $x, H(x)$;

$\forall x[H(x)] \equiv$ for every $x, H(x) \equiv$ for all $x, H(x)$.

The sentences $\exists x[H(x)]$ and $\forall x[H(x)]$ are called, respectively, **the existential quantification** and **the universal quantification** of the sentence $H(x)$.
For instance, if the sentences:

$L \equiv$ (the number) 3 is less then (the number) 4;

$M \equiv$ (the number) 4 is prime;

$N(x) \equiv$ (the number) x is divisible by (the number) 2,

then:

$\neg(L) \equiv$ non $L \equiv$ (the number) 3 is not less than (the number) 4;

$\neg(M) \equiv$ non $M \equiv$ (the number) 4 is not prime;

and so on:

$(L) \lor (M) \equiv$ (the number) 3 is less than (the number) 4 or (the number) 4 is prime;

$(L) \land (M) \equiv$ (the number) 3 is less than (the number) 4 and (the number) 4 is prime;

$(L) \Rightarrow (M) \equiv$ if (the number) 3 is less than (the number) 4, then (the number) 4 is prime;

$(L) \Leftrightarrow (M) \equiv$ (the number) 3 is less than (the number) 4 if and only if (the number) 4 is prime.

Of course, some of these sentences are true, others are false. For instance, the proposition L is true and the proposition M is false, the proposition $\neg(M)$ is true and so on. But it is important to notice that we have in view only the expression capability of the logical language and, for the moment, we do not concentrate our attention on the truth values of the sentences under consideration.

In what concerns quantifications, one can construct the following two sentences:

$\exists x[N(x)] \equiv$ for some x, x is divisible by 2;
$\forall x[N(x)] \equiv$ for every x, x is divisible by 2.

The first quantification means that there are even numbers (which is obviously true) while the second asserts that all numbers are even (which is obviously false).

(5) Language of naive mathematical logic

Let S be a set of given (initial) sentences. The set of all sentences which can be obtained from the sentences belonging to S by exclusively using the operators $\neg, \lor, \land, \Rightarrow, \Leftrightarrow$ is called the language of propositional logic generated by S and is denoted by $\mathcal{L}_0(S) = \mathcal{L}_0$.

The set of all sentences which can be obtained from S by using both operators and quantifiers is called the language of predicate logic generated by S and is denoted by $\mathcal{L}(S) = \mathcal{L}$. It is clear that the language $\mathcal{L}$ is larger and more expressive than the language $\mathcal{L}_0$ ($\mathcal{L}_0 \subset \mathcal{L}$).

(6) Truth values and logical deduction

Logic has two components: a logical language $\mathcal{L}$ and a truth structure $\mathcal{S}$ on $\mathcal{L}$. The language $\mathcal{L}$ was sketched above. The truth structure $\mathcal{S}$ can be induced on $\mathcal{L}$ in a semantic way by means of so-called interpretations and in a syntactic way by means of logical axioms and inference rules. Classical logic is concerned exclusively with sentences which can have only two truth values: true and false. That is why classical logic is called bivalent.

In this preliminary section we shall limit ourselves to discussing only some aspects of the semantic version for the truth structure on $\mathcal{L}$. As we already have seen in the semantic approach, every sentence F has a truth value which can be identified with an element of the set $\Gamma = \{0, 1\}$; let us denote this truth value by $v(F)$. If $v(F) = 1$, then we say that F is true, and if $v(F) = 0$ we say that F is false. Using this convention we can formulate the following basic rules concerning the behavior of truth values in bivalent logic:

$$(*) \quad \begin{cases} v(\neg F) & = \overline{v(F)} = 1 - v(F); \\ v(F \vee G) & = \max\{v(F), v(G)\}; \\ v(F \wedge G) & = \min\{v(F), v(G)\}; \\ v(F \Rightarrow G) & = \max\{v(\neg F), v(G)\}; \\ v(F \Leftrightarrow G) & = \min\{\max\{v(\neg F), v(G)\}, \max\{v(\neg G), v(F)\}\}. \end{cases}$$

In the above equalities the following (natural) definitions concerning the set $\Gamma = \{0, 1\}$ are used: $0 < 1$, $\max\{0, 0\} = 0$, $\max\{0, 1\} = \max\{1, 0\} = 1$, $\max\{1, 1\} = 1$, $\min\{0, 0\} = 0$, $\min\{1, 0\} = \min\{0, 1\} = 0$, $\min\{1, 1\} = 1$.

The equalities $(*)$ can be expressed equivalently by means of the following "truth tables":

(a) Negation

F	$\neg(F)$
1	0
0	1

(b) Disjunction

F	G	$(F) \vee (G)$
1	1	1
0	1	1
1	0	1
0	0	0

(c) Conjunction

F	G	$(F) \wedge (G)$
1	1	1
0	1	0
1	0	0
0	0	0

(d) Implication

F	G	$(F) \Rightarrow (G)$
1	1	1
0	1	1
1	0	0
0	0	1

(e) Equivalence

F	G	$(F) \Leftrightarrow (G)$
1	1	1
0	1	0
1	0	0
0	0	1

From these tables one can see that:

(a): $\neg(F)$ is true if F is false; $\neg(F)$ is false if F is true;

(b): $(F) \vee (G)$ is true if at least one of sentences F, G is true; $(F) \vee (G)$ is false if both F and G are false;

(c): $(F) \wedge (G)$ is true if both sentences are true; $(F) \wedge (G)$ is false if at least one of the sentences F, G is false;

(d): $(F) \Rightarrow (G)$ is false only when F is true and G is false; in all other cases $(F) \Rightarrow (G)$ is true;

(e): $(F) \Leftrightarrow (G)$ is true if sentences F and G have the same truth value; $(F) \Leftrightarrow (G)$ is false if sentences F, G have different truth values.

Using truth tables one can compute the truth values of compound sentences. For instance, if $K_1 \equiv (F \Rightarrow G) \Leftrightarrow (\neg F \wedge \neg G)$, then the "truth table" below furnishes the truth values of K_1:

F	G	$(F) \Rightarrow (G)$	$\neg F$	$\neg G$	$\neg F \wedge \neg G$	K_1
1	1	1	0	0	0	0
0	1	1	1	0	0	0
1	0	0	0	1	0	1
0	0	1	1	1	1	1

In the same way we can obtain the truth values of the sentence

$$K_2 \equiv (F \Rightarrow G) \Rightarrow [(H \vee F) \Rightarrow (H \vee G)] :$$

F	G	H	F $\Rightarrow$ G	H $\vee$ F	H $\vee$ G	(H $\vee$ F) $\Rightarrow$ (H $\vee$ G)	K_2
1	1	1	1	1	1	1	1
0	1	1	1	1	1	1	1
1	0	1	0	1	1	1	1
0	0	1	1	1	1	1	1
1	1	0	1	1	1	1	1
0	1	0	1	0	1	1	1
1	0	0	0	1	0	0	1
0	0	0	1	0	0	1	1

In the two previous tables we have computed the truth values of the compound sentences K_1 and K_2 giving all possible truth values to the components. We shall say that truth values of K_1 in the first table are computed for all possible interpretations of F and G; similarly, in the second table the truth values of K_2 are computed for all possible interpretations of F, G, H. Each line of these tables corresponds to some interpretation of the component sentences. For instance, in the second line in the table of K_1 we have $v(F) = 0$, $v(G) = 1$ and we obtain the resulting value $v(K_1) = 0$. In the third line of the same table we have $v(F) = 1$, $v(G) = 0$ and $v(K_1) = 1$. Therefore, one can say that in the interpretation $v(F) = 0$ and $v(G) = 1$ of the sentences F, G one obtains $v(K_1) = 0$, while in the interpretation $v(F) = 1$ and $v(G) = 0$ one gets $v(K_1) = 1$.

In summary one can say that in the first two interpretations of F, G the sentence K_1 is false and in the last two interpretations of F, G, the sentence K_1 is true. A similar analysis shows that K_2 is true in all interpretations of the components F, G, H. In more specific terms we call **interpretation of the language** $\mathcal{L}_0$ any function $v: \mathcal{L}_0 \to \Gamma : F \mapsto v(F)$ having properties (∗), p.6. Given an interpretation v of $\mathcal{L}_0$, the sentence $S \in \mathcal{L}_0$ is said to be: v-true if $v(S) = 1$ and v-false if $v(S) = 0$; the sentence S is said to be **universally true** or **a tautology** (respectively, **universally false** or **antilogy**) if for any interpretation v we have $v(S) = 1$ (respectively, $v(S) = 0$). The sentence S is called contingent if it is neither a tautology nor an antilogy. Using this terminology the sentence K_2 is obviously a tautology while K_1 is a contingent sentence. Taking into account that the notion of interpretation intended in the above sense provides some (very simple) "meaning" to the sentences (the "meaning" of each sentence being its truth value), we shall say that the resulting truth structure on the language is obtained by *semantic* means. The tautologies form a remarkable class of sentences which express the laws of logical thinking. The set of tautologies is infinite but some of the most important ones will be presented below.

(A) "Classical" tautologies

(A1) : $F \Rightarrow F$; (the principle of identity)

(A2) : $(F \Rightarrow G) \Leftrightarrow (\neg G \Rightarrow \neg F)$; (the principle of contraposition)

(A3) : $F \Leftrightarrow \neg\neg F$; (the principle of double negation)

(A4) : $(F \Rightarrow G) \Rightarrow [(G \Rightarrow H) \Rightarrow (F \Rightarrow H)]$; (the principle of syllogism)

(A5) : $[F \wedge (F \Rightarrow G)] \Rightarrow G$; (the principle of "modus ponens")

(A6) : $[(F \Rightarrow G) \wedge \neg G] \Rightarrow \neg F$; (the principle of "modus tollens")

(A7) : $F \vee \neg F$; (the principle of "tertium non datur" or the principle of "third excluded")

(A8) : $[F \Rightarrow (G \Rightarrow H)] \Rightarrow [G \Rightarrow (F \Rightarrow H)]$; (the principle of permutation of the antecedents)

(A9) : $\begin{cases} \neg(F \vee G) \Leftrightarrow \neg F \wedge \neg G; \\ \neg(F \wedge G) \Leftrightarrow \neg F \vee \neg G \end{cases}$; (De Morgan laws)

(B) "Algebraic" tautologies

(B1) : $\begin{cases} F \vee G \Leftrightarrow G \vee F; \\ F \wedge G \Leftrightarrow G \wedge F; \end{cases}$ (commutativity laws)

(B2) : $\begin{cases} F \vee (G \vee H) \Leftrightarrow (F \vee G) \vee H; \\ F \wedge (G \wedge H) \Leftrightarrow (F \wedge G) \wedge H; \end{cases}$ (associativity laws)

(B3) : $\begin{cases} F \vee (G \wedge H) \Leftrightarrow (F \vee G) \wedge (F \vee H); \\ F \wedge (G \vee H) \Leftrightarrow (F \wedge G) \vee (F \wedge H); \end{cases}$ (distributivity laws)

(C) Additional remarkable tautologies

(C1) : $(F \vee F) \Rightarrow F$;

(C2) : $F \Rightarrow (F \vee G)$;

(C3) : $(F \vee G) \Rightarrow (G \vee F)$;

(C4) : $(F \Rightarrow G) \Rightarrow [(H \vee F) \Rightarrow (H \vee G)]$;

(C5) : $F \Rightarrow (G \Rightarrow F)$;

(C6) : $[F \Rightarrow (G \Rightarrow H)] \Rightarrow [(F \Rightarrow G) \Rightarrow (F \Rightarrow H)]$;

(C7) : $(\neg F \Rightarrow \neg G) \Rightarrow [(\neg F \Rightarrow G) \Rightarrow F]$;

(C8) : $[(F \Rightarrow G) \Rightarrow G] \Rightarrow [(G \Rightarrow F) \Rightarrow F]$.

The reader is invited to prove (as an exercise) using the above definition of interpretation and of tautology, that all sentences (A), (B), (C) are tautologies.

REMARK 0.4   We highlighted the group C of tautologies because in what follows they will play a special role. Namely, the sentences $(C1)-(C4)$ will represent the axioms of Hilbert's propositional logic, while the last four axioms will be adopted as axioms for fuzzy propositional logic.   ∎

Using tautologies one can sometimes obtain important information about truth values of sentences without tedious computations or truth tables. For instance, given sentences S,T in $\mathcal{L}_0$, the following assertion holds: The sentence $S \Rightarrow T$ is a tautology if and only if $v(S) \leq v(T)$ for all interpretations v of $\mathcal{L}_0$. The proof is not difficult. Indeed, suppose that $S \Rightarrow T$ is a tautology and let v be an interpretation of $\mathcal{L}_0$. From the definition of tautologies and from the fourth equality in (∗) we get:

$$1 = v(S \Rightarrow T) = \max\{v(\neg S), v(T)\} = \max\{1 - v(S), v(T)\};$$

if $v(S) = 1$, then $1 - v(S) = 0$ and $\max\{1 - v(S), v(T)\} = v(T)$ so that $v(T) = 1$; if $v(S) = 0$, then $v(S) \leq v(T)$. It follows that $v(S) \leq v(T)$ in any case. The "only if" part of the assertion is proved. Conversely, suppose that $v(S) \leq v(T)$ for any interpretation v. If $v(S) = 0$, then $1 - v(S) = 1$ and $\max\{1 - v(S), v(T)\} = 1$; if $v(S) = 1$, then $v(T) = 1$ and $\max\{1 - v(S), v(T)\} = 1$. It follows that in any case $v(S \Rightarrow T) = \max\{1 - v(S), v(T)\} = 1$, which means that $S \Rightarrow T$ is a tautology.

From this result it follows immediately that $S \Leftrightarrow T$ is a tautology if and only if $v(S) = v(T)$ for all interpretations v. Indeed, $v(S \Leftrightarrow T) = \min\{v(S \Rightarrow T), v(T \Rightarrow S)\}$ and $v(S \Leftrightarrow T) = 1$ if and only if $v(S \Rightarrow T) = 1$ and $v(T \Rightarrow S) = 1$, or if and only if $v(S) \leq v(T)$ and $v(T) \leq v(S)$. Therefore, $S \Leftrightarrow T$ is a tautology if and only if $v(S) = v(T)$ for all v.

Another application of tautologies is the following: If sentences S and $S \Rightarrow T$ are true in some interpretation, then the sentence T is also true in the same interpretation. In particular, if S and $S \Rightarrow T$ are tautologies, then T is also a tautology. Indeed, suppose that for an interpretation v we have $v(S) = 1$ and $v(S \Rightarrow T) = 1$. Taking into account the considerations above (concerning the implication $S \Rightarrow T$), from $v(S \Rightarrow T) = 1$ it follows $v(S) \leq v(T)$, hence $v(T) = 1$ (because $v(S) = 1$) and the assertion is proved. This result can by formulated in simplified form as follows:

if sentences S and $S \Rightarrow T$ are true, then the sentence T is also true.

This assertion represents the so-called MP-rule ("Modus Ponens"-rule). Generally, any rule permitting the derivation of true sentences from some other true sentences is called an **inference rule** or a **deduction rule**. Each process of correct reasoning (proof, deduction) is based on some explicitly given inference rules.

The notion of interpretation for the language $\mathcal{L}$ of predicate logic is more sophisticated than that of propositional logic. This is due to the existence of individual variables and of predicates. Intuitively, individual variables and constants are names of objects in the universe of discourse while sentences in $\mathcal{L}$ (propositions or predicates) represent assertions about objects represented by individual variables and constants. It was found that for the needs of predicate bivalent logic (as in the case of propositional logic) it is sufficient to interpret propositions and predicates, by elements of $\Gamma = \{0, 1\}$ and by Γ-valued functions, respectively. Thus any interpretation of $\mathcal{L}$ is roughly speaking a compound entity $I = (\mathcal{D}, w, v)$, where $\mathcal{D}$ is an abstract set called the **domain of interpretation**, w is a function which associates each individual variable or constant z with an element $w(z) \in \mathcal{D}$, while v is a function which associates each proposition $P \in \mathcal{L}$ with an element $v(P) \in \Gamma$ and each n-ary predicator π with a Γ-valued function $v(\pi)$ defined on $\mathcal{D}^n$ so that, the predicate $\pi(z_1, z_2, ..., z_n)$ is associated with the element $v(\pi)(w(z_1), w(z_2), ..., w(z_n)) \in \Gamma$. In other words, an interpretation I is defined by three elements: a set $\mathcal{D}$ of "objects", an interpretation w of subjects (of individual variables and constants) and an interpretation v of predicators so that each sentence $S \in \mathcal{L}$ is interpreted as being an element $I[S] \in \Gamma$ such that:

$$I[S] = \begin{cases} v(P), \text{ if } S \equiv P, \text{ where P is a proposition;} \\ v(\pi)(w(z_1), w(z_2), ..., w(z_n)), \text{ if } S \equiv \pi(z_1, z_2, ..., z_n), \\ \text{where } \pi \text{ is an n-ary predicator and } z_1, z_2, ..., z_n \text{ are individual} \\ \text{variables or constants.} \end{cases}$$

By means of interpretations of the language $\mathcal{L}$ one can define I-true sentences, universally true sentences, I-false sentences and so on. For instance, given an interpretation $I = (\mathcal{D}, w, v)$ of $\mathcal{L}$ and some predicate $H(x)$, where H is a unary predicator and x is a given individual variable, we shall say that the predicate $H(x)$ is: I-true if $I[H(x)] = v(H)(w(x)) = 1$ and I-false if $I[H(x)] = v(H)(w(x)) = 0$. The predicate $H(x)$ is said to be universally true (respectively, universally false) if for all interpretations $I, I[H(x)] = 1$ (respectively, $I[H(x)] = 0$). Of course, these notions need more detailed explanation. If the above "definitions" are formulated with more accuracy, then one gets some important properties of interpretations. For instance, if S, S_1, S_2 are arbitrary sentences (propositions or predicates) and $\pi(x)$ is a unary predicate containing the individual variable x, then

$$I[\neg S] = \overline{I[S]}, \quad I[S_1 \vee S_2] = \max\{I[S_1], I[S_2]\}$$

$$I[S_1 \wedge S_2] = \min\{I[S_1], I[S_2]\}, \text{ etc.}$$

and

$$I[\exists x(\pi(x))] = \sup_{u \in \mathcal{D}} v(\pi)(u),$$

$$I[\forall x(\pi(x))] = \inf_{u \in \mathcal{D}} v(\pi)(u),$$

where $\mathcal{D}$ is the domain of the considered interpretation. From the two last equalities it follows easily, for instance, that sentences

$$\pi(z) \Rightarrow \exists x(\pi(x)) \quad \text{and} \quad \forall x(\pi(x)) \Rightarrow \pi(z)$$

(where z is any individual variable or constant) are universally true. Moreover, as in the case of propositional logic, one can establish new inference rules and define the process of deduction in predicate logic. These topics will be treated in the second part of the book.

Exercises

1. Let P be "It is cold" and let Q be "It is raining". Express in the usual (natural) language the following sentences:

   (a) $\neg P$ (b) $P \wedge Q$ (c) $P \vee P$ (d) $Q \vee \neg P$ (e) $\neg P \wedge \neg Q$

2. Let P be "Alain is rich" and Q be "Alain is handsome". Write each of the following sentences in symbolic form:

   (a) Alain is not rich but handsome.

   (b) Alain is neither rich nor handsome.

   (c) Alain is either rich or not handsome.

   (d) Alain is not rich or else he is both rich and not handsome.

3. Let P_1, P_2, P_3, P_4 and P_5 be the following sentences:
   P_1: "He needs a doctor."
   P_2: "He needs a lawyer."
   P_3: "He has an accident."
   P_4: "He is sick."
   P_5: "He is injured."
   State the following formulas in English:

   (a) $(P_4 \Rightarrow P_1) \wedge (P_3 \Rightarrow P_2)$

   (b) $P_1 \Rightarrow (P_4 \vee P_5)$

   (c) $(P_1 \wedge P_2) \Rightarrow P_3$

   (d) $(P_1 \wedge P_2) \Leftrightarrow (P_4 \wedge P_5)$

   (e) $\neg(P_4 \vee P_5) \Rightarrow \neg P_1$.

4. Let A, B and C be the following sentences:
   A: "Roses are red."
   B: "Violets are blue."
   C: "Sugar is sweet."
   Translate the following compound sentences into symbolic notation:

   (a) Roses are red and violets are blue.

   (b) Roses are red and either violets are blue or sugar is sweet.

   (c) Whenever violets are blue, roses are red and sugar is sweet.

   (d) Roses are red only if violets aren't blue or sugar is sour.

   (e) Roses are red and if sugar is sour, then either violets aren't blue or sugar is sweet.

   Translate the following formulas in English:

(a) $B \vee \neg C$

(b) $\neg B \vee (A \Rightarrow C)$

(c) $(C \wedge \neg A) \Leftrightarrow B$

(d) $C \wedge (\neg A \Leftrightarrow B)$

(e) $\neg (B \wedge \neg C) \Rightarrow A$

(f) $A \vee ((B \wedge \neg C)$

(g) $(A \vee B) \wedge \neg C.$

5. Let us consider the following two predicates: $A(x) \equiv$ "(the animal) x is a cat" and $B(x) \equiv$ "(the animal) x is a pet". Using quantifiers and logical operators write in symbolic form the following six sentences:

   (a) All cats are pets.

   (b) No cats are pets.

   (c) There does not exist a cat which is a pet.

   (d) Some cats are not pets.

   (e) Not all cats are pets.

   (f) Some cats are pets.

6. Suppose that A, B and C represent conditions that will be true or false when a certain computer program is executed. Suppose further that you want the program to carry out a certain task only when A or B is true (but not both) and C is false. Using A, B and C and the connectives AND, OR and NOT, write a statement (sentence) that will be true only under these conditions.

7. Denoting by A and B the propositions "I take an umbrella" and "It rains", express in logical language the sentences below using A and B. What is the relationship between them?

   (a) If I do not take an umbrella, then it rains.

   (b) It rains if I do not take an umbrella.

   (c) It rains unless I take an umbrella.

   (d) If it doesn't rain, I take an umbrella.

   (e) It doesn't rain only if I take an umbrella.

8. Verify that
$$(F \Rightarrow G) \Rightarrow ((F \vee H) \Rightarrow (G \vee H))$$
is a tautology.

9. Show that

   (a) $(P \wedge Q) \Rightarrow P$ is a tautology;

   (b) $(P \vee Q) \Rightarrow P$ is not a tautology.

Part I

PROPOSITIONAL LOGIC

Part I of this book is concerned with classical (bivalent) logic (Chapters 1-5), with some basic aspects of propositional fuzzy logic (Chapter 6) and some applications (Chapter 7).

Logic means a logical language $\mathcal{L}$ and a truth structure $\mathcal{S}$ on $\mathcal{L}$. From a formal point of view one can say that classical and fuzzy propositional logic have essentially the same language, which will be designated by $\mathcal{L}_0$. The language $\mathcal{L}_0$ is presented in Chapter 1 in abstract form. This means that all sentences of $\mathcal{L}_0$ are constructed recursively by means of formal rules starting from some set of initial (atomic) sentences and using a finite system of operator symbols (connectives). The initial sentences will be called propositional variables. The set of propositional variables is supposed to be finite, so that the variables will be designated by $X_1, \ldots, X_N$ (where $N \geq 1$ is an arbitrary natural number). The notation of logical operators used in this book is a variant (adopted also by a great number of authors) of the notations introduced in the classical works of G. Frege, B. Russell, P. Bernays, W. Ackermann, D. Hilbert and others. More precisely, we shall use the following notations: $\neg$ for the negation, $\vee$ for the disjunction, $\wedge$ for the conjunction, $\Rightarrow$ for implication and $\Leftrightarrow$ for equivalence. So that, if P and Q are sentences, then:

$\neg(P)$ is the negation of P and we read it "not P";

$(P) \vee (Q)$ is the disjunction of P, Q and we read it "P or Q";

$(P) \wedge (Q)$ is the conjunction of P, Q and we read it "P and Q";

$(P) \Rightarrow (Q)$ is the implication of P, Q and we read it "P implies Q" or "if P, then Q";

$(P) \Leftrightarrow (Q)$ is the equivalence of P, Q and we read it "P is equivalent to Q" or "P if and only if Q".

There are many other versions for the notation of operators. For instance, in various books one can find the following notations:

for $\neg(P) : \sim P$ or $\overline{P}$ or P';

for $(P) \vee (Q) : (P) \cup (Q)$;

for $(P) \wedge (Q) : (P) \cap (Q)$ or $(P)\&(Q)$;

for $(P) \Rightarrow (Q) : (P) \supset (Q)$ or $(P) \rightarrow (Q)$;

for $(P) \Leftrightarrow (Q) : (P) \leftrightarrow (Q)$ or $(P) \sim (Q)$ or $(P) \equiv (Q)$.

It is important to notice that operators $\neg, \vee, \wedge, \Rightarrow, \Leftrightarrow$ are not logically in-dependent, because some of them can be expressed by means of others. For instance:

(a) the operators $\wedge, \Rightarrow, \Leftrightarrow$ can be expressed in terms of $\neg$ and $\vee$;

(b) the operators $\vee, \Rightarrow, \Leftrightarrow$ can be expressed in terms of $\neg$ and $\wedge$;

(c) the operators $\vee, \wedge, \Leftrightarrow$ can be expressed in terms of $\neg$ and $\Rightarrow$.

In order to make assertions (a), (b), (c) more explicit, let us introduce the following convention: if $\mathcal{O}$ is a formal object (a formal sentence or an arbitrary sequence of formal symbols) and A is a letter which does not belong to the (formal) language $\mathcal{L}_0$, then $A \equiv \mathcal{O}$ means that A is simply a notation for $\mathcal{O}$ (or an abbreviation of $\mathcal{O}$); at the same time, $A \equiv \mathcal{O}$ means that by A we understand the formal object $\mathcal{O}$.
In the case (a), one can define:

$$(P) \wedge (Q) \equiv \neg(\neg P \vee \neg Q);$$
$$(P) \Rightarrow (Q) \equiv \neg P \vee Q;$$
$$(P) \Leftrightarrow (Q) \equiv (P \Rightarrow Q) \wedge (Q \Rightarrow P).$$

In case (b), one can define:

$$(P) \vee (Q) \equiv \neg(\neg P \wedge \neg Q);$$
$$(P) \Rightarrow (Q) \equiv \neg(P \wedge \neg Q);$$
$$(P) \Leftrightarrow (Q) \equiv (P \Rightarrow Q) \wedge (Q \Rightarrow P).$$

In case (c), one can define:

$$(P) \vee (Q) \equiv (P \Rightarrow Q) \Rightarrow Q;$$
$$(P) \wedge (Q) \equiv \neg((P \Rightarrow Q) \Rightarrow \neg P);$$
$$(P) \Leftrightarrow (Q) \equiv (P \Rightarrow Q) \wedge (Q \Rightarrow P).$$

For the sake of simplicity, in the right side of the above definitions some parentheses were dropped; such "abusive" omissions will be permitted when no confusion can arise.

As we already outlined, propositional logic has two components: the language $\mathcal{L}_0$ and the truth structure $\mathcal{S}_0$ on $\mathcal{L}_0$, so that when we talk of propositional logic we refer to the pair $[\mathcal{L}_0; \mathcal{S}_0]$. Some ideas about the language $\mathcal{L}_0$ were briefly discussed above. The truth structure $\mathcal{S}_0$ provides the "logical character" of the logic. The structure $\mathcal{S}_0$ allows us, roughly speaking, to distinguish among the sentences of $\mathcal{L}_0$ the so-called valid (true) sentences. At the same time, by means of $\mathcal{S}_0$, we can introduce the formal notion corresponding to the intuitive concept of "correct reasoning".

The truth structure $\mathcal{S}_0$ can be approached in two ways, i.e., in two different versions: the semantic version and the syntactic one. Intuitively, propositions of natural language reflect connections between objects or properties of objects belonging to a certain material or ideal "world" $\mathcal{U}$ called the universe of discourse. Therefore, we can say that, when the universe $\mathcal{U}$ is given, then the propositions concerning $\mathcal{U}$ have certain "meanings".

The formal language $\mathcal{L}_0$, being an abstract one, it is not supposed to have such a property, although it is constructed with the possibility of very general interpretations in mind. Being abstract, the formal language $\mathcal{L}_0$ provides propositions (sentences) without "meanings". However, for the needs of classical propositional logic one can consider a very simple universe Γ containing only two elements called truth values and denoted, for instance, by 1 (true) and 0 (false). In this case, the "meaning" of each sentence P in $\mathcal{L}_0$ will be an element of Γ. Indeed, one can prove that there are Γ-valued functions (in a mathematical sense) φ defined on $\mathcal{L}_0$

$$\varphi : \mathcal{L}_0 \to \Gamma, \quad \Gamma = \{0, 1\}$$

having the following two "good" properties:

$(*)$ $\qquad \begin{cases} \varphi(\neg P) = 1 - \varphi(P) \\ \varphi(P \vee Q) = \max\{\varphi(P), \varphi(Q)\} \end{cases}$

Each such function φ will be called a (Boolean) interpretation of $\mathcal{L}_0$. If φ is an interpretation of $\mathcal{L}_0$, then for each sentence $P \in \mathcal{L}_0$, we shall have $\varphi(P) = 1$ or $\varphi(P) = 0$. So that, for a given φ, each sentence is φ-true (if $\varphi(P) = 1$) or φ- false (if $\varphi(P) = 0$). Taking into account that for any interpretation φ each sentence P is necessarily true or false (a third possibility being excluded), the propositional logic having the truth structure $\mathcal{S}_0$ based on the above concept of interpretation is said to be *bivalent*. The equalities $(*)$ allow us to compute the truth value $\varphi(R)$ of any compound sentence $R \equiv R(P_1, \cdots, P_r) \in \mathcal{L}_0$ once we know the truth values $\varphi(P_1), \ldots, \varphi(P_r)$ of the propositions $P_1, \ldots, P_r \in \mathcal{L}_0$. In this way, propositional logic becomes an authentic "propositional calculus". In Chapter 2 we shall see that there are many sentences in $\mathcal{L}_0$ which are φ-true for all interpretations φ. Such (compound) sentences will be called universally true or tautologies. For instance all sentences of the form

$$P \vee \neg P$$

$$P \Leftrightarrow \neg\neg P$$

$$(P \wedge (P \Rightarrow Q)) \Rightarrow Q$$

are tautologies. The tautologies, being true for any interpretation, express formally the "logical laws" of usual human thinking. Taking into account that this propositional logic is obtained by means of interpretations (providing "meanings" to the sentences), we shall say that its truth structure is defined in the semantic version.

The truth structure in a syntactic version is defined in a completely different way. The "logical character" induced on $\mathcal{L}_0$ in a syntactic way leads to the formalization of logical reasoning which deals with concepts such as axioms, rules of inference, deduction, proof and theorem. By contrast with the semantic version, the syntactic one is not algorithmic, but it is much more similar to common human reasoning, particularly to mathematical reasoning. In the syntactic version of S_0 we do not obtain tautologies (as valid sentences), but theorems. Theorems are (compound) sentences furnished by deduction processes, step by step, from a given set of axioms using so-called rules of inference.

In Chapter 3 the syntactic version of S_0 is precisely described following Hilbert's approach. This version will be called in what follows the deductive system of Hilbert and denoted by **H**. The system **H** contains four axioms and a unique rule of inference called "MODUS PONENS", briefly denoted by MP. The axioms of **H** are given by the following schemes of formulas:

H1: $(F \vee F) \Rightarrow F$;

H2: $F \Rightarrow (F \vee G)$;

H3: $(F \vee G) \Rightarrow (G \vee F)$;

H4: $(F \Rightarrow G) \Rightarrow [(H \vee F) \Rightarrow (H \vee G)]$.

Axioms **H1** – **H4** are called schemes of axioms because F, G, H, are arbitrary formulas in $\mathcal{L}_0$. The MP rule essentially states that starting from P and $P \Rightarrow Q$ one can deduce Q. A finite sequence

$$(\Delta) \qquad F_1, F_2, \ldots, F_r \quad (r \geq 1)$$

will be called a proof in $(\mathcal{L}_0, \mathbf{H})$, i.e., in the system of Hilbert, if for each i $(1 \leq i \leq r)$, the formula F_i either is an axiom or can be obtained by the MP rule from formulas F_h, F_k preceding F_i in (Δ). Each formula F which can be inserted in such a sequence (Δ) is called a theorem in $(\mathcal{L}_0, \mathbf{H})$; in such a case, we write $\vdash F$. In particular, the formula F_r in the sequence (Δ) is a theorem and the sequence (Δ) is called a proof of F_r. From the above definitions it follows clearly that axioms **H1** – **H4** are theorems and the formula F_1 in any proof (Δ) is necessarily an axiom.

On the basis of axioms, inference rules and proofs one can develop a whole theory about formal theorems and formal proofs. At this junction the natural question arises: what connections are there between theorems and tautologies? More generally, what connections are there between the syntactic and the semantic structures of truth on $\mathcal{L}_0$? A complete answer to this question is given in Chapter 4. First, it is not so difficult to verify that any theorem is a tautology; conversely, one can also prove that any tautology is a theorem. The latter assertion is not trivial and constitutes the content of the so-called completeness metatheorem of Hilbert. The completeness metatheorem and its converse have a certain philosophical significance. The identity between theorems and tautologies shows that axioms and the inference rule are well chosen, because they allow proof of all intuitive logical laws (tautologies) and only them. Consequently, depending on our needs, we can use either the syntactic or the semantic point of view in propositional logic.

In the literature, there are many (equivalent or not) syntactic versions for the truth structure on $\mathcal{L}_0$. Some of them are presented in Chapter 5.

In Chapter 6 some elements of fuzzy propositional logic are exposed. Our basic idea in the construction of fuzzy propositional logic consists of the systematic use of fuzzy sets having a fixed (arbitrary) finite number M of membership degrees. In such an option, we sketch the construction of fuzzy propositional logic in which the set of truth values has M elements ($M \geq 2$). In the case $M > 2$, the set of truth values becomes an MV-algebra and the propositional logic is not bivalent anymore. Some aspects of the connection between the semantic and the syntactic versions of the fuzzy truth structure on $\mathcal{L}_0$ are also discussed.

Chapter 7 is dedicated to some applications of these ideas within computer science.

1

The Formal Language of Propositional Logic

The formal language of propositional logic is the set of all formal sentences which can be constructed by means of the operator symbols $\neg, \vee, \wedge, \Rightarrow, \Leftrightarrow$ (called connectives), starting from a set of initial (atomic) formal sentences $X_1, X_2, X_3, ..., X_N$ called propositional variables. The formal language is conceived prior to any interpretation or "meaning" attached to the involved symbols. This means that in the construction of the formal language we deal with some abstract set of symbols $\mathbf{X}_0 = \{X_1, X_2, X_3 ..., X_N, \neg, \vee, \wedge, \Rightarrow, \Leftrightarrow, (,)\}$ called the *alphabet* of the language and with some finite sequences $\sigma_1 \sigma_2 ... \sigma_m$ of symbols belonging to $\mathbf{X}_0$; such sequences will be called *formal words*. In our setting, the set of propositional variables will be finite. Among formal words we have to distinguish those which are "well-formed", i.e., which represent in formal language the usual sentences. Such "good", well-formed formal words will be called *formulas*. Consequently, formulas are sequences of symbols in $\mathbf{X}_0$ which correspond to the common sentences of a non-formalized logical language. The distinction between arbitrary formal words and formulas is a non-trivial problem. The theory of the formal language of propositional logic presented below gives a precise answer to this problem. The solution is based on certain key concepts such as: formal alphabet, formal word, formal construction, formula, etc., and is due to the work of many generations of logicians (G. Frege, B. Russell, G. Peano, D. Hilbert, J. Lukasiewicz, A. Tarski, A. Church and others).

1.1 The Formal Language $\mathcal{L}_0$ of Propositional Logic Using Parentheses

Let us consider a natural number $N \geq 1$. The finite set

$$\mathbf{X}_0 = \{X_1, X_2, ..., X_N, \neg, \vee, (,)\}$$

is called **the alphabet** of the (formal) language of propositional logic. The elements of $\mathbf{X}_0$ are called **symbols**. Symbols $X_1, X_2, ..., X_N$ are called **propositional variables**. The symbols $\neg, \vee$ are called (logical) **operators** or **connectives** and the symbols (,) are called **parentheses** (left parenthesis and right parenthesis respectively).

DEFINITION 1.1   *Any finite string*

$$A \equiv \sigma_1 \sigma_2 \ldots \sigma_m \quad (m \geq 1)$$

*of symbols $\sigma_1, \sigma_2, \ldots, \sigma_m$ belonging to $\mathbf{X}_0$ is called a **formal word** (in $\mathbf{X}_0$). The number $\lambda(A) = m$ is called the **length** of the formal word A.*

REMARK 1.1   Formal words are arbitrary finite strings of symbols in $\mathbf{X}_0$, regardless of their intuitive meanings. For instance, the strings

$B_1 \equiv X_2(X_4\neg$
$B_2 \equiv \neg\neg\neg\neg$
$B_3 \equiv (X_3) \vee (X_1)$
$B_4 \equiv \neg(X_1) \vee (X_2)$

are formal words. The lengths of these formal words are $\lambda(B_1) = 4$, $\lambda(B_2) = 4$, $\lambda(B_3) = 7$ and $\lambda(B_4) = 8$, respectively.   ∎

DEFINITION 1.2   *(a) The formal words*

$$A_1 \equiv \sigma_1 \sigma_2 \ldots \sigma_{m_1}, \quad A_2 \equiv \tau_1 \tau_2 \ldots \tau_{m_2}$$

*are said to be **equal** if they are identical, i.e., if $m_1 = m_2 = m$ ($\lambda(A_1) = \lambda(A_2) = m$) and $\sigma_i \equiv \tau_i$ for all i ($1 \leq i \leq m$). In such a case we shall use the notation*

$$A_1 \equiv A_2;$$

if A_1, A_2 are not equal, we shall write $A_1 \not\equiv A_2$.
(b) If $A \equiv \sigma_1 \sigma_2 \ldots \sigma_p$ and $B \equiv \tau_1 \tau_2 \ldots \tau_q$ are formal words, then one can define the formal words $\neg(A), (A) \vee (B)$ by

$$\neg(A) \equiv \neg(\sigma_1 \sigma_2 \ldots \sigma_p);$$
$$(A) \vee (B) \equiv (\sigma_1 \sigma_2 \ldots \sigma_p) \vee (\tau_1 \tau_2 \ldots \tau_q).$$

(c) For arbitrary formal words $A \equiv \sigma_1 \sigma_2 \ldots \sigma_p$ and $B \equiv \tau_1 \tau_2 \ldots \tau_q$ we can also define the formal words $(A) \wedge (B)$, $(A) \Rightarrow (B)$ and $(A) \Leftrightarrow (B)$ by the

following conventions (abbreviations):

$$(A) \wedge (B) \equiv \neg((\neg(A)) \vee (\neg(B)));$$
$$(A) \Rightarrow (B) \equiv (\neg(A)) \vee (B);$$
$$(A) \Leftrightarrow (B) \equiv ((A) \Rightarrow (B)) \wedge ((B) \Rightarrow (A)).$$

REMARK 1.2   The correct use of parentheses is important. However, in order to simplify the writing of some formal words, we shall sometimes drop certain parentheses if no confusion is possible. For instance, we shall often use notations like $\neg A, A \vee B, A \Rightarrow B$, etc. instead of $\neg(A), (A) \vee (B), (A) \Rightarrow (B)$, etc., provided the sense of such "simplified" writings is clear.   ∎

We are now ready to define the distinction between arbitrary formal words and formulas (well-formed words). This distinction will be based on the notion of formal construction (of formulas).

DEFINITION 1.3   *A finite sequence of formal words (in the alphabet* $\mathbf{X_0}$*)*

$$(fc): \quad A_1, A_2, ..., A_r \ (r \geq 1)$$

is called a **formal construction** *(in* $\mathbf{X_0}$*) or a* **construction of formulas** *if for each index* i *(*$1 \leq i \leq r$*) the formal word* A_i *in* *(fc) satisfies one of three conditions below:*

(a) A_i *coincides with a propositional variable* $(A_i \equiv X)$*;*

(b) there is some index j *(*$j < i$*) such that:*

$$A_i \equiv \neg(A_j);$$

(c) there are some indices h, k *(*$h, k < i$*) such that*

$$A_i \equiv (A_h) \vee (A_k).$$

Any formal word in $\mathbf{X_0}$ *which can be inserted in a formal construction is called a* **formula** *(in* $\mathbf{X_0}$*). In particular, the formal word* A_r *in (fc) is a formula and (fc) is often called a formal construction of the formula* A_r*.*

REMARK 1.3   In each formal construction *(fc)*, the first term A_1 is necessarily a propositional variable; it is clear also that variables X_1,

$X_2, \ldots, X_N$ are formulas. Moreover, variables are the only formulas of length 1. ∎

From Definition 1.3 one can deduce the following almost obvious assertion:
if

$$(fc)_1 : \quad A_1, A_2, \ldots, A_r$$

and

$$(fc)_2 : \quad B_1, B_2, \ldots, B_s$$

are formal constructions (in $\mathbf{X}_0$), then any sequence of formal words obtained by "mixing" the constructions $(fc)_1$ and $(fc)_2$ and keeping the mutual order in each construction, is also a formal construction. For instance, from the constructions $(fc)_1$ and $(fc)_2$ one obtains formal constructions such as:

$$(\alpha) : \quad A_1, A_2, \ldots, A_r, B_1, B_2, \ldots, B_s;$$
$$(\beta) : \quad A_1, B_1, A_2, B_2, \ldots.$$

REMARK 1.4
(1) In a formal construction of a given formula $F \equiv F_r$ the number of terms is not necessarily minimal because it may contain repeated formulas or arbitrary variables, etc. Of course, it is often interesting to have a minimal formal construction for a given formula. This problem can be easily solved.
(2) Each formal construction can be extended to a new one by adding to the given construction an arbitrary variable or any formal word of the form $\neg(B)$ or $(B) \vee (C)$, where B, C are terms in the initial construction. ∎

DEFINITION 1.4   *The set of all formulas (in $\mathbf{X}_0$) is denoted by* $\mathcal{L}(\mathbf{X}_0)$ *or* $\mathcal{L}(X_1, X_2, \ldots, X_N)$ *or simply by* $\mathcal{L}_0$ *and is called* **the language of propositional logic (with parentheses)** *generated by the propositional variables* $X_1, X_2, \ldots, X_N$.

REMARK 1.5   The language $\mathcal{L}_0$ contains abstract strings of symbols which correspond to the common (natural) compound sentences, constructed from some initially given sentences $X_1, X_2, \ldots, X_N$. ∎

It is easy to see that if A, B are formulas, then $\neg(A), (A) \vee (B)$ are also formulas and consequently $(A) \wedge (B), (A) \Rightarrow (B), (A) \Leftrightarrow (B)$ are formulas, too. From Definition 1.3 and Definition 1.4 it follows that for any formula F in $\mathcal{L}_0$ one of the following must hold:

(a) $F \equiv X$ or (b) $F \equiv \neg(B)$ or (c) $F \equiv (C) \vee (D)$, where B, C, D are formulas in $\mathcal{L}_0$.

1.2 The Formal Language of Propositional Logic without Parentheses (Polish Notation)

The Polish school of mathematical logic (J. Łukasiewicz, A. Mostowski, A. Tarski, H. Rasiowa, R. Sikorski and others), who became famous after 1920, discovered, among other remarkable contributions, the possibility of avoiding parentheses in the formal language of logic. Thus, Łukasiewicz created the so-called Polish notation for the logical formal language.

The Polish language is equivalent (has the same capacity of expression) to the language $\mathcal{L}_0$ presented in the previous section and will be denoted by $\mathcal{L}_{00}$. Although less intuitive than $\mathcal{L}_0$, it possesses certain advantages, such as brevity and facilitates the arithmetical approach of the language.

In the Polish notation the operators $\neg, \vee, \wedge, \Rightarrow, \Leftrightarrow$ are respectively denoted by N, A, K, C, E. However, for many reasons, we shall keep the notations already adopted in Section 1.1.

Let us consider a natural number $N \geq 1$; the finite set

$$\mathbf{X}_{00} = \{X_1, X_2, ..., X_N, \neg, \vee\}$$

is called **the (formal) alphabet of propositional logic in Polish notation**. Symbols $X_1, X_2, ..., X_N$ are still called **propositional variables** and $\neg, \vee$ are the symbols of negation and of disjunction. Any finite string

$$A \equiv \sigma_1\sigma_2...\sigma_m \quad (m \geq 1)$$

of symbols in $\mathbf{X}_{00}$ will be called a **formal word (in $\mathbf{X}_{00}$)**. The length of the formal word A is the number $\lambda(A) = m$. The equality of formal words is defined as in Definition 1.2 (a). If $A \equiv \sigma_1\sigma_2...\sigma_p, B \equiv \tau_1\tau_2...\tau_q$ are formal words (in $\mathbf{X}_{00}$), then we define the formal words

$$\neg A \equiv \neg\sigma_1\sigma_2...\sigma_p$$

and

$$\vee AB \equiv \vee\sigma_1\sigma_2...\sigma_p\tau_1\tau_2...\tau_q.$$

The formal words $\neg A$ and $\vee AB$ in $\mathbf{X}_{00}$ obviously correspond to the formal words $\neg(A)$ and $(A) \vee (B)$ in $\mathbf{X}_0$. Following the same idea one may introduce the formal words $\wedge AB, \Rightarrow AB, \Leftrightarrow AB$ by the following notational abbreviations:

$$\wedge AB \equiv \neg \vee \neg A \neg B;$$

$$\Rightarrow AB \equiv \vee \neg AB;$$

$$\Leftrightarrow AB \equiv \wedge \Rightarrow AB \Rightarrow BA.$$

It is clear that the formal words $\wedge AB,\ \Rightarrow AB,\ \Leftrightarrow AB$ in $\mathbf{X}_{00}$ correspond respectively to the formal words $(A) \wedge (B), (A) \Rightarrow (B), (A) \Leftrightarrow (B)$ in $\mathbf{X}_0$.
The notion of formula is defined in a similar way as in Section 1.1.

DEFINITION 1.5   *A finite sequence of formal words in* $\mathbf{X}_{00}$

$$(fc): \quad A_1, A_2, ..., A_r \quad (r \geq 1)$$

is called **a formal construction (in** $\mathbf{X}_{00}$**)** *if for each index i $(1 \leq i \leq r)$ the formal word A_i satisfies one of three conditions below:*

(a) A_i coincides with a propositional variable $(A_i \equiv X)$;

(b) there is some index j $(j < i)$ such that

$$A_i \equiv \neg A_j;$$

(c) there are some indices h, k $(h, k < i)$ such that

$$A_i \equiv \vee A_h A_k.$$

Any formal word in $\mathbf{X}_{00}$ *which can be inserted in a formal construction is called* **a formula (in** $\mathbf{X}_{00}$**)**. *In particular, the formal word A_r in (fc) is a formula (in* $\mathbf{X}_{00}$*) and (fc) is often called* **a formal construction of the formula** A_r.

From the above definition it follows that formal words $A_1, ..., A_r$ in (fc) are formulas (in $\mathbf{X}_{00}$). In each formal construction (fc), the first term A_1 is necessarily a propositional variable. Moreover, all propositional variables are formulas and they are the only formulas (in $\mathbf{X}_{00}$) having their length equal to 1.

It is clear that all remarks in Section 1.1 concerning the "mixing" and extension of formal constructions in $\mathbf{X}_0$ remain valid for the formal constructions in $\mathbf{X}_{00}$. It is also easy to see that if A, B, C are formulas in $\mathbf{X}_{00}$, then $\neg A, \vee AB, \wedge AB, \Rightarrow AB$ and $\Leftrightarrow AB$ are also formulas (in $\mathbf{X}_{00}$).

DEFINITION 1.6   *The set of all formulas (in* $\mathbf{X}_{00}$*) denoted by $\mathcal{L}(\mathbf{X}_{00})$ or simply by $\mathcal{L}_{00}$ is called* **the language of propositional logic (in Polish version)** *generated by the propositional variables $X_1, X_2, ..., X_N$.*

In order to study the structure of formulas in $\mathcal{L}_{00}$ and to facilitate the translation from $\mathcal{L}_{00}$ to $\mathcal{L}_{0}$, we have to introduce some new notions.

DEFINITION 1.7   Let $A \equiv \sigma_1\sigma_2...\sigma_m$ be a formal word (in $\mathbf{X}_{00}$) and suppose that $\lambda(A) = m > 1$. The formal words

$$S_1 \equiv \sigma_1,$$
$$S_2 \equiv \sigma_1\sigma_2,$$
$$\vdots$$
$$S_{m-1} \equiv \sigma_1\sigma_2...\sigma_{m-1}$$

are called **segments** of A. The fact that a formal word S is a segment of formal word A $(\lambda(A) > 1)$ will be denoted (conventionally) by $S \subset A$. Formal words of length 1 do not have segments.

DEFINITION 1.8   Each symbol σ in $\mathbf{X}_{00}$ has a **weight** $\pi(\sigma)$, which is the number defined by

$$\pi(\sigma) = \begin{cases} 0, & \text{if} \quad \sigma \equiv X \\ 1, & \text{if} \quad \sigma \equiv \neg \\ 2, & \text{if} \quad \sigma \equiv \vee. \end{cases}$$

The weight $\pi(A)$ of any formal word $A \equiv \sigma_1\sigma_2...\sigma_m$ (in $\mathbf{X}_{00}$) is the number $\pi(A) = \sum_{i=1}^{m} \pi(\sigma_i)$.

REMARK 1.6   If $A \equiv \sigma_1\sigma_2...\sigma_p, B \equiv \tau_1\tau_2...\tau_q$ are formal words (in $\mathbf{X}_{00}$) and if we denote $AB \equiv \sigma_1\sigma_2...\sigma_p\tau_1\tau_2...\tau_q$, then $\pi(AB) = \pi(A) + \pi(B)$.
∎

In what follows, our aim will be to characterize formulas in arithmetic terms. For this purpose we introduce the notion of a **balanced formal word**, which will be designated by **bfw** for brevity's sake.

DEFINITION 1.9   A formal word A (in $\mathbf{X}_{00}$) is called **bfw** if one of the two conditions below is satisfied:

(a) $\lambda(A) = 1$ and $A \equiv X$;

(b) $\lambda(A) > 1$ and

    (b_1) $\lambda(A) = \pi(A) + 1$;

(b₂) $\lambda(S) \leq \pi(S)$ for every $S \subset A$.

The set of all bfw (in $\mathbf{X}_{00}$) will be denoted by $\mathcal{E}$.

REMARK 1.7

(1) The notion of bfw is appropriate for arithmetic computation and can be programmed on computers.

(2) From Definition 1.9 it follows immediately that all variables are bfw, i.e., $X_1, ..., X_N \in \mathcal{E}$. From the same definition it follows that any segment S of a bfw A *is not* a bfw; in other words, if $A \in \mathcal{E}$ and $S \subset A$, then $S \notin \mathcal{E}$. That is because the segment S of $A \in \mathcal{E}$ satisfies the condition (b_2) $\lambda(S) \leq \pi(S)$, therefore $\lambda(S) \neq \pi(S) + 1$ and hence condition (b_1) is not satisfied by S. By similar arguments, one obtains that $\neg, \vee \notin \mathcal{E}$. ∎

PROPOSITION 1.1

If A, B are bfw, then $\neg A, \vee AB, \wedge AB, \Rightarrow AB, \Leftrightarrow AB$ are also bfw.

PROOF   It is sufficient to prove that the formal words $\neg A$ and $\vee AB$ are bfw. Consider first the formal word $\neg A$ and suppose that A is bfw. It is obvious that $\lambda(\neg A) = \lambda(\neg) + \lambda(A) = 1 + \lambda(A) > 1$, hence we have to verify conditions $(b_1), (b_2)$ in Definition 1.9.

$$\lambda(\neg A) = 1 + \lambda(A) = 1 + (\pi(A) + 1) = 1 + \pi(A) + \pi(\neg) = \pi(\neg A) + 1,$$

consequently $\neg A$ satisfies (b_1). On the other hand, if $T \subset \neg A$ (T is a segment of $\neg A$), then $T \equiv \neg$ or $T \equiv \neg S$ where $S \subset A$. In the first case, $\lambda(T) = 1 = \pi(T)$; in the second case, we have

$$\lambda(T) = \lambda(\neg S) = \lambda(\neg) + \lambda(S) = 1 + \lambda(S) \leq 1 + \pi(S) =$$

$$= \pi(\neg) + \pi(S) = \pi(\neg S) = \pi(T);$$

here we have used the fact that A is bfw. Thus, $\lambda(S) \leq \pi(S)$ for any $S \subset A$, hence T satisfies condition (b_2). Therefore, $\neg A$ is a bfw.

Let us verify now that $\vee AB$ is also a bfw, supposing that A, B are bfw. We have to verify only conditions $(b_1), (b_2)$, because

$$\lambda(\vee AB) = \lambda(\vee) + \lambda(A) + \lambda(B) \geq 3.$$

Taking into account that A, B are bfw, we obtain successively:

$$\lambda(\vee AB) = 1 + \lambda(A) + \lambda(B) = 1 + (\pi(A) + 1) + (\pi(B) + 1) =$$
$$= (2 + \pi(A) + \pi(B)) + 1 = \pi(\vee AB) + 1;$$

hence the formal word $\vee AB$ satisfies condition (b_1). Now let U be a segment of $\vee AB$ $(U \subset \vee AB)$. We have to analyze four cases:

$$(i) \ \ U \equiv \vee; \ \ \ (ii) \ \ \ U \equiv \vee S, \ \ \ \text{where} \ \ S \subset A;$$

$$(iii) \ \ U \equiv \vee A \ \ \text{and} \ \ (iv) \ \ U \equiv \vee AT, \ \ \text{where} \ \ T \subset B.$$

In the case (i),

$$\lambda(U) = \lambda(\vee) = 1 < 2 = \pi(\vee) = \pi(U).$$

In the case (ii),

$$\lambda(U) = \lambda(\vee S) = 1 + \lambda(S) \leq 1 + \pi(S) < 2 + \pi(S) =$$

$$= \pi(\vee) + \pi(S) = \pi(\vee S) = \pi(U).$$

In the case (iii),

$$\lambda(U) = \lambda(\vee A) = 1 + \lambda(A) = 1 + (\pi(A) + 1) =$$

$$= 2 + \pi(A) = \pi(\vee A) = \pi(U).$$

In the case (iv),

$$\lambda(U) = \lambda(\vee AT) = 1 + \lambda(A) + \lambda(T) = 1 + (\pi(A) + 1) + \lambda(T) \leq$$

$$\leq (1 + \pi(A) + 1) + \pi(T) = 2 + \pi(A) + \pi(T) =$$

$$= \pi(\vee) + \pi(A) + \pi(T) = \pi(\vee AT) = \pi(U).$$

∎

PROPOSITION 1.2
If $A \in \mathcal{E}$ and $S \subset A$, then there is a unique $T \in \mathcal{E}$ such that $ST \subset A$ or $ST \equiv A$.

PROOF   Let $A \equiv \sigma_1 \sigma_2 ... \sigma_k \sigma_{k+1} ... \sigma_r$ a bfw and $S \equiv \sigma_1 \sigma_2 ... \sigma_k$, where $1 \leq k < r$. Consider the following set of indices

$$\mathcal{I} = \{\, i \mid k+1 \leq i \leq r \ \text{and} \ \pi(\sigma_{k+1}\sigma_{k+2}...\sigma_i) < \lambda(\sigma_{k+1}\sigma_{k+2}...\sigma_i) \,\}.$$

It is easy to see that the set $\mathcal{I}$ is not empty. Indeed, taking the formal word $U \equiv \sigma_{k+1}...\sigma_r$, we have: $\lambda(U) = \lambda(A) - \lambda(S) = (\pi(A)+1) - \lambda(S) \geq (\pi(A)+1) - \pi(S) = (\pi(A) - \pi(S)) + 1 = \pi(U) + 1 > \pi(U)$, hence $\lambda(\sigma_{k+1}...\sigma_r) > \pi(\sigma_{k+1}...\sigma_r)$; this means that $r \in \mathcal{I}$ and $\mathcal{I} \neq \emptyset$.

Now let us denote by p the smallest element of $\mathcal{I}$, $p = \min \mathcal{I}$. From the definition of set $\mathcal{I}$ it follows that $p \geq k + 1$. There are two possible cases:

$$p = k + 1 \quad \text{and} \quad p > k + 1.$$

In the first case, by the definition of $\mathcal{I}$ we have $\lambda(\sigma_{k+1}) > \pi(\sigma_{k+1})$ or $1 > \pi(\sigma_{k+1})$, hence $\pi(\sigma_{k+1}) = 0$. It follows that $\sigma_{k+1} \equiv X$, hence the formal word $T \equiv \sigma_{k+1}$ is a bfw and $ST \equiv A$ or $ST \subset A$.

In the second case, the formal word $T \equiv \sigma_{k+1}...\sigma_p$ is necessarily a bfw. Indeed, first $\lambda(\sigma_{k+1}...\sigma_{p-1}) \leq \pi(\sigma_{k+1}...\sigma_{p-1})$, because $p - 1 \notin \mathcal{I}$, hence

$$\lambda(\sigma_{k+1}...\sigma_p) = \lambda(\sigma_{k+1}...\sigma_{p-1}) + 1 \leq \pi(\sigma_{k+1}...\sigma_{p-1}) + 1 \leq \pi(\sigma_{k+1}...\sigma_p) + 1$$

or $\lambda(T) \leq \pi(T) + 1$. Taking into account that $p \in \mathcal{I}$, we have $\lambda(\sigma_{k+1}...\sigma_p) > \pi(\sigma_{k+1}...\sigma_p)$, hence $\lambda(T) > \pi(T)$ that is $\lambda(T) \geq \pi(T) + 1$. From the two obtained inequalities we get $\lambda(T) = \pi(T) + 1$ (condition (b_1)).

Consider now a segment W of T. This segment has the form $W \equiv \sigma_{k+1}...\sigma_l$, where $k + 1 \leq l < p$, hence $l \notin \mathcal{I}$. From the definition of $\mathcal{I}$ it follows that

$$\lambda(W) = \lambda(\sigma_{k+1}...\sigma_l) \leq \pi(\sigma_{k+1}...\sigma_l) = \pi(W) \quad (\text{condition}(b_2)),$$

hence $T \in \mathcal{E}$ and $ST \equiv A$ or $ST \subset A$. The existence of the requested $T \in \mathcal{E}$ is thus proved.

The uniqueness of T follows almost immediately. Indeed, suppose that there are two bfw $T_1, T_2, T_1 \not\equiv T_2$ such that $ST_1 \equiv A$ or $ST_1 \subset A$ and $ST_2 \equiv A$ or $ST_2 \subset A$. Then necessarily $T_1 \subset T_2$ or $T_2 \subset T_1$, which is a contradiction because T_1, T_2 are bfw and a segment of a bfw cannot be a bfw. ∎

PROPOSITION 1.3

Let be $A \in \mathcal{E}$, $A \equiv \sigma_1 \sigma_2...\sigma_m$.

(a) If $\sigma_1 \not\equiv \neg, \vee$, then $\lambda(A) = 1$ and $A \equiv X$;

(b) If $\sigma_1 \equiv \neg$, then there is a unique $B \in \mathcal{E}$ such that $A \equiv \neg B$;

(c) If $\sigma_1 \equiv \vee$, then there is a unique pair $C, D \in \mathcal{E}$ such that $A \equiv \vee CD$.

PROOF   (a) If σ_1 does not coincide with $\neg$ or with $\vee$, then σ_1 is necessarily a propositional variable, i.e., $\sigma_1 \equiv X$. If we suppose that $\lambda(A) > 1$, then X is a segment in A, which is impossible, X being a bfw. Hence, $\lambda(A) = 1$ and $A \equiv X$.

(b) If $\sigma_1 \equiv \neg$, then $\lambda(A) > 1$ because $A \in \mathcal{E}$ and $\neg \notin \mathcal{E}$. Therefore, $\neg$ is a segment of A. By Proposition 1.2 there is a unique $B \in \mathcal{E}$ such that

$$\neg B \equiv A \quad \text{or} \quad \neg B \subset A.$$

The second situation is impossible, because $\neg B \in \mathcal{E}$ (by Proposition 1.1) and consequently $\neg B$ cannot be a segment of A. So that $\neg B \equiv A$.

(c) As in case (b) it follows that $\lambda(A) > 1$, hence $\vee$ is a segment of A. By Proposition 1.2 there is a unique $C \in \mathcal{E}$ such that

$$\vee C \equiv A \quad \text{or} \quad \vee C \subset A.$$

The first situation cannot arise because $\vee C \notin \mathcal{E}$. Indeed, $\lambda(\vee C) = 1 + \lambda(C)$ and taking into account that $C \in \mathcal{E}$ we have $\lambda(C) = \pi(C) + 1$, therefore

$$\lambda(\vee C) = 1 + (\pi(C) + 1) = 2 + \pi(C) = \pi(\vee C) < \pi(\vee C) + 1$$

and $\vee C$ does not fulfill condition (b_1) of Definition 1.9. It follows necessarily that $\vee C \subset A$. By the same Proposition 1.2 there is a unique $D \in \mathcal{E}$ such that

$$\vee C D \equiv A \quad \text{or} \quad \vee C D \subset A.$$

But the second situation is impossible because $\vee C D \in \mathcal{E}$ (by Proposition 1.1), hence $\vee C D$ cannot be a segment of A. ∎

REMARK 1.8   Proposition 1.3 is relevant because it furnishes the structure of bfw. ∎

In the remaining part of this section we shall show that a formal word is a formula (in $\mathbf{X}_{00}$) if and only if it is a bfw, i.e.,

$$\mathcal{L}_{00} = \mathcal{E}.$$

We shall prove this equality by showing that $\mathcal{E} \subseteq \mathcal{L}_{00}$ and $\mathcal{L}_{00} \subseteq \mathcal{E}$.

PROPOSITION 1.4
Each bfw is a formula (in $\mathbf{X}_{00}$), i.e., $\mathcal{E} \subseteq \mathcal{L}_{00}$.

PROOF   Let be $A \in \mathcal{E}$. We shall prove by induction on the length $\lambda(A) \geq 1$ that $A \in \mathcal{L}_{00}$. Suppose $\lambda(A) = 1$; we know that the only bfw of length 1 is the propositional variables. Therefore, $A \equiv X$, hence $A \in \mathcal{L}_{00}$. Suppose now that $\lambda(A) = m > 1$ and that any bfw A', with $\lambda(A') < m$, belongs to $\mathcal{L}_{00}$. By Proposition 1.3, A cannot be a variable, hence $A \equiv \neg B$ or $A \equiv \vee C D$, where B, C, D are bfw with $\lambda(B), \lambda(C), \lambda(D) < m$. By the inductive assumption we have $B \in \mathcal{L}_{00}$, $C, D \in \mathcal{L}_{00}$, respectively. So that we have $A \equiv \neg B \in \mathcal{L}_{00}$, $A \equiv \vee CD \in \mathcal{L}_{00}$, respectively (see considerations following Definition 1.5). Therefore, in any case, $A \in \mathcal{L}_{00}$. This proves (by the principle of mathematical induction) that all *bfw* are formulas; in other words, $\mathcal{E} \subseteq \mathcal{L}_{00}$. ∎

PROPOSITION 1.5

Each formula (in $\mathbf{X}_{00}$*) is a bfw, i.e.,* $\mathcal{L}_{00} \subseteq \mathcal{E}$.

PROOF   Let $F \in \mathcal{L}_{00}$ and let us prove that $F \in \mathcal{E}$. F being a formula (in $\mathbf{X}_{00}$), there is a formal construction

$$(fc): \ A_1, A_2, ...A_r \ (r \geq 1)$$

such that $A_r \equiv F$. We shall prove by induction that for all i, $\ 1 \leq i \leq r$

$$A_i \in \mathcal{E}.$$

For $i = 1$ in (fc), the formal word A_1 is necessarily a propositional variable, hence $A_1 \in \mathcal{E}$.

Suppose now that $i > 1$ and assume that all formal words $A_1, A_2, ...,$ A_{i-1} belong to $\mathcal{E}$. From Definition 1.5, formula A_i satisfies one of three conditions:

(a) $A_i \equiv X$;

(b)   there is $j < i$ such that $A_i \equiv \neg A_j$;

(c)   there are $h, k < i$ such that $A_i \equiv \vee A_h A_k$.

In case (a), A_i being a propositional variable belongs to $\mathcal{E}$. In case (b), by the inductive assumption, $A_j \in \mathcal{E}$ so that (by Proposition 1.1) $A_i \equiv \neg A_j \in \mathcal{E}$. Case (c) is similar to case (b). Consequently, in all cases, $A_i \in \mathcal{E}$. From the principle of mathematical induction, it follows that $A_i \in \mathcal{E}$ for all i ($1 \leq i \leq r$). In particular, $F \equiv A_r \in \mathcal{E}$ and the inclusion $\mathcal{L}_{00} \subset \mathcal{E}$ is thus proved. ∎

From Proposition 1.4 and Proposition 1.5, one obtains the following final result:

PROPOSITION 1.6

$\mathcal{L}_{00} = \mathcal{E}$*, i.e., a formal word is a formula (in* $\mathbf{X}_{00}$*) if and only if it is a balanced formal word (bfw).*

REMARK 1.9

(1) The identity between formulas and balanced formal words furnishes an arithmetical tool for deciding if a formal word (in $\mathbf{X}_{00}$) is a formula or not.
(2) Proposition 1.6 can be used in computer programming to describe precisely the structure of each formula (in $\mathbf{X}_{00}$) and consequently gives the possibility of automatically translating the language $\mathcal{L}_{00}$ into the language $\mathcal{L}_0$. The converse translation from $\mathcal{L}_0$ to $\mathcal{L}_{00}$ is also possible, but it needs some additional considerations about the distribution of parentheses in formulas of $\mathcal{L}_0$. ∎

2

The Truth Structure on $\mathcal{L}_0$ in Semantic Version

In this chapter we shall define the Boolean interpretations of $\mathcal{L}_0$ and we shall prove the existence of such interpretations. Based on Boolean interpretations we shall define the tautologies and construct the so-called Lindenbaum algebra of propositional logic. We shall systematically use some fundamental results about Boolean algebras; the reader can find all the necessary elements on this subject in APPENDIX A at the end of the book.

2.1 Boolean Interpretations of the Language $\mathcal{L}_0$

Let $\Gamma = \{0, 1\}$ be the Boolean algebra containing only two elements endowed with one internal unary operation $x \mapsto \bar{x}$ ($\bar{x}$ being called the complement of x) and two internal binary operations $(x, y) \mapsto x \vee y$ and $(x, y) \mapsto x \wedge y$, where

$$\bar{x} = 1 - x;$$

$$x \vee y = \max\{x, y\};$$

$$x \wedge y = \min\{x, y\}.$$

It is supposed that Γ is a set ordered by the natural order of numbers; that is, $0 < 1$, and $x \leq y$ if $x = y$ or $x < y$. It follows that $\max\{0, 0\} = 0$, $\max\{0, 1\} = \max\{1, 0\} = 1$, $\max\{1, 1\} = 1$ and $\min\{0, 0\} = 0$, $\min\{0, 1\} = \min\{1, 0\} = 0$, $\min\{1, 1\} = 1$.

DEFINITION 2.1 *A function $\varphi : \mathcal{L}_0 \longrightarrow \Gamma : F \longmapsto \varphi(F)$ is called a* **Boolean interpretation of** $\mathcal{L}_0$ *if it satisfies the following two conditions:*

(i) $\varphi(\neg(F)) = \overline{\varphi(F)}$;

(ii) $\varphi((F) \vee (G)) = \varphi(F) \vee \varphi(G)$,

where F, G are arbitrary formulas in $\mathcal{L}_0$.

REMARK 2.1   Taking into account that in any Boolean algebra (see APPENDIX A), particularly in Γ, one can prove the equalities

$\bar{\bar{x}} = x$ (involution property of $^-$) and

$\overline{x \vee y} = \bar{x} \wedge \bar{y}$,

$\overline{x \wedge y} = \bar{x} \vee \bar{y}$ (De Morgan's laws),

it follows that

$$\varphi((F) \wedge (G)) = \varphi(\neg(\neg F \vee \neg G)) = \overline{\varphi(\neg F \vee \neg G)} = \overline{\varphi(\neg F) \vee \varphi(\neg G)} =$$

$$= \overline{\overline{\varphi(F)} \vee \overline{\varphi(G)}} = \overline{\overline{\varphi(F)}} \wedge \overline{\overline{\varphi(G)}} = \varphi(F) \wedge \varphi(G);$$

$$\varphi((F) \Rightarrow (G)) = \varphi(\neg F \vee G) = \varphi(\neg F) \vee \varphi(G) = \overline{\varphi(F)} \vee \varphi(G);$$

$$\varphi((F) \Leftrightarrow (G)) = \varphi[(F \Rightarrow G) \wedge (G \Rightarrow F)] = \varphi(F \Rightarrow G) \wedge \varphi(G \Rightarrow F).$$

Consequently, for the (derived) operators $\wedge, \Rightarrow, \Leftrightarrow$ in $\mathcal{L}_0$ we have finally

$$\varphi(F \wedge G) = \varphi(F) \wedge \varphi(G);$$
$$\varphi(F \Rightarrow G) = \overline{\varphi(F)} \vee \varphi(G);$$
$$\varphi(F \Leftrightarrow G) = (\overline{\varphi(F)} \vee \varphi(G)) \wedge (\overline{\varphi(G)} \vee \varphi(F)).$$

∎

    If φ is an interpretation of $\mathcal{L}_0$, then the formula $F \in \mathcal{L}_0$ is called φ-**true** if $\varphi(F) = 1$, and it is called φ-**false** if $\varphi(F) = 0$. For this reason Γ is called **the set of truth values** for $\mathcal{L}_0$. Let φ be an arbitrary fixed interpretation of $\mathcal{L}_0$ and let F, G be formulas in $\mathcal{L}_0$. The φ-truth values of formulas $\neg(F), (F) \vee (G), (F) \wedge (G), (F) \Rightarrow (G), (F) \Leftrightarrow (G)$ can be expressed by means of the "truth tables" given in "Preliminaries".

    Now, the notion of Boolean interpretation having been defined, two natural questions arise: Is there at least one Boolean interpretation of $\mathcal{L}_0$? If yes, how many are there? In other words, denoting by $\mathcal{V}$ the set of all interpretations of $\mathcal{L}_0$, the two questions become: Is the set $\mathcal{V}$ nonempty ($\mathcal{V} \neq \emptyset$)? If yes, then what is the number of these interpretations (card $(\mathcal{V}) = ?$)? The following result yields a complete answer to both these questions.

    Let Γ^N be the set of all N - tuples $(\alpha_1, \alpha_2, ...\alpha_N)$, where $\alpha_i \in \Gamma$ $(1 \leq i \leq N)$, i.e.,

$$\Gamma^N = \underbrace{\Gamma \times \Gamma \times ... \times \Gamma}_{N \text{ times}}.$$

PROPOSITION 2.1

(a) *For any* $\alpha = (\alpha_1, \alpha_2, ..., \alpha_N) \in \Gamma^N$ *there exists a unique function* $\varphi_\alpha : \mathcal{L}_0 \to \Gamma$ *having the following properties:*

    (i) $\varphi_\alpha(\neg F) = \overline{\varphi_\alpha(F)}$ *for all* $F \in \mathcal{L}_0$;

    (ii) $\varphi_\alpha(F \vee G) = \varphi_\alpha(F) \vee \varphi_\alpha(G)$ *for all* $F, G \in \mathcal{L}_0$;

    (iii) $\varphi_\alpha(X_1) = \alpha_1, \varphi_\alpha(X_2) = \alpha_2, ..., \varphi_\alpha(X_N) = \alpha_N$.

(b) *The correspondence* $\alpha \mapsto \varphi_\alpha$ *is a bijective map from* Γ^N *into* $\mathcal{V}$, *hence card* $(\mathcal{V}) =$ *card* $(\Gamma^N) = 2^N$; *in particular,* $\mathcal{V} \neq \emptyset$.

PROOF   Let us fix an element $\alpha \in \Gamma^N$, $\alpha = (\alpha_1, \alpha_2, ..., \alpha_N)$.

(a) Consider the set $\mathcal{K}$ of all subsets $K \subseteq \mathcal{L}_0 \times \Gamma = \{(F, \gamma) | F \in \mathcal{L}_0, \gamma \in \Gamma\}$ having the following properties:

(j) If $(F, \gamma) \in K$, then $(\neg F, \bar{\gamma}) \in K$;

(jj) If $(F_1, \gamma_1) \in K$ and $(F_2, \gamma_2) \in K$ then $(F_1 \vee F_2, \gamma_1 \vee \gamma_2) \in K$;

(jjj) $(X_i, \alpha_i) \in K$, for all i $(1 \leq i \leq N)$.

First, let us notice that $\mathcal{K} \neq \emptyset$, i.e., there are subsets $K \subseteq \mathcal{L}_0 \times \Gamma$ having properties (j), (jj), (jjj). Indeed, such a subset K is, for instance, the whole set $\mathcal{L}_0 \times \Gamma$, i.e., $\mathcal{L}_0 \times \Gamma \in \mathcal{K}$. Now let us consider the set

$$K_0 = \bigcap_{K \in \mathcal{K}} K.$$

It is clear that $K_0 \subseteq \mathcal{L}_0 \times \Gamma$ and K_0 have properties (j), (jj), (jjj), hence $K_0 \in \mathcal{K}$ and for each $K \in \mathcal{K}$ we have $K \supseteq K_0$. This means that K_0 is the "smallest" set in $\mathcal{K}$. It is not difficult to see that, for any $K \in \mathcal{K}$, the set $pr_1(K)$ of all formulas F such that $(F, \gamma) \in K$ for some $\gamma \in \Gamma$, coincides with $\mathcal{L}_0$. It is obvious that $pr_1(K) \subseteq \mathcal{L}_0$. For the converse inclusion, let $F \in \mathcal{L}_0$ be an arbitrary formula. By definition, there is a formal construction of F

$$\text{(fc)}: \quad A_1, A_2, ..., A_r \equiv F.$$

We shall prove by induction along (fc) that for an arbitrary $K \in \mathcal{K}$, $A_i \in pr_1(K)$ for all i $(1 \leq i \leq r)$. The formal word A_1 is necessarily a propositional variable, i.e., there exists some index l $(1 \leq l \leq N)$ such that $A_1 \equiv X_l$. By the definition of sets K one gets $(X_l, \alpha_l) \in K$, hence $A_1 \equiv X_l \in pr_1(K)$. Let be $i > 1$ $(i \leq r)$ and assume that $A_1, A_2, ..., A_{i-1} \in pr_1(K)$. If $A_i \equiv X$, then as above it results that $A_i \in pr_1(K)$. If $A_i \equiv \neg(A_h)$ $(h < i)$, then $A_h \in pr_1(K)$ by the inductive assumption. From the definition of set $pr_1(K)$, it follows that there exists $\gamma_h \in \Gamma$ such that $(A_h, \gamma_h) \in K$; hence by (j) one gets $(\neg(A_h), \bar{\gamma}_h)) \in K$, so that $A_i \equiv \neg(A_h) \in pr_1(K)$. By similar arguments it follows that if $A_i \equiv (A_p) \vee (A_q)$, with $p, q < i$, then $A_i \in pr_1(K)$. From the principle of induction it follows that $A_i \in pr_1(K)$ for all i ($1 \leq i \leq r$); in particular, $F \equiv A_r \in pr_1(K)$. So we obtain the converse inclusion $\mathcal{L}_0 \subseteq pr_1(K)$ and, therefore, the equality $pr_1(K) = \mathcal{L}_0$ holds for any $K \in \mathcal{K}$.

Taking into account that $K_0 \in \mathcal{K}$ we also get

(*) $$pr_1(K_0) = \mathcal{L}_0.$$

In the remaining part of the proof we shall show that the set K_0 is the graph of a certain Γ-valued mapping φ defined on $\mathcal{L}_0$. For this purpose we have to show that

(**) if $(F, \gamma_1) \in K_0$ and $(F, \gamma_2) \in K_0$, then $\gamma_1 = \gamma_2$.

The proof of the implication (**) is not too difficult, but is tedious enough. The idea of the proof consists of the following. Suppose that there exists a formula $F \in \mathcal{L}_0$ and elements $\gamma_1, \gamma_2 \in \Gamma, \gamma_1 \neq \gamma_2$, such that $(F, \gamma_1) \in K_0$ and $(F, \gamma_2) \in K_0$. We shall say that such a formula has the property π. We shall prove that formulas having property π do not exist. Let us suppose that such formulas exist.

Let F_0 be a formula with minimal length having property π. If $\lambda(F_0) = 1$, then $F_0 \equiv X_l$ for some l ($1 \leq l \leq N$). By (jjj) we have $(F_0, \alpha_l) = (X_l, \alpha_l) \in K_0$. On the other hand, F_0 has property π, hence for F_0 there is some $\gamma \in \Gamma, \gamma \neq \alpha_l$, such that $(F_0, \gamma) \equiv (X_l, \gamma) \in K$. Defining the set

$$K_0^* = K_0 \setminus \{(X_l, \gamma)\}$$

it is obvious that $K_0^* \subseteq K_0$ and $K_0^* \neq K_0$. On the other hand, it is not difficult to verify that K_0^* has properties (j), (jj), (jjj); hence $K_0^* \in \mathcal{K}$ and $K_0^* \supseteq \bigcap_{K \in \mathcal{K}} K = K_0$, which is impossible. Therefore, each formula F having the property π must have a length greater than 1 ($\lambda(F_0) > 1$). If $\lambda(F_0) > 1$, then $F_0 \equiv \neg(G_0)$ or $F_0 \equiv (P_0) \vee (Q_0)$. In the first case, we have $\lambda(G_0) < \lambda(F_0)$, so that from the minimality of $\lambda(F_0)$ there is a unique γ_0 such that $(G_0, \gamma_0) \in K_0$. By (j) it follows $(F_0, \bar{\gamma}_0) = (\neg(G_0), \bar{\gamma}_0) \in K_0$. Taking into account that F_0 is supposed to have property π, there exists $\gamma_1 \in \Gamma, \gamma_1 \neq \bar{\gamma}_0$, such that $(F_0, \gamma_1) \in K_0$. If we consider the set

$$K_0^* = K_0 \setminus \{(F_0, \gamma_1)\},$$

then one can verify that K_0^* has properties (j), (jj), (jjj); hence $K_0^* \in \mathcal{K}$ and as above we get a contradiction because $K_0^* \neq K_0$. In the case where $F_0 \equiv (P_0) \vee (Q_0)$ similar arguments lead to a contradiction. The above reasoning proves that the implication (∗∗) is true for any formula $F \in \mathcal{L}_0$, i.e., for any $F \in \mathcal{L}_0$ there is a unique $\gamma \in \Gamma$ such that $(F, \gamma) \in K_0$. Finally, we define the mapping

$$\varphi_\alpha : \mathcal{L}_0 \to \Gamma : F \mapsto \varphi_\alpha(F)$$

in the following way: if F is an arbitrary formula, then $\varphi_\alpha(F) = \gamma$, where γ is the unique element of Γ such that $(F, \gamma) \in K_0$. Now from properties (j), (jj), (jjj) it easily follows that φ_α satisfies conditions (i), (ii), (iii) stated in the proposition to be proved.

For the assertion (a), we have to also prove the uniqueness of φ_α. For this purpose, let φ be a Γ-valued function defined on $\mathcal{L}_0$ having properties (i), (ii), (iii). Therefore,

$$(* * *) \qquad \varphi(X_i) = \varphi_\alpha(X_i) = \alpha_i \quad \text{for all} \quad i \ (1 \leq i \leq N).$$

We shall establish by induction on the length $\lambda(F) \geq 1$ that $\varphi(F) = \varphi_\alpha(F)$ for all $F \in \mathcal{L}_0$.

In the case $\lambda(F) = 1$, we have $F \equiv X_l$ for some l $(1 \leq l \leq N)$ so that, by $(* * *)$,

$$\varphi(F) = \varphi(X_l) = \alpha_l = \varphi_\alpha(X_l) = \varphi_\alpha(F).$$

In the case $\lambda(F) > 1$, assume that $\varphi(F') = \varphi_\alpha(F')$ for any $F' \in \mathcal{L}_0$ with $\lambda(F') < \lambda(F)$. It is clear that F is not a propositional variable; hence there are two possibilities:

$$F \equiv \neg(G) \quad \text{or} \quad F \equiv (P) \vee (Q), \quad \text{where } G, P, Q \in \mathcal{L}_0.$$

In the case $F \equiv \neg(G)$, we have $\lambda(G) < \lambda(F)$, hence by the inductive assumption $\varphi(G) = \varphi_\alpha(G)$. Therefore, from (i), one gets

$$\varphi(F) = \varphi(\neg G) = \overline{\varphi(G)} = \overline{\varphi_\alpha(G)} = \varphi_\alpha(\neg G) = \varphi_\alpha(F).$$

In the case $F \equiv (P) \vee (Q)$, by similar arguments we get also $\varphi(F) = \varphi_\alpha(F)$. From the principle of mathematical induction it follows that $\varphi(F) = \varphi_\alpha(F)$ for any formula $F \in \mathcal{L}_0$, that is $\varphi = \varphi_\alpha$. Consequently, we have proved the uniqueness of φ_α, hence assertion (a).

(b) Let us notice that from (a) it follows that the correspondence $\alpha \mapsto \varphi_\alpha$ is a well-defined mapping from Γ^N to $\mathcal{V}$. It follows, in particular, that $\mathcal{V}$ is nonempty ($\mathcal{V} \neq \emptyset$). In order to prove that the mapping $\alpha \mapsto \varphi_\alpha$ is bijective we have to show that it is injective and surjective. Let $\alpha, \beta \in \Gamma^N$ be two arbitrary elements in $\Gamma^N, \alpha = (\alpha_1, \alpha_2, ..., \alpha_N), \beta = (\beta_1, \beta_2, ..., \beta_N)$. Suppose $\alpha \neq \beta$ there is some l $(1 \leq l \leq N)$ such that $\alpha_l \neq \beta_l$. Taking into account that φ_α and φ_β satisfy condition (iii) we get $\varphi_\alpha(X_l) = \alpha_l \neq \beta_l = \varphi_\beta(X_l)$. Therefore, $\varphi_\alpha \neq \varphi_\beta$, hence the mapping $\alpha \mapsto \varphi_\alpha$ is injective.

This mapping is also surjective. Indeed, take an arbitrary $\varphi \in \mathcal{V}$ and denote $\alpha = (\alpha_1, \alpha_2, ..., \alpha_N)$, where $\alpha_i = \varphi(X_i)$ $(1 \leq i \leq N)$. By (a) it follows that there is $\varphi_\alpha \in \mathcal{V}$ satisfying (i), (ii), (iii). By the uniqueness property we get $\varphi = \varphi_\alpha$. So that the mapping $\alpha \mapsto \varphi_\alpha$ is surjective and hence bijective.

The mapping $\alpha \mapsto \varphi_\alpha : \Gamma^N \to \mathcal{V}$ being bijective one obtains that

$$\text{card} \, (\mathcal{V}) = \text{card} \, (\Gamma^N) = 2^N.$$

Assertion (b), hence the entire Proposition, is now completely proved.   ∎

REMARK 2.2   From the above proposition it follows that the set of all Boolean interpretations of $\mathcal{L}_0$ is nonempty. Moreover, if the number of propositional variables is finite and equals N, then there are exactly 2^N Boolean interpretations of $\mathcal{L}_0$.   ∎

As was already mentioned in the introduction to this section, by means of Boolean interpretations we can introduce truth values for formulas.

DEFINITION 2.2   *A Boolean interpretation $\varphi \in \mathcal{V}$ being given we shall say that formula $F \in \mathcal{L}_0$ is φ-true if $\varphi(F) = 1$ and is φ-false if $\varphi(F) = 0$.*

DEFINITION 2.3

(a) *A formula F which is φ-true for all interpretations $\varphi \in \mathcal{V}$ is called* **universally true** *or a* **tautology.**

(b) *A formula F which is φ-false for all interpretations $\varphi \in \mathcal{V}$* *called* **universally false** *or an* **antilogy.**

(c) *A formula F for which there are interpretations $\varphi', \varphi'' \in \mathcal{V}$ such that $\varphi'(F) = 1$ and $\varphi''(F) = 0$ is called* **contingent.**

REMARK 2.3   In the "Preliminaries" chapter of the book a quite extensive list of tautologies is given.   ∎

In the following, in addition to the above considerations, we shall need to introduce the concept of "functional interpretation" of $\mathcal{L}_0$. For this purpose we denote by $B(N, \Gamma)$ the set of all Γ-valued functions defined on Γ^N, where N is the number of propositional variables in $\mathcal{L}_0$; hence

$$B(N, \Gamma) = \{f : \Gamma^N \to \Gamma\}.$$

In $B(N, \Gamma)$ there are N remarkable functions $p_1, p_2, ..., p_N$ called **projectors** defined by

$$p_i(x_1, ..., x_i, ..., x_N) = x_i \quad (1 \le i \le N).$$

The set $B(N, \Gamma)$ endowed with internal operations $\bar{f}, f \vee g, f \wedge g$ defined by

$$\bar{f}(x_1, ..., x_N) = \overline{f(x_1, ..., x_N)};$$
$$(f \vee g)(x_1, ..., x_N) = f(x_1, ..., x_N) \vee g(x_1, ..., x_N);$$
$$(f \wedge g)(x_1, ..., x_N) = f(x_1, ..., x_N) \wedge g(x_1, ..., x_N)$$

is a Boolean algebra with 2^{2^N} elements. Details about Boolean algebras $B(N, \Gamma)$ can be found in APPENDIX A. In the same appendix it is proved that any function $f \in B(N, \Gamma)$ can be represented as a Boolean polynomial. More precisely if for any $f \in B(N, \Gamma)$ one defines the function $f^{\sigma} \in B(N, \Gamma)$ by

$$f^{\sigma} = \begin{cases} f, & \text{if } \sigma = 1 \\ \bar{f}, & \text{if } \sigma = 0, \end{cases}$$

then, for any $f \ne 0$, the following equality holds:

$$f = \bigvee_{f(\sigma_1, ..., \sigma_N) = 1} (p_1^{\sigma_1} \wedge p_2^{\sigma_2} \wedge ... \wedge p_N^{\sigma_N}).$$

In equivalent form this equality can be written

$$f(x_1, ..., x_N) = \bigvee_{f(\sigma_1, ..., \sigma_N) = 1} (x_1^{\sigma_1} \wedge ... \wedge x_N^{\sigma_N})$$

where for any $x \in \Gamma$, $x^{\sigma} = x$ if $\sigma = 1$ and $x^{\sigma} = \bar{x}$ if $\sigma = 0$. The polynomial $f = \bigvee_{f(\sigma_1, ..., \sigma_N) = 1} (p_1^{\sigma_1} \wedge ... \wedge p_N^{\sigma_N})$ is said to be in disjunctive form. It is clear that using de Morgan laws any function $f \in B(N, \Gamma)$, $f \ne 1$, can be represented also by a polynomial in a conjunctive form:

$$f = \bigwedge_{f(\sigma_1, ..., \sigma_N) = 0} (p_1^{\bar{\sigma}_1} \vee ... \vee p_N^{\bar{\sigma}_N}).$$

PROPOSITION 2.2
There exists a unique mapping

$$\Phi : \mathcal{L}_0 \to B(N, \Gamma) : F \mapsto \Phi(F)$$

having the following properties:

(i) $\Phi(\neg(F)) = \overline{\Phi(F)}$ *for all $F \in \mathcal{L}_0$;*

(ii) $\Phi((F) \vee (G)) = \Phi(F) \vee \Phi(G)$　*for all* $F, G, \in \mathcal{L}_0$;

(iii) $\Phi(X_l) = p_l$ *for all* l $(1 \leq l \leq N)$.

Moreover, the mapping Φ *is surjective.*

PROOF　For any fixed $F \in \mathcal{L}_0$, let us define a function $\Phi(F) \in B(N, \Gamma)$ by the following equality:

$$(*) \qquad \Phi(F)(\alpha) = \varphi_\alpha(F), \quad \text{for each } \alpha \in \Gamma^N.$$

Therefore, $\Phi(F)$ is an element of $B(N, \Gamma)$ and we can define the mapping

$$\Phi : \mathcal{L}_0 \to B(N, \Gamma) : F \mapsto \Phi(F)$$

where the function $\Phi(F)$ is given by $(*)$. The properties (i), (ii), (iii) of Φ easily follow from the corresponding properties (i), (ii), (iii) (Definition 2.1) of the Boolean interpretations. For instance, by $(*)$ one gets

$$\Phi(X_l)(\alpha) = \varphi_\alpha(X_l) = \alpha_l$$

for each $\alpha = (\alpha_1, \alpha_2, ...\alpha_N) \in \Gamma^N$. But $\alpha_l = p_l(\alpha)$, hence

$$\Phi(X_l)(\alpha) = p_l(\alpha) \quad \text{for all } \alpha \in \Gamma^N$$

therefore $\Phi(X_l) = p_l$ and property (iii) is proved. Using property (i) of Boolean interpretations we can also directly deduce property (i) for Φ. Indeed, for all $\alpha \in \Gamma^N$,

$$\Phi(\neg(F))(\alpha) = \varphi_\alpha(\neg(F)) = \overline{\varphi_\alpha(F)} = \overline{\Phi(F)(\alpha)} = \overline{\Phi(F)}(\alpha).$$

Therefore $\Phi(\neg(F)) = \overline{\Phi(F)}$, and property (i) is proved. Property (ii) can be proved in a similar way.

　　The uniqueness of a mapping Φ having properties (i), (ii), (iii) can easily be established by induction. Indeed, if $\Psi : \mathcal{L}_0 \to B(N, \Gamma)$ is an arbitrary map having properties (i), (ii), (iii), then for any formula $F \in \mathcal{L}_0$ with $\lambda(F) = 1$ there is some l $(1 \leq l \leq N)$ such that $F \equiv X_l$ and by (iii) $\Psi(F) = \Psi(X_l) = p_l = \Phi(X_l) = \Phi(F)$. So that $\Psi(F) = \Phi(F)$ for any formula of length 1. Now let F be a formula with $\lambda(F) > 1$ and assume that $\Psi(F') = \Phi(F')$ for any $F' \in \mathcal{L}_0$ with $\lambda(F') < \lambda(F)$. Two cases have to be considered:

$$(j) \quad F \equiv \neg(G) \quad \text{and} \quad (jj) \quad F \equiv (P) \vee (Q).$$

In the case (j), from the inductive assumption (and taking into account that Ψ has property (i)), one obtains successively:

$$\Psi(F) = \Psi(\neg(G)) = \overline{\Psi(G)} = \overline{\Phi(G)} = \Phi(\neg(G)) = \Phi(F).$$

In the case (jj), by the same induction assumption and taking into account that Ψ has property (ii) we obtain:

$$\Psi(F) = \Psi((P)\vee(Q)) = \Psi(P)\vee\Psi(Q) = \Phi(P)\vee\Phi(Q) = \Phi((P)\vee(Q)) = \Phi(F).$$

Consequently, in all cases $\Psi(F) = \Phi(F)$. Applying the principle of mathematical induction we deduce that $\Psi(F) = \Phi(F)$ for all $F \in \mathcal{L}_0$, hence $\Psi = \Phi$ and the uniqueness of Φ is proved.

In order to prove the surjectivity of the mapping Φ, consider an arbitrary function $f \in B(N, \Gamma)$. By the result mentioned before, Proposition 2.2, the function f, $f \neq 0$, can be represented in disjunctive polynomial form, i.e.,

$$f = \bigvee_{f(\sigma_1,...,\sigma_N)=1} (p_1^{\sigma_1} \wedge ... \wedge p_N^{\sigma_N}).$$

Let $\mathcal{C}$ be the finite set of all terms $p_1^{\sigma_1} \wedge ... \wedge p_N^{\sigma_N}$ for which $f(\sigma_1, \sigma_2, ..., \sigma_N) = 1$ and consider an arbitrary fixed ordering of these terms. Therefore one can write

$$\mathcal{C} = \{c_1, c_2, ..., c_L\}$$

and $f = \bigvee_{s=1}^{L} c_s$. Taking into account that for each s ($1 \leq s \leq L$) there exists a unique N-tuple $(\sigma_1, ..., \sigma_N) \in \Gamma^N$ with $f(\sigma_1, ..., \sigma_N) = 1$ such that $c_s = p_1^{\sigma_1} \wedge ... \wedge p_N^{\sigma_N}$, we can uniquely associate to c_s the formula

$$C_s \equiv X_1^{\sigma_1} \wedge X_2^{\sigma_2} \wedge X_3^{\sigma_3} \wedge ... \wedge X_N^{\sigma_N},$$

where the distribution of parentheses is not marked (it is easy to see that this distribution is irrelevant).

It is clear that

$$\Phi(C_s) = c_s \quad \text{for all} \quad s \ (1 \leq s \leq L).$$

Finally, if we consider the formula $F \in \mathcal{L}_0$ defined by

$$F \equiv C_1 \vee C_2 \vee C_3 \vee ... \vee C_L,$$

then taking into account the properties of Φ, we get

$$\Phi(F) = \bigvee_{s=1}^{L} \Phi(C_s) = \bigvee_{s=1}^{L} c_s = \bigvee_{f(\sigma_1,...\sigma_n)=1} (p_1^{\sigma_1} \wedge ... \wedge p_N^{\sigma_N}) = f.$$

Thus we have proved that for any $f \in B(N, \Gamma)$ there is some $F \in \mathcal{L}_0$ such that $\Phi(F) = f$. This means that mapping Φ is surjective. ∎

Proposition 2.2 provides another concept of interpretation for the language $\mathcal{L}_0$. In contrast with Boolean interpretations φ_α, which assign to each formula $F \in \mathcal{L}_0$ an element of Γ, the new interpretation furnished by Φ associates with any formula a certain Boolean function f belonging to $B(N, \Gamma)$. This remark is formulated in the definition below.

DEFINITION 2.4   *The unique mapping $\Phi : \mathcal{L}_0 \to B(N, \Gamma)$, having the properties stated in Proposition 2.2 will be called the $B(N, \Gamma)$- interpretation or simply the* **functional interpretation** *of $\mathcal{L}_0$.*

Using the functional interpretation Φ one can classify the sentences of $\mathcal{L}_0$ into valid, inconsistent and contingent sentences. We shall see that this classification is equivalent to the one obtained by means of Boolean interpretations.

DEFINITION 2.5   *A formula $F \in \mathcal{L}_0$ is said to be:*

   (a) **valid** *if $\Phi(F) = 1$ (the constant function 1);*

   (b) **inconsistent** *if $\Phi(F) = 0$ (the constant function 0);*

   (c) **contingent** *if the function $\Phi(F)$ is not constant;*

   (d) **non valid** *if it is inconsistent or contingent;*

   (e) **consistent** *if it is valid or contingent.*

From the connection between the mapping Φ and the interpretations $\varphi_\alpha, \alpha \in \Gamma^N$, given by the equality

$$\Phi(F)(\alpha) = \varphi_\alpha(F)$$

(for all $\alpha \in \Gamma^N$ and each $F \in \mathcal{L}_0$) one immediately deduces the following conclusion.

PROPOSITION 2.3
A formula $F \in \mathcal{L}_0$ is

   (a) valid if and only if it is universally true (tautology);

   (b) inconsistent if and only if it is universally false (antilogy).

REMARK 2.4   The above proposition shows that universally true formulas (tautologies) coincide with valid formulas, universally false formulas (antilogies) coincide with inconsistent formulas and so on. In the following we shall use both terminologies without additional specifications. ∎

From now on we adopt the following notation: the assertion "F is a tautology" will be denoted by $\models F$.

The notions of consistent and inconsistent formulas can be immediately extended to arbitrary sets of formulas. From Definition 2.5 it easily follows that a formula $F \in \mathcal{L}_0$ is (semantically) consistent if and only if there is an interpretation $\varphi \in \mathcal{V}$ such that $\varphi(F) = 1$.

DEFINITION 2.6   *(a) A set $\mathcal{H}$ of formulas in $\mathcal{L}_0$ is said to be* **semantically consistent** *or simply,* **consistent** *if there is an interpretation $\varphi \in \mathcal{V}$ such that*

$$\varphi(H) = 1 \quad for\ all\ \ H \in \mathcal{H};$$

in such a case the interpretation φ is called a **model** *of (for) $\mathcal{H}$ and we will say that $\mathcal{H}$ has a model.*
(b) If for each interpretation $\varphi \in \mathcal{V}$ there is at least one formula $H \in \mathcal{H}$ such that $\varphi(H) = 0$, then the set $\mathcal{H}$ is called **semantically inconsistent**, *or simply,* **inconsistent**; *in other words, $\mathcal{H}$ is inconsistent if $\mathcal{H}$ has no model.*

DEFINITION 2.7   *A set $\mathcal{H}$ of formulas in $\mathcal{L}_0$ is called* **finitely consistent** *if every finite subset of $\mathcal{H}$ is consistent; in other words, $\mathcal{H}$ is finitely consistent if every finite subset of $\mathcal{H}$ has a model.*

It is clear that each consistent set $\mathcal{H}$ of formulas in $\mathcal{L}_0$ is finitely consistent. The converse of this remark asserts that each finitely consistent set of formulas is consistent. This result is also true (but less trivial) and is called the compactness (or compacity) theorem of propositional logic. The proof of the compactness theorem will be given below after some preliminaries.

DEFINITION 2.8   *A set $\mathcal{H}$ in $\mathcal{L}_0$ is said to be* **maximal finitely consistent** *if it is finitely consistent and if it has the following* **maximality** *property: if $\mathcal{K}$ is a finitely consistent set and $\mathcal{K} \supseteq \mathcal{H}$, then $\mathcal{K} = \mathcal{H}$.*

The following proposition emphasizes the connection between the class of maximal finitely consistent sets and the Boolean interpretations $\varphi \in \mathcal{V}$ of $\mathcal{L}_0$.

PROPOSITION 2.4

For each Boolean interpretation φ of $\mathcal{L}_0$ the set $\mathcal{H}_\varphi$ of all formulas $H \in \mathcal{L}_0$ such that $\varphi(H) = 1$ is a maximal finitely consistent set; moreover, the set $\mathcal{H}_\varphi$ is even consistent in $\mathcal{L}_0$.

PROOF The set $\mathcal{H}_\varphi$ is consistent, hence finitely consistent, because φ is obviously a model for $\mathcal{H}_\varphi$. Let us prove that $\mathcal{H}_\varphi$ is a maximal finitely consistent set. In order to do this, consider a finitely consistent set $\mathcal{K}$ such that $\mathcal{K} \supseteq \mathcal{H}_\varphi$ and take an arbitrary formula $F \in \mathcal{K}$. Suppose that $F \notin \mathcal{H}_\varphi$, hence $\varphi(F) = 0$; it follows that $\varphi(\neg F) = 1$ and consequently (by the definition of $\mathcal{H}_\varphi$), $\neg F \in \mathcal{H}$. It follows that $\neg F \in \mathcal{K}$ and (by hypothesis) $F \in \mathcal{K}$. But this is impossible because $\mathcal{K}$ is finitely consistent and the finite subset $\{F, \neg F\} \subset \mathcal{K}$ has no model. Therefore, $F \in \mathcal{H}_\varphi$ and $\mathcal{K} = \mathcal{H}_\varphi$. ∎

PROPOSITION 2.5

The set $\mathcal{H}$ in $\mathcal{L}_0$ is maximal finitely consistent if and only if it is finitely consistent and has the following property: if F is a formula in $\mathcal{L}_0$, then either $F \in \mathcal{H}$ or $\neg F \in \mathcal{H}$.

PROOF Suppose that $\mathcal{H}$ is a maximal finitely consistent set in $\mathcal{L}_0$ and take an arbitrary formula $F \in \mathcal{L}_0$. We will prove that either $\mathcal{K}' = \mathcal{H} \cup \{F\}$ or $\mathcal{K}'' = \mathcal{H} \cup \{\neg F\}$ is finitely consistent. If $\mathcal{K}'$ is finitely consistent the assertion is true. Suppose that $\mathcal{K}'$ is not finitely consistent. By definition, in this case, there is a finite subset $\mathcal{K}'_0 \subset \mathcal{K}'$, $\mathcal{K}'_0 = \{K'_1, K'_2, ..., K'_r, K'_{r+1}\}$, $(r \geq 0)$ which is inconsistent. This means that, for every $\varphi \in \mathcal{V}$, we have $\varphi(K'_1 \wedge K'_2 \wedge ... \wedge K'_{r+1}) = 0$. Taking into account that $\mathcal{H}$ is finitely consistent, the formula F necessarily belongs to $\mathcal{K}'_0$, say $K'_{r+1} \equiv F$. It follows that, for each $\varphi \in \mathcal{V}$, we have $\varphi(K'_1 \wedge K'_2 \wedge ... \wedge K'_r \wedge F) = 0$; in particular, $\varphi(F) = 0$ for each model φ of $\mathcal{K}'_0 \setminus \{F\} \subset \mathcal{H}$. Consequently, $\varphi(\neg F) = 1$ for each model φ of $\mathcal{K}'_0 \setminus \{F\}$.

Now take an arbitrary finite subset $\mathcal{K}''_0 \subset \mathcal{K}''$, $\mathcal{K}''_0 = \{K''_1, K''_2, ..., K''_s, K''_{s+1}\}$, $(s \geq 0)$; we have to prove that $\mathcal{K}''_0$ is consistent. If $\neg F \notin \mathcal{K}''_0$, then $\mathcal{K}''_0 \subset \mathcal{H}$ and $\mathcal{K}''_0$ is consistent. If $\neg F \in \mathcal{K}''_0$, then $\mathcal{K}''_0 \setminus \{\neg F\} \subset \mathcal{H}$; on the other hand, we also have $\mathcal{K}'_0 \setminus \{F\} \subset \mathcal{H}$. Consequently, the finite set $\mathcal{K}_0 = (\mathcal{K}'_0 \setminus \{F\}) \cup (\mathcal{K}''_0 \setminus \{\neg F\})$ is contained in $\mathcal{H}$, hence $\mathcal{K}_0$ has a model $\psi \in \mathcal{V}$. The interpretation ψ is obviously a model for $\mathcal{K}'_0 \setminus \{F\}$, hence, from the above considerations, $\psi(\neg F) = 1$; at the same time ψ is also a model for $\mathcal{K}''_0 \setminus \{\neg F\}$ and consequently ψ is a model for $\mathcal{K}''_0 = (\mathcal{K}''_0 \setminus \{\neg F\}) \cup \{\neg F\}$, i.e., $\mathcal{K}''_0$ is consistent. The finite set $\mathcal{K}''_0$ being an arbitrary finite subset of $\mathcal{K}''$, it follows that $\mathcal{K}''$ is finitely consistent.

So, we have proved that, for every $F \in \mathcal{L}_0$, either $\mathcal{K}' = \mathcal{H} \cup \{F\}$ or $\mathcal{K}'' = \mathcal{H} \cup \{\neg F\}$ is finitely consistent. Let us note that $\mathcal{K}' \supseteq \mathcal{H}$ and

$\mathcal{K}'' \supseteq \mathcal{H}$. If $\mathcal{K}'$ is finitely consistent, then $\mathcal{K}' = \mathcal{H}$ and $F \in \mathcal{H}$; if $\mathcal{K}''$ is finitely consistent, then $\mathcal{K}'' = \mathcal{H}$ and $\neg F \in \mathcal{H}$. ∎

PROPOSITION 2.6

Each maximal finitely consistent set $\mathcal{H}$ has a model; and this model is unique.

PROOF   Consider the Boolean interpretation φ of $\mathcal{L}_0$ which has the property: $\varphi(X) = 1$ for all propositional variables belonging to $\mathcal{H}$ and $\varphi(X) = 0$ for the propositional variables which do not belong to $\mathcal{H}$. By Proposition 2.1 such a Boolean interpretation φ exists and is unique.

We will prove by induction on the length $\lambda(F) = n$ $(n \geq 1)$ of the formula $F \in \mathcal{L}_0$, that

$$(*) \qquad F \in \mathcal{H} \text{ if and only if } \varphi(F) = 1.$$

If $\lambda(F) = 1$, then F coincides with a propositional variable and by the definition of φ, $F \in \mathcal{H}$ if and only if $\varphi(F) = 1$. Now suppose $n > 1$ and assume that, for all formulas $F' \in \mathcal{L}_0$, with $\lambda(F') < n$, the assertion $(*)$ is true. Taking a formula $F \in \mathcal{L}_0$, with $\lambda(F) = n$, we have to analyze two possibilities:
(a) $F \equiv \neg G$, where $G \in \mathcal{L}_0$ and $\lambda(G) < n$;
(b) $F \equiv (H) \vee (K)$, where $H, K \in \mathcal{L}_0$ and $\lambda(H) < n, \lambda(K) < n$.
In case (a), if $F \in \mathcal{H}$, then $G \notin \mathcal{H}$ because $\mathcal{H}$ is finitely consistent. From the inductive assumption it follows that $\varphi(G) = 0$, hence $\varphi(F) = \varphi(\neg G) = 1$. Conversely, if $\varphi(F) = 1$, then $\varphi(G) = 0$ and from the inductive assumption it follows that $G \notin \mathcal{H}$. The set $\mathcal{H}$ being maximal finitely consistent, it follows (from Proposition 2.5) that $\neg G \in \mathcal{H}$, hence $F \in \mathcal{H}$. So, in case (a), the assertion $(*)$ is proved.

In case (b), if $F \in \mathcal{H}$, then we will show that $\varphi(F) = 1$. Indeed, if we suppose that $\varphi(F) = 0$, then $\varphi(H) = \varphi(K) = 0$ and from the inductive assumption it follows that $H \notin \mathcal{H}$ and $K \notin \mathcal{H}$. The set $\mathcal{H}$ being maximal finitely consistent, by Proposition 2.5 we get $\neg H \in \mathcal{H}$ and $\neg K \in \mathcal{H}$. Therefore the finite subset $\{H \vee K, \neg H, \neg K\}$ of $\mathcal{H}$ must be consistent which is impossible; it follows that $\varphi(F) = 1$. Conversely, suppose that $\varphi(F) = 1$, hence $\varphi(H) = 1$ or $\varphi(K) = 1$. From the inductive assumption it follows $H \in \mathcal{H}$ or $K \in \mathcal{H}$. If we suppose that $F \notin \mathcal{H}$, then from the maximality of $\mathcal{H}$ it follows that $\neg F \in \mathcal{H}$. Consequently, at least one of the subsets $\{\neg F, H\}$, $\{\neg F, K\}$ must be consistent which is impossible. Therefore, $F \in \mathcal{H}$. So, the assertion $(*)$ is true in the case (b), too. Now applying the principle of mathematical induction we deduce that $F \in \mathcal{H}$ if and only if $\varphi(F) = 1$. In particular, this means that φ is a model for $\mathcal{H}$. In order to prove that the model φ is unique, let us consider an arbitrary model $\psi \in \mathcal{V}$ of $\mathcal{H}$. This means that $\psi(F) = 1$ for all $F \in \mathcal{H}$. On the other hand, by Proposition

2.4, the set $\mathcal{H}_\psi$ of all formulas in $\mathcal{L}_0$, with $\psi(F) = 1$, is finitely consistent and obviously contains the set $\mathcal{H}$. From the maximality of $\mathcal{H}$ it follows that $\mathcal{H}_\psi = \mathcal{H}$, hence $\psi(F) = 1$ if and only if $\varphi(F) = 1$; therefore $\psi = \varphi$ and the uniqueness of the model is proved. ∎

PROPOSITION 2.7
(The compactness theorem of propositional logic)
Each finitely consistent set $\mathcal{H}$ in $\mathcal{L}_0$ has a model, hence is consistent.

PROOF   It is not difficult to see that the set of all formulas in $\mathcal{L}_0$ is countable and consequently it can be represented (in many ways) as an infinite sequence:

$$(*) \quad F_1, F_2, ..., F_n, ...$$

We will define inductively an infinite sequence of sets in $\mathcal{L}_0$,

$$(**) \quad \mathcal{H}_0 \subseteq \mathcal{H}_1 \subseteq ... \subseteq \mathcal{H}_n \subseteq ...$$

by the following definition:

$$\mathcal{H}_0 = \mathcal{H},$$

and supposing that $\mathcal{H}_0, \mathcal{H}_1, ..., \mathcal{H}_{n-1}$, $(n \geq 1)$ are already defined, we put

$$\mathcal{H}_n = \begin{cases} \mathcal{H}_{n-1} \cup \{F_n\}, & \text{if } \mathcal{H}_{n-1} \cup \{F_n\} \text{ is finitely consistent,} \\ \mathcal{H}_{n-1} \cup \{\neg F_n\}, & \text{if } \mathcal{H}_{n-1} \cup \{F_n\} \text{ is not finitely consistent.} \end{cases}$$

Let us prove by induction that the sets in $(**)$ are finitely consistent.

For $n = 0$ we have $\mathcal{H}_0 = \mathcal{H}$, hence $\mathcal{H}_0$ is finitely consistent by hypothesis. Now suppose $n > 1$ and assume that $\mathcal{H}_0, \mathcal{H}_1, ..., \mathcal{H}_{n-1}$ are finitely consistent. If $\mathcal{H}_{n-1} \cup \{F_n\}$ is finitely consistent, then $\mathcal{H}_n = \mathcal{H}_{n-1} \cup \{F_n\}$ is finitely consistent. If $\mathcal{H}_{n-1} \cup \{F_n\}$ is not finitely consistent, then there is a finite subset $\mathcal{K}_0' = \{K_1', K_2', ..., K_r', K_{r+1}'\}$ of $\mathcal{H}_{n-1} \cup \{F_n\}$ $(r \geq 0)$ which is inconsistent. The set $\mathcal{H}_{n-1}$ being finitely consistent (by the inductive assumption), the formula F_n belongs necessarily to $\mathcal{K}_0'$, say $K_{r+1}' \equiv F$. From the inconsistency of $\mathcal{K}_0'$ it follows that $\varphi(K_1' \wedge K_2' \wedge ... \wedge K_r' \wedge F_n) = 0$ for each interpretation $\varphi \in \mathcal{V}$. In particular, for each model $\varphi \in \mathcal{V}$ of $\mathcal{K}_0' \setminus \{F_n\}$ we have $\varphi(F_n) = 0$. Let us prove now that $\mathcal{H}_{n-1} \cup \{\neg F_n\}$ is finitely consistent. For this purpose, take an arbitrary finite subset $\mathcal{K}_0''$ of $\mathcal{H}_n = \mathcal{H}_{n-1} \cup \{\neg F_n\}$ and let us show that $\mathcal{K}_0''$ is consistent. If $\neg F_n$ does not belong to $\mathcal{K}_0''$, then $\mathcal{K}_0'' \subset \mathcal{H}_{n-1}$ and by the inductive assumption $\mathcal{K}_0''$ is consistent. If $\neg F_n$ belongs to $\mathcal{K}_0''$, then $\mathcal{K}_0'' \setminus \{\neg F_n\}$ is consistent. It follows that the set $\mathcal{K}_0 = (\mathcal{K}_0' \setminus \{F_n\}) \cup (\mathcal{K}_0'' \setminus \{\neg F_n\})$ is contained in $\mathcal{H}_{n-1}$, hence is consistent; therefore, there is a model $\varphi \in \mathcal{V}$ for $\mathcal{K}_0$, hence φ is a model for $\mathcal{K}_0' \setminus \{F_n\}$. From the above considerations we have $\varphi(F_n) = 0$ and $\varphi(\neg F_n) = 1$. On the other hand, the interpretation φ is a model for

$\mathcal{K}_0'' \setminus \{\neg F_n\}$, too. Consequently, φ is a model for $(\mathcal{K}_0'' \setminus \{\neg F_n\}) \cup (\neg F_n) = \mathcal{K}_0''$, hence $\mathcal{K}_0''$ is a consistent set. So, we have proved that if $\mathcal{H}_{n-1} \cup \{F_n\}$ is not finitely consistent, then the set $\mathcal{H}_n = \mathcal{H}_{n-1} \cup \{\neg F_n\}$ is finitely consistent. From the induction principle we deduce that all sets $\mathcal{H}_n$ ($n \geq 0$) are finitely consistent. Let us consider the set $\mathcal{K} = \bigcup_{n \geq 0} \mathcal{H}_n$; taking into account that each finite subset of $\mathcal{K}$ is contained in some set $\mathcal{H}_n$ for a sufficiently large index n and that all sets $\mathcal{H}_n$ are finitely consistent, it follows that $\mathcal{K}$ is finitely consistent. We will prove that $\mathcal{K}$ is a maximal finitely consistent set. In order to do this, consider an arbitrary formula $F_1 \mathcal{L}_0$; this formula belongs necessarily to the sequence $(*)$, hence $F \equiv F_n$ for some $n \geq 1$. If $\mathcal{H}_n = \mathcal{H}_{n-1} \cup \{F_n\}$, then $F \equiv F_n \in \mathcal{H}_n$; if $\mathcal{H}_n = \mathcal{H}_{n-1} \cup \{\neg F_n\}$, then $\neg F \equiv \neg F_n \in \mathcal{H}_n$. Therefore, for each formula F in $\mathcal{L}_0$ we have either $F \in \mathcal{K}$ or $\neg F \in \mathcal{K}$. From Proposition 2.5 the set $\mathcal{K}$ is a maximal finitely consistent set and by Proposition 2.6 $\mathcal{K}$ has a model; therefore, $\mathcal{K}$ is consistent. ∎

REMARK 2.5   From the compactness theorem it follows that a set $\mathcal{H}$ is consistent if and only if each finite subset of $\mathcal{H}$ is consistent. In other words, the set $\mathcal{H}$ is inconsistent if and only if it contains a finite inconsistent set. We remark here a certain analogy with the compactness property in topological spaces. This analogy has deep reasons and the reader can find more details about it in the book by J. L. Bell and M. Machover (*A Course in Mathematical Logic* edited in 1977 by North-Holland publishers, Amsterdam). ∎

2.2   Semantic Deduction

Let $\mathcal{H}$ be an arbitrary (finite or not) and possibly empty, subset of formulas in $\mathcal{L}_0$.

DEFINITION 2.9   *A formula* $F \in \mathcal{L}_0$ *will be called* **a semantic consequence** *or* **a logical consequence** *of* $\mathcal{H}$ *and we shall write*

$$\mathcal{H} \models F$$

if F is φ-true in all interpretations $\varphi \in \mathcal{V}$ for which all formulas belonging to $\mathcal{H}$ are φ-true. In other words, $\mathcal{H} \models F$ means that

$$\varphi(H) = 1 \ \text{for all} \ \ H \in \mathcal{H} \ \ \text{implies} \ \ \varphi(F) = 1.$$

In particular, F is a semantic consequence (or logical consequence) of $\emptyset$ and *we shall write*

$$\emptyset \models F.$$

if F is φ-true in all interpretations $\varphi \in \mathcal{V}$. In other words, $\emptyset \models F$ means that

$$\varphi(F) = 1, \text{ for all } \varphi \in \mathcal{V}.$$

The formulas H belonging to $\mathcal{H}$ will be called **hypotheses**.

REMARK 2.6

(1) If $\mathcal{H}$ is a finite nonempty set,

$$\mathcal{H} = \{H_1, H_2, ..., H_p\},$$

then instead of $\mathcal{H} \models F$ we shall often write

$$H_1, H_2, ..., H_p \models F$$

In particular, if $\mathcal{H}$ is reduced to a single formula H, ,i.e., $\mathcal{H} = \{H\}$, then instead of $\mathcal{H} \models F$ we shall also write

$$H \models F.$$

The assertion $\mathcal{H} \cup \mathcal{K} \models F$ will be also denoted by

$$\mathcal{H}, \mathcal{K} \models F.$$

In particular, if $\mathcal{K}$ consists of a single element K, i.e., $\mathcal{K} = \{K\}$, then instead of $\mathcal{H}, \{K\} \models F$ we shall often write

$$\mathcal{H}, K \models F.$$

It is also clear that for any two subsets $\mathcal{H}_1, \mathcal{H}_2$ of hypotheses with $\mathcal{H}_1 \subseteq \mathcal{H}_2$ one deduces that

$$\text{if } \mathcal{H}_1 \models F, \text{ then } \mathcal{H}_2 \models F;$$

in particular, if $\models F$, then for any $\mathcal{H}$ one also has $\mathcal{H} \models F$.

(2) It is obvious, from Definition 2.9, that the notation $\emptyset \models F$ means that F is a tautology, hence $\models F$. Consequently, instead of $\emptyset \models F$ we shall usually write $\models F$.

(3) The set of all Boolean interpretations of $\mathcal{L}_0$ was denoted by $\mathcal{V}$ and the Boolean interpretations themselves were denoted by φ. In Proposition 2.1 we proved that there exists a bijective correspondence $\alpha \mapsto \varphi_\alpha$ from Γ^N to $\mathcal{V}$ such that for any $F \in \mathcal{L}_0$, $\varphi_\alpha(F) = \Phi(F)(\alpha)$. Therefore, for each $\varphi \in \mathcal{V}$ there is some (unique) $\alpha \in \Gamma^N$ such that $\varphi = \varphi_\alpha$. For this reason, in what follows the Boolean interpretations of $\mathcal{L}_0$ will be denoted equivalently by φ or by φ_α depending on our needs. ∎

The set of all semantic consequences of $\mathcal{H}$, i.e., the set of all $F \in \mathcal{L}_0$ such that $\mathcal{H} \models F$, will be denoted by $C^{sem}(\mathcal{H})$; that is,

$$C^{sem}(\mathcal{H}) = \{F \in \mathcal{L}_0 \mid \mathcal{H} \models F\}.$$

It follows that the assertion $\mathcal{H} \models F$ is equivalent to the assertion that $F \in C^{sem}(\mathcal{H})$.

If $\mathfrak{S}$ is the set of all tautologies, then it is clear that

$$C^{sem}(\emptyset) = \mathfrak{S}.$$

In Definition 2.9 the set $\mathcal{H}$ was supposed to be an arbitrary finite or infinite subset of $\mathcal{L}_0$; we shall show that, in this definition, the set $\mathcal{H}$ can always be supposed to be finite. In order to do this we shall first introduce a binary relation in $\mathcal{L}_0$ which will also be useful in other cases.

DEFINITION 2.10   *The formulas $F, G \in \mathcal{L}_0$ are said to be* **logically** *or* **semantically equivalent** *and we write*

$$F \approx G$$

if

$$\varphi(F) = \varphi(G) \quad \text{for all} \quad \varphi \in V$$

(or, which is the same, $\varphi_\alpha(F) = \varphi_\alpha(G)$ for all $\alpha \in \Gamma^N$).

The following three propositions concerning the binary relation "$\approx$" are almost obvious.

PROPOSITION 2.8
 (a) $F \approx G$ iff $\Phi(F) = \Phi(G)$;   (b) $F \approx G$ iff $\models F \Leftrightarrow G$.

PROPOSITION 2.9

(a) *The binary relation "$\approx$" is an equivalence relation in the set $\mathcal{L}_0$; this means that for any formulas $F, G \in \mathcal{L}_0$ the following assertions hold: (i) $F \approx F$;   (ii) if $F \approx G$, then $G \approx F$;   (iii) if $F \approx G$ and $G \approx H$, then $F \approx H$. It follows that the relation "$\approx$" induces a decomposition of $\mathcal{L}_0$ in disjoint and nonempty classes of equivalent formulas. The set $\mathcal{L}_0/_\approx$ of all these classes will be denoted by $\hat{\mathcal{L}}_0$;*

(b) *If $\xi \in \hat{\mathcal{L}}_0$ is a class of equivalent formulas, then there exists a unique function $f_\xi \in B(N, \Gamma)$ such that $\Phi(F) = f_\xi$ for all $F \in \xi$;*

(c) *The correspondence $\xi \longmapsto f_\xi$ is an injective and surjective (hence bijective) mapping from $\hat{\mathcal{L}}_0$ onto $B(N, \Gamma)$.*

PROPOSITION 2.10
If $H \approx K$ and $F \approx G$, then $H \models F$ if and only if $K \models G$.

Now we are ready to prove that the set $\mathcal{H}$ can be considered finite.

PROPOSITION 2.11
For each infinite subset $\mathcal{H}$ of $\mathcal{L}_0$ there is some finite subset $\mathcal{H}_0$ of $\mathcal{H}$ such that

$$\mathcal{H} \models F \quad \text{if and only if} \quad \mathcal{H}_0 \models F;$$

therefore, $\quad \mathcal{C}^{sem}(\mathcal{H}) = \mathcal{C}^{sem}(\mathcal{H}_0).$

PROOF   Each formula $H \in \mathcal{H}$ belongs to some (unique) class $\xi_H \in \hat{\mathcal{L}}_0$. Let us denote by $\hat{\mathcal{H}}$ the image of $\mathcal{H}$ in $\hat{\mathcal{L}}_0$ by the mapping $F \mapsto \xi_F$, where $\xi_F \ni F$. In other words, $\hat{\mathcal{H}}$ is the set of all classes ξ_H with $H \in \mathcal{H}$. Taking into account that the mapping $\xi \mapsto f_\xi : \hat{\mathcal{L}}_0 \to B(N, \Gamma)$ is bijective it follows that $\hat{\mathcal{L}}_0$ is a finite set ($B(N, \Gamma)$ being finite), resulting that the subset $\hat{\mathcal{H}}$ of $\hat{\mathcal{L}}_0$ is also finite. Now, in each class $\xi \in \hat{\mathcal{H}}$, choose an arbitrary (but fixed) formula $H_\xi \in \xi \cap \mathcal{H}$ and denote by $\mathcal{H}_0$ the set of all these formulas H_ξ. It is clear that the correspondence $\xi \mapsto H_\xi$ is a bijective mapping from $\hat{\mathcal{H}}$ onto $\mathcal{H}_0$. It follows that $\mathcal{H}_0$ is a finite subset of $\mathcal{H}$ having the property that for each $H \in \mathcal{H}$ there exists a unique $H_0 \in \mathcal{H}_0$ such that $H \approx H_0$. We shall show that $\mathcal{H}_0$ is the requested finite subset of $\mathcal{H}$, i.e.,

$$\mathcal{H} \models F \quad \text{if and only if} \quad \mathcal{H}_0 \models F.$$

It is obvious that if $\mathcal{H}_0 \models F$, then, taking into account that $\mathcal{H}_0 \subseteq \mathcal{H}$, one has $\mathcal{H} \models F$. For the converse implication, suppose that $\mathcal{H} \models F$ and let $\varphi \in \mathcal{V}$ be such that $\varphi(H) = 1$ for all $H \in \mathcal{H}_0$. We have to prove that $\varphi(F) = 1$. For this purpose consider an arbitrary formula $H \in \mathcal{H}$. From the definition of $\mathcal{H}_0$ there exists $H_0 \in \mathcal{H}_0$ such that $H_0 \approx H$, hence $\varphi(H) = \varphi(H_0) = 1$. It follows that $\varphi(H) = 1$ for all $H \in \mathcal{H}$. Taking into account the assumption that $\mathcal{H} \models F$ it follows that $\varphi(F) = 1$. Therefore, $\mathcal{H}_0 \models F$.   ∎

On the basis of the above proposition, in the following we shall make exclusive use of finite and possibly empty sets of hypotheses.

PROPOSITION 2.12
Let $\mathcal{H}$ be a nonempty set of hypotheses $\mathcal{H} = \{H_1, H_2, ..., H_p\}$ and denote $H_1 \wedge H_2 \wedge ... \wedge H_p \equiv H$. A formula $F \in \mathcal{L}_0$ is a semantic consequence of $\mathcal{H}$

if and only if it is a semantic consequence of H; in other words,

$$\mathcal{H} \models F \quad \text{if and only if} \quad H \models F.$$

PROOF   Suppose that $\mathcal{H} \models F$. This means that $\varphi(F) = 1$ for all $\varphi \in \mathcal{V}$ such that $\varphi(H_1) = \varphi(H_2) = ... = \varphi(H_p) = 1$. Now consider an interpretation $\varphi \in \mathcal{V}$ with $\varphi(H) = 1$. It follows that $1 = \varphi(H) = \varphi(H_1) \wedge \varphi(H_2) \wedge ... \wedge \varphi(H_p)$. Taking into account that the operation "$\wedge$" in Γ means "min" we obtain $\varphi(H_1) = \varphi(H_2) = ... = \varphi(H_p) = 1$, thus $\varphi(F) = 1$ (because $\mathcal{H} \models F$). Therefore, $\varphi(F) = 1$ for any $\varphi \in \mathcal{V}$ with $\varphi(H) = 1$, hence $H \models F$.

Conversely, suppose that $H \models F$ and take $\varphi \in \mathcal{V}$ with $\varphi(H_1) = \varphi(H_2) = ... = \varphi(H_p) = 1$. We obtain $\varphi(H) = \varphi(H_1) \wedge \varphi(H_2) \wedge ... \wedge \varphi(H_p) = 1$, hence, by assumption, $\varphi(F) = 1$. Therefore, for any $\varphi \in \mathcal{V}$ with $\varphi(H_1) = \varphi(H_2) = ... = \varphi(H_p) = 1$, it follows $\varphi(F) = 1$; this means that $\mathcal{H} \models F$, which concludes the proof.   ∎

PROPOSITION 2.13
Keeping the notations adopted in Proposition 2.12

$$H \models F \quad \text{if and only if} \quad \Phi(H) \le \Phi(F).$$

PROOF   Suppose that $H \models F$. This means $\varphi_\alpha(F) = 1$ for all φ_α such that $\varphi_\alpha(H) = 1$. On the other hand, by Proposition 2.2, we know that $\Phi(F)(\alpha) = \varphi_\alpha(F)$ and $\Phi(H)(\alpha) = \varphi_\alpha(H)$. Thus,

$$\Phi(H)(\alpha) = \Phi(F)(\alpha) = 1$$

for all α such that $\varphi_\alpha(H) = 1$. For the other $\alpha \in \Gamma^N$ we have $\varphi_\alpha(H) = 0$, hence $\Phi(H)(\alpha) = 0$. It follows that $\Phi(H)(\alpha) \le \Phi(F)(\alpha)$, for all $\alpha \in \Gamma^N$, that is $\Phi(H) \le \Phi(F)$.

Conversely, suppose that $\Phi(H) \le \Phi(F)$. Taking into account the definition of Φ one obtains

$$\varphi_\alpha(H) = \Phi(H)(\alpha) \le \Phi(F)(\alpha) = \varphi_\alpha(F),$$

for all $\alpha \in \Gamma^N$. From the inequality $\varphi_\alpha(H) \le \varphi_\alpha(F)$ it follows that for all φ_α with $\varphi_\alpha(H) = 1$ we have $\varphi_\alpha(F) = 1$. This means that $H \models F$.   ∎

The concept of semantic consequence can be presented in an equivalent form which is similar to intuitive deduction.

DEFINITION 2.11   *Let $\mathcal{H}$ be an arbitrary finite set, possibly empty of formulas in $\mathcal{L}_0$, called* **set of hypotheses.** *The finite sequence of formulas*

$$(d) \; : \quad F_1, F_2, ..., F_r \quad (r \geq 1)$$

is called a **semantic deduction from** $\mathcal{H}$ *if for each index i $(1 \leq i \leq r)$ the formula F_i satisfies one of the conditions:*

(i) F_i is a tautology;

(ii) F_i is a hypothesis $(F_i \in \mathcal{H})$;

(iii) there are indices h, k $(1 \leq h, k < i)$ such that $F_k \equiv (F_h) \Rightarrow (F_i)$.

The sequence (d) will also be called a semantic deduction of its last term F_r. Any formula F which can be inserted in a sequence (d) will be called **semantically deducible from** $\mathcal{H}$.

REMARK 2.7   It is clear that in any semantic deduction from $\mathcal{H}$ the first term is necessarily a tautology or an element of $\mathcal{H}$. The condition (iii) represents the inference rule called "modus ponens" (the MP-rule). This rule states that if $F_h \equiv A$, $F_k \equiv A \Rightarrow B$, then for any $i > h$, k one can put $F_i \equiv B$. It is also clear that from two semantic deductions one can obtain new semantic deductions by "mixing" the terms of the initial deductions provided the mutual order in each of the given sequences remains unchanged (as in the case of formal constructions).   ∎

PROPOSITION 2.14
    Let $\mathcal{H}$ be an arbitrary finite set (possibly empty) of formulas in $\mathcal{L}_0$. A formula F in $\mathcal{L}_0$ is a semantic consequence of $\mathcal{H}$, i.e.,

$$\mathcal{H} \models F$$

if and only if there exists a semantic deduction from $\mathcal{H}$

$$(d) : \quad F_1, F_2, ..., F_r \quad (r \geq 1)$$

such that $F \equiv F_r$.

PROOF   Let be $\mathcal{H} = \{H_1, H_2, ...H_p\}$. For the "if" part, suppose that there is a semantic deduction from $\mathcal{H}$

$$(*) \quad F_1, F_2, ..., F_r$$

such that $F_r \equiv F$. We shall prove that in this case we have $\mathcal{H} \models F$. Taking into account Proposition 2.13 we have to prove that $\Phi(H) \leq \Phi(F)$, where

$H = H_1 \wedge H_2 \wedge ... \wedge H_p$. We shall proceed by induction, proving that for all i $(1 \leq i \leq r)$

$$\Phi(H) \leq \Phi(F_i).$$

Let $i = 1$. The formula F_1 can be a tautology or an element of $\mathcal{H}$. If F_1 is a tautology, then $\Phi(F_1) = 1 \geq \Phi(H)$; if $F_1 \in \mathcal{H}$, say $F_1 \equiv H_l$, then $\Phi(H) = \Phi(H_1 \wedge H_2 \wedge ... \wedge H_p) = \Phi(H_1) \wedge \Phi(H_2) \wedge ... \wedge \Phi(H_p) \leq \Phi(H_l) = \Phi(F_1)$. Therefore, $\Phi(H) \leq \Phi(F_1)$.

Let $i > 1$ and assume that

$$\Phi(H) \leq \Phi(F_1), ..., \Phi(H) \leq \Phi(F_{i-1}).$$

If F_i is a tautology or a member of $\mathcal{H}$, then as in case $i = 1$ it follows that $\Phi(H) \leq \Phi(F_i)$. Now suppose that there are indices $h, k < i$ such that F_i is obtained by MP from F_h and F_k. In other words, $F_h \equiv A$, $F_k \equiv A \Rightarrow B$ and $F_i \equiv B$. By the inductive assumption

$\Phi(H) \leq \Phi(A)$;

$\Phi(H) \leq \Phi(A \Rightarrow B) = \overline{\Phi(A)} \vee \Phi(B)$.

Denoting $\Phi(H) = h, \Phi(A) = a, \Phi(B) = b$ we have $h, a, b \in B(N, \Gamma)$ and the above inequalities become $h \leq a$; $h \leq \bar{a} \vee b$. Taking into account that $B(N, \Gamma)$ is a Boolean algebra, we get

$$h \leq a \wedge (\bar{a} \vee b) = (a \wedge \bar{a}) \vee (a \wedge b) = 0 \vee (a \wedge b) = a \wedge b \leq b.$$

Therefore, $\Phi(H) \leq \Phi(F_i)$. From the principle of mathematical induction we obtain

$$\Phi(H) \leq \Phi(F_i) \quad \text{for all} \quad i \quad (1 \leq i \leq r).$$

In particular, $\Phi(H) \leq \Phi(F_r) = \Phi(F)$. This means that $\mathcal{H} \models F$.

For the "only if" part, consider conversely that $\mathcal{H} \models F$ and prove that for F there is some semantic deduction from $\mathcal{H}$. By Proposition 2.9, $\Phi(H) \leq \Phi(F)$. In order to simplify the writing let us denote $\Phi(H_1) = h_1$, $\Phi(H_2) = h_2, ...,$ $\Phi(H_p) = h_p$, $\Phi(H) = h$ and $\Phi(F) = f$. Our assumption becomes $h \leq f$. Let us now define the function $f_0 \in B(N, \Gamma)$ as follows:

$$f_0(x_1, x_2, ..., x_N) = \begin{cases} 0 & \text{if } f(x_1, x_2, ..., x_N) = 0 \\ h(x_1, x_2, ..., x_N) & \text{if } f(x_1, x_2, ..., x_N) = 1. \end{cases}$$

Taking into account that $h \leq f$ it is easy to see that

$$f(x_1, x_2, ..., x_N) \wedge f_0(x_1, x_2, ..., x_N) = h(x_1, x_2, ..., x_N)$$

for all $(x_1, x_2, ..., x_N) \in \Gamma^N$ hence

$$(**) \qquad\qquad f \wedge f_0 = h.$$

The mapping $\Phi : \mathcal{L}_0 \rightarrow B(N, \Gamma)$ being surjective there exists $F_0 \in \mathcal{L}_0$ such that $\Phi(F_0) = f_0$. Therefore,

$$\Phi(F \wedge F_0) = \Phi(F) \wedge \Phi(F_0) = f \wedge f_0 = h = \Phi(H)$$

hence $\models (F \wedge F_0) \Leftrightarrow H$, in particular $\models H \Rightarrow (F \wedge F_0)$. Consider now the following semantic deduction from H:

(1) H ; hypothesis;

(2) $H \Rightarrow (F \wedge F_0)$; tautology;

(3) $F \wedge F_0; (MP) : (1), (2)$;

(4) $F \wedge F_0 \Rightarrow F$; tautology;

(5) $F; (MP) : (3), (4)$.

It follows that F is semantically deducible from H, hence from $\mathcal{H}$. Consequently, the converse assertion is also proved. In the case $\mathcal{H} = \emptyset$ we have $\models F$ and there is nothing to prove. ∎

REMARK 2.8   If $F_1, F_2, ..., F_r \equiv F$ is a semantic deduction from $\mathcal{H}$, we shall also say that the sequence $F_1, F_2, ...F_r$ is a semantic deduction of F from $\mathcal{H}$. The above proposition states that F is a semantic consequence of $\mathcal{H}$ if and only if there is some semantic deduction of F from $\mathcal{H}$. ∎

We shall prove now an additional property of semantic deduction.

METATHEOREM 2.1 (Metatheorem of semantic deduction)
Let $F, G \in \mathcal{L}_0$ and $\mathcal{H}$ be an arbitrary finite set of hypotheses in $\mathcal{L}_0$. With these notations,
$$\mathcal{H}, F \models G \quad \text{if and only if} \quad \mathcal{H} \models F \Rightarrow G.$$

PROOF   Consider first the case $\mathcal{H} \neq \emptyset$, $\mathcal{H} = \{H_1, H_2, ...H_p\}$ and put $H = H_1 \wedge H_2 \wedge ... \wedge H_p$. Suppose that $\mathcal{H}, F \models G$ that is

$$(*) \quad H, F \models G.$$

By Proposition 2.13, $(*)$ means that $\Phi(H \wedge F) \leq \Phi(G)$ or $\Phi(H) \wedge \Phi(F) \leq \Phi(G)$. It follows that

$$\Phi(F \Rightarrow G) = \overline{\Phi(F)} \vee \Phi(G) \geq \overline{\Phi(F)} \vee (\Phi(H) \wedge \Phi(F)) =$$

$$= (\overline{\Phi(F)} \vee \Phi(H)) \wedge (\overline{\Phi(F)} \vee \Phi(F)) = \overline{\Phi(F)} \vee \Phi(H) \geq \Phi(H),$$

hence
$$\Phi(H) \leq \Phi(F \Rightarrow G)$$

therefore, $H \models F \Rightarrow G$ (or $\mathcal{H} \models F \Rightarrow G$).

Conversely, suppose that $\mathcal{H} \models F \Rightarrow G$ or, which is the same, $H \models F \Rightarrow G$. By Proposition 2.9, this means that $\Phi(H) \leq \Phi(F \Rightarrow G)$ or

$$\Phi(H) \leq \overline{\Phi(F)} \vee \Phi(G).$$

From this inequality we can obtain successively:

$$\Phi(H \wedge F) = \Phi(H) \wedge \Phi(F) \leq (\overline{\Phi(F)} \vee \Phi(G)) \wedge \Phi(F) =$$

$$= (\overline{\Phi(F)} \wedge \Phi(F)) \vee (\Phi(G) \wedge \Phi(F)) = \Phi(G) \wedge \Phi(F) \leq \Phi(G).$$

Therefore, $\Phi(H \wedge F) \leq \Phi(G)$, hence $H, F \models G$ or $\mathcal{H}, F \models G$; our metatheorem is now completely proved. ∎

REMARK 2.9 The metatheorem of semantic deduction has a remarkable analogue in the syntactic version (Chapter 3) called the metatheorem of syntactic deduction. The proof of this last result is less trivial and is due to J. Herbrand. ∎

Let us recall that for any Boolean algebra B (in particular for Γ and $B(N, \Gamma)$) we have introduced the following notation: if $x \in B$ and $\sigma \in \Gamma$, then

$$x^\sigma = \begin{cases} x \,, & \text{if } \sigma = 1 \\ \bar{x} \,, & \text{if } \sigma = 0. \end{cases}$$

In a similar way for $F \in \mathcal{L}_0$ and $\sigma \in \Gamma$ we introduce the notation :

$$F^\sigma = \begin{cases} F \,, & \text{if } \sigma = 1 \\ \neg(F) \,, & \text{if } \sigma = 0. \end{cases}$$

It is easy to see that $\Phi(F^\sigma) = \Phi(F)^\sigma$.

PROPOSITION 2.15
Let $H_1, H_2 \in \mathcal{L}_0$ and let $\varphi \in \mathcal{V}$ be an arbitrary Boolean interpretation of $\mathcal{L}_0$.

$$\text{If } \models H_1 \Leftrightarrow H_2, \text{ then } \models H_1^{\varphi(H_1)} \Leftrightarrow H_2^{\varphi(H_2)}.$$

PROOF Suppose that $\models H_1 \Leftrightarrow H_2$. From Proposition 2.8 it follows that $H_1 \approx H_2$ hence $\Phi(H_1) = \Phi(H_2)$. If $\varphi_\alpha \in \mathcal{V}$, then

$$\varphi_\alpha(H_1) = \Phi(H_1)(\alpha) = \Phi(H_2)(\alpha) = \varphi_\alpha(H_2);$$

hence,

$$\Phi(H_1^{\varphi(H_1)}) = \Phi(H_1^{\varphi(H_2)}) = \Phi(H_1)^{\varphi(H_2)} = \Phi(H_2)^{\varphi(H_2)} = \Phi(H_2^{\varphi(H_2)}).$$

This means that $H_1^{\varphi(H_1)} \approx H_2^{\varphi(H_2)}$ or, by the same proposition, that

$$\models H_1^{\varphi(H_1)} \Leftrightarrow H_2^{\varphi(H_2)}. \quad \blacksquare$$

REMARK 2.10   The converse of Proposition 2.15 is not true. Indeed, let Δ_1, Δ_2 be two subsets of Γ^N such that $\Delta_1 \neq \emptyset, \Delta_2 \neq \emptyset, \Delta_1 \cap \Delta_2 = \emptyset$ and $\Delta_1 \cup \Delta_2 = \Gamma^N$ and consider the functions $h_1, h_2 \in B(N, \Gamma)$ defined as follows: $h_2 = \bar{h}_1$ and

$$h_1(x_1, x_2, ..., x_N) = \begin{cases} 1, & \text{if } (x_1, x_2, ..., x_N) \in \Delta_1, \\ 0, & \text{if } (x_1, x_2, ..., x_N) \in \Delta_2; \end{cases}$$

it follows that $h_1 \neq h_2$. It is also easy to verify that

$$(*) \quad h_1^{h_1(\beta)} = h_2^{h_2(\beta)}, \quad \text{for all} \quad \beta \in \Gamma^N.$$

Now let $H_1, H_2 \in \mathcal{L}_0$ be two formulas such that $\Phi(H_1) = h_1, \Phi(H_2) = h_2$ and consider an arbitrary Boolean interpretation φ_β of $\mathcal{L}_0$. It follows successively:

$$\varphi_\beta(H_1) = \Phi(H_1)(\beta) = h_1(\beta), \quad \varphi_\beta(H_2) = \Phi(H_2)(\beta) = h_2(\beta)$$

therefore, for any fixed $\beta \in \Gamma^N$,

$$\Phi(H_1^{\varphi_\beta(H_1)}) = (\Phi(H_1))^{\varphi_\beta(H_1)} = h_1^{h_1(\beta)}$$

and

$$\Phi(H_2^{\varphi_\beta(H_2)}) = (\Phi(H_2))^{\varphi_\beta(H_2)} = h_2^{h_2(\beta)}.$$

The functions $h_1^{h_1(\beta)}, h_2^{h_2(\beta)} \in B(N, \Gamma)$ are equal because of $(*)$, hence

$$H_1^{\varphi_\beta(H_1)} \approx H_2^{\varphi_\beta(H_2)}, \quad \text{for all} \quad \beta \in \Gamma^N.$$

In other words,

$$\models H_1^{\varphi(H_1)} \Leftrightarrow H_2^{\varphi(H_2)}, \quad \text{for all} \quad \varphi \in \mathcal{V}.$$

On the other hand, $\Phi(H_1) = h_1 \neq h_2 = \Phi(H_2)$, hence $H_1 \Leftrightarrow H_2$ is not a tautology. $\quad \blacksquare$

The last part of this section is dedicated to the description of the set $\mathcal{C}^{sem}(\mathcal{H})$ of all formulas F which are semantic consequences from $\mathcal{H}$. In what follows we shall consider only the case $\mathcal{H} \neq \emptyset, \mathcal{H} = \{H_1, H_2, ..., H_p\}$ and we shall denote $H_1 \wedge H_2 \wedge ... \wedge H_p \equiv H$. We can deduce that

$$C^{sem}(\mathcal{H}) = C^{sem}(H)$$

and

$$F \in C^{sem}(H) \text{ if and only if } \varphi(H) \leq \varphi(F) \text{ for all } \varphi \in \mathcal{V}.$$

Let φ be a given Boolean interpretation of $\mathcal{L}_0$ and let U_φ be the set of all formulas $F \in \mathcal{L}_0$ with $\varphi(H) \leq \varphi(F)$, i.e.,

$$U_\varphi = \{F \in \mathcal{L}_0 \mid \varphi(H) \leq \varphi(F)\}.$$

From the definition of $C^{sem}(H)$ we get

$$C^{sem}(H) = \bigcap_{\varphi \in \mathcal{V}} U_\varphi.$$

DEFINITION 2.12 *A subset U in $\mathcal{L}_0$ is said to be* **closed under the** *MP-rule or* **MP-closed** *if*

$$A \in U \text{ and } A \Rightarrow B \in U \text{ imply } B \in U.$$

The set of all subsets U of $\mathcal{L}_0$ having the properties

(i) U is MP-closed;

(ii) U contains all tautologies (briefly $U \supseteq \mathfrak{S}$);

(iii) $U \ni H$

will be denoted by $\mathcal{U}_\mathcal{H}$.

PROPOSITION 2.16

(a) If S is a subset in $\mathcal{U}_\mathcal{H}$, then $\bigcap_{U \in S} U \in \mathcal{U}_\mathcal{H}$;

(b) $U_\varphi \in \mathcal{U}_\mathcal{H}$ for all $\varphi \in \mathcal{V}$;

(c) $\bigcap_{\varphi \in \mathcal{V}} U_\varphi = \bigcap_{U \in \mathcal{U}_\mathcal{H}} U$;

(d) $C^{sem}(\mathcal{H}) = \bigcap_{U \in \mathcal{U}_\mathcal{H}} U$.

PROOF   The assertion (a) is obvious.

(b) Let φ be an arbitrary Boolean interpretation of $\mathcal{L}_0$. In order to prove that U_φ is MP-closed (condition (i)) let A and $A \Rightarrow B$ be formulas in U_φ. This means that

$$\varphi(H) \leq \varphi(A) \quad \text{and} \quad \varphi(H) \leq \varphi(A \Rightarrow B).$$

We have to show that $\varphi(H) \leq \varphi(B)$. From the above inequalities we deduce

$$\varphi(H) \leq \varphi(A) \wedge \varphi(A \Rightarrow B) = \varphi(A) \wedge (\overline{\varphi(A)} \vee \varphi(B)) =$$

$$= (\varphi(A) \wedge \overline{\varphi(A)}) \vee (\varphi(A) \wedge \varphi(B)) = \varphi(A) \wedge \varphi(B) \leq \varphi(B),$$

hence $B \in U_\varphi$. Therefore, U_φ is MP-closed (condition (i)). If $F \in \mathfrak{F}$, then $\varphi(F) = 1 \geq \varphi(H)$, hence $F \in U_\varphi$; therefore, $\mathfrak{F} \subset U_\varphi$ (condition (ii)). It is clear that condition (iii) is also fulfilled. So $U_\varphi \in \mathcal{U}_\mathcal{H}$ and assertion (b) is proved.

(c) The set $\mathcal{S}$ of all U_φ with $\varphi \in \mathcal{V}$ is a subset of $\mathcal{U}_\mathcal{H}$, hence

$$\bigcap_{\varphi \in \mathcal{V}} U_\varphi = \bigcap_{U \in \mathcal{S}} U \supseteq \bigcap_{U \in \mathcal{U}_\mathcal{H}} U.$$

It remains to prove the inverse inclusion.

Let $F \in \bigcap_{\varphi \in \mathcal{V}} U_\varphi = \mathcal{C}^{sem}(\mathcal{H})$; from Proposition 2.14 there exists some semantic deduction from H

$$F_1, F_2, ..., F_r \quad (r \geq 1)$$

such that $F_r \equiv F$. Now take an arbitrary element $U \in \mathcal{U}_\mathcal{H}$ and let us prove by (finite) induction that:

$$F_i \in U, \text{ for all } i \ (1 \leq i \leq r).$$

For $i = 1, F_1$ coincides with H or $F_1 \in \mathfrak{F}$, hence $F_1 \in U$. For $i > 1$ suppose that $F_1, ...F_2, ..., F_{i-1} \in U$. If $F_i \equiv H$ or $F_i \in \mathfrak{F}$, it follows as above that $F_i \in U$. If there are some $h, k < i$ such that $F_h \equiv A, F_k \equiv A \Rightarrow B$ and $F_i \equiv B$, then by the inductive assumption $A \in U$ and $A \Rightarrow B \in U$. But U is MP-closed, hence $F_i \equiv B \in U$. From the induction principle one gets that $F_i \in U$ for all i $(1 \leq i \leq r)$. In particular, $F \equiv F_r \in U$. Therefore, if $F \in \bigcap_{\varphi \in \mathcal{V}} U_\varphi$, then $F \in U$ for all $U \in \mathcal{U}_\mathcal{H}$, that is $F \in \bigcap_{U \in \mathcal{U}_\mathcal{H}} U$; so we get the second inclusion $\mathcal{C}^{sem}(\mathcal{H}) \subset \bigcap_{U \in \mathcal{U}_\mathcal{H}} U$ which together with the first one provides the requested equality (c).

The assertion (d) follows directly from the remarks preceding Proposition 2.16 and from (c).   ∎

Comments. Equality (d) means that

$$F \in C^{sem}(\mathcal{H}) \text{ if and only if } F \in \bigcap_{U \in \mathcal{U}_{\mathcal{H}}} U.$$

Recall that $\mathfrak{F}$ is the set of all tautologies, $\mathcal{H} = \{H_1, H_2, ... H_p\}$ and $H = H_1 \wedge H_2 \wedge ... \wedge H_p$. Furthermore, if we denote by χ_Z the characteristic function of any subset $Z \subset \mathcal{L}_0$, i.e.,

$$\chi_Z(F) = \begin{cases} 1, & \text{if } F \in Z \\ 0, & \text{if } F \notin Z, \end{cases}$$

then equality (d) from Proposition 2.16 can be expressed equivalently in terms of characteristic functions:

$$\chi_{C^{sem}(\mathcal{H})} = \bigwedge_{U \in \mathcal{U}_{\mathcal{H}}} \chi_U$$

where $\displaystyle\bigwedge_{U \in \mathcal{U}_{\mathcal{H}}} \chi_U : \mathcal{L}_0 \to \Gamma$ is the function defined by

$$(\bigwedge_{U \in \mathcal{U}_{\mathcal{H}}} \chi_U)(F) = \bigwedge_{U \in \mathcal{U}_{\mathcal{H}}} \chi_U(F) = \inf_{U \in \mathcal{U}_{\mathcal{H}}} \chi_U(F).$$

In other words,

$$F \in C^{sem}(\mathcal{H}) \text{ if and only if } \chi_{C^{sem}(\mathcal{H})}(F) = 1$$

or

$$F \in C^{sem}(\mathcal{H}) \text{ if and only if } \chi_U(F) = 1 \text{ for all } U \in U_{\mathcal{H}}.$$

In Chapter 6 we shall see that the concept of semantic fuzzy deduction will be introduced by similar equalities, replacing crisp sets by fuzzy sets. That is why in the last part of this section we have introduced the set $C^{sem}(\mathcal{H})$ and described it more explicitly by the equality (d) and by means of characteristic functions.

2.3   The Semantic Lindenbaum Algebra of $\mathcal{L}_0$

In Section 2.1 we introduced the Boolean interpretations φ of $\mathcal{L}_0$; the set of all these interpretations φ was denoted by $\mathcal{V}$. In the same section it was proved that there is a bijective mapping $\alpha \mapsto \varphi_\alpha$ from Γ^N onto $\mathcal{V}$. By means of Boolean interpretations we have also defined the functional interpretation of $\mathcal{L}_0$ as being the unique mapping $\Phi : \mathcal{L}_0 \to B(N, \Gamma) : F \mapsto \Phi(F)$ having the properties:

(i) $\Phi(\neg F) = \overline{\Phi(F)}$;

(ii) $\Phi(F \vee G) = \Phi(F) \vee \Phi(G)$;

(iii) $\Phi(X_j) = p_j$, for all j $(1 \leq j \leq N)$.

In the above equalities $X_1, X_2, ..., X_N$ are the propositional variables of $\mathcal{L}_0$ and F, G are arbitrary formulas in $\mathcal{L}_0$; the functions $p_j \in B(N, \Gamma)$ called projectors are defined by:

$$p_j(x_1, x_2, ..., x_N) = x_j.$$

The mapping Φ is surjective (that is for any $f \in B(N, \Gamma)$, there is some $F \in \mathcal{L}_0$ such that $\Phi(F) = f$) and for any $F \in \mathcal{L}_0$ the function $\Phi(F) \in B(N, \Gamma)$ is defined by $\Phi(F)(\alpha) = \varphi_\alpha(F)$ for all $\alpha \in \Gamma^N$.

In Section 2.2 we introduced in $\mathcal{L}_0$ the binary relation "$\approx$" by the following definition: two formulas $F, G \in \mathcal{L}_0$ are said to be semantically equivalent, or briefly $F \approx G$, if $\varphi(F) = \varphi(G)$ for all $\varphi \in \mathcal{V}$. In other words,

$$F \approx G \text{ if and only if } F \Leftrightarrow G \text{ is a tautology}$$

or

$$F \approx G \text{ if and only if } \Phi(F) = \Phi(G).$$

From the last characterization of "$\approx$" it immediately follows that for any formulas $F, G, H \in \mathcal{L}_0$ the following hold:

(a) $F \approx F$;

(b)   if $F \approx G$, then $G \approx F$;

(c)   if $F \approx G$ and $G \approx H$, then $F \approx H$.

The properties (a), (b), (c) show that "$\approx$" is an equivalence relation in $\mathcal{L}_0$, hence $\mathcal{L}_0$ can be decomposed in nonempty, disjoint classes of equivalent formulas. The set $\mathcal{L}_0/\approx$ of all these classes was denoted by $\hat{\mathcal{L}}_0$ and became a Boolean algebra by endowing it with appropriate internal operations.

PROPOSITION 2.17
If F, F', G, G' are formulas in $\mathcal{L}_0$, then the following assertions hold:

(a) If $F \approx F'$, then $\neg F \approx \neg F'$;

(b) If $F \approx F'$ and $G \approx G'$, then $F \vee G \approx F' \vee G'$;

(c) If $F \approx F'$ and $G \approx G'$, then $F \wedge G \approx F' \wedge G'$.

PROOF   The assertions (a), (b) follow directly from the definition of "$\approx$ " and the properties of Φ. In the proof of (c) we additionally have to take into account the property of Φ (proved in Section 2.1):

$$\Phi(F \wedge G) = \Phi(F) \wedge \Phi(G).$$

Let us prove the assertion (c). Suppose that $F \approx F'$ and $G \approx G'$ or equivalently $\Phi(F) = \Phi(F')$ and $\Phi(G) = \Phi(G')$. It follows that

$$\Phi(F \wedge G) = \Phi(F) \wedge \Phi(G) = \Phi(F') \wedge \Phi(G') = \Phi(F' \wedge G'),$$

hence $F \wedge G \approx F' \wedge G'$.   ∎

For any $F \in \mathcal{L}_0$ denote by $\xi_F \in \hat{\mathcal{L}}_0$ the unique class containing F. From the definition of "$\approx$ " it follows that

$$\xi_F = \xi_{F'} \quad \text{if and only if} \quad F \approx F'.$$

So we obtain the mapping:

$$\delta : \mathcal{L}_0 \to \hat{\mathcal{L}}_0 : F \mapsto \delta(F); \quad \delta(F) = \xi_F.$$

It is clear that δ is a surjective mapping, i.e., for any $\xi \in \hat{\mathcal{L}}_0$ there is a formula $F \in \mathcal{L}_0$ (actually an infinite set of formulas) such that $\xi = \delta(F)$ or

$$\xi = \xi_F.$$

Now consider two arbitrary classes $\xi_F, \xi_G \in \hat{\mathcal{L}}_0$ and define:

$$\bar{\xi}_F = \xi_{\neg F}; \quad \xi_F \vee \xi_G = \xi_{F \vee G}; \quad \xi_F \wedge \xi_G = \xi_{F \wedge G}.$$

By Proposition 2.17 these definitions are correct.

Let us notice that in $\hat{\mathcal{L}}_0$ there are two remarkable classes, namely the class $\Im$ of all tautologies and the class $\mathcal{O}$ of all antilogies.

PROPOSITION 2.18
The set $\hat{\mathcal{L}}_0$ endowed with operations $\bar{\xi}, \xi_1 \vee \xi_2, \xi_1 \wedge \xi_2$ is a Boolean algebra, in which the unit elements are $\mathcal{O}$ and $\Im$.

PROOF   The commutativity, associativity and distributivity laws in $\hat{\mathcal{L}}_0$ follow immediately from the definition of the operations in $\hat{\mathcal{L}}_0$.

For instance, if ξ_1, ξ_2, are two arbitrary classes in $\hat{\mathcal{L}}_0$, then there are $F_1, F_2 \in \mathcal{L}_0$ such that $\xi_1 = \xi_{F_1}$ and $\xi_2 = \xi_{F_2}$.
Therefore,

$$\xi_1 \vee \xi_2 = \xi_{F_1} \vee \xi_{F_2} = \xi_{F_1 \vee F_2};$$

$$\xi_2 \vee \xi_1 = \xi_{F_2} \vee \xi_{F_1} = \xi_{F_2 \vee F_1}.$$

Taking into account that $F_1 \vee F_2 \approx F_2 \vee F_1$, hence $\xi_{F_1 \vee F_2} = \xi_{F_2 \vee F_1}$, we get $\xi_1 \vee \xi_2 = \xi_2 \vee \xi_1$.

The properties involving the operations $\bar{\xi}$ follow in the same way. For instance, if $F \in \hat{\mathcal{L}}_0, \xi = \xi_F$, then $\xi \vee \bar{\xi} = \xi_F \vee \xi_{\neg F} = \xi_{F \vee \neg F}$. Taking into account that $F \vee \neg F \in \Im$ it follows that $\xi_{F \vee \neg F} = \Im$, hence $\xi \vee \bar{\xi} = \Im$. Similarly one deduces that $\xi_1 \vee (\xi_1 \wedge \xi_2) = \xi_1$ and so on. ∎

In the Boolean algebra $(\hat{\mathcal{L}}_0, {}^-, \vee, \wedge)$, still denoted by $\hat{\mathcal{L}}_0$, the order relation "$\leq$" is introduced in a usual way, namely

$$\xi_1 \leq \xi_2 \quad \text{if} \quad \xi_1 = \xi_1 \wedge \xi_2$$

or equivalently,

$$\xi_1 \leq \xi_2 \quad \text{if} \quad \xi_2 = \xi_1 \vee \xi_2.$$

If $\xi_1 \leq \xi_2$ and $\xi_1 \neq \xi_2$, we shall write $\xi_1 < \xi_2$. It is clear that $\mathcal{O} < \Im$ and for each $\xi \in \hat{\mathcal{L}}_0$, $\mathcal{O} \leq \xi \leq \Im$. From the last relation it follows that $\mathcal{O}$ and $\Im$ are, respectively, the smallest and the greatest elements of $\hat{\mathcal{L}}_0$.

DEFINITION 2.13   *The Boolean algebra* $(\hat{\mathcal{L}}_0, {}^-, \vee, \wedge)$ *denoted simply by* $\hat{\mathcal{L}}_0$ *is called* **the (semantic) Lindenbaum algebra** *associated to the language* $\mathcal{L}_0$.

Recall now that in Section 2.2 it was noted that there is a bijective mapping $\xi \mapsto f_\xi$ from $\hat{\mathcal{L}}_0$ onto $B(N, \Gamma)$. The precise definition of this mapping will be given now and some properties of it will be emphasized. As we know, if $\xi \in \hat{\mathcal{L}}_0$ and $F', F'' \in \xi$, then $\Phi(F') = \Phi(F'')$, hence Φ is a constant on each class ξ. This means that there is a unique Boolean function $f_\xi \in B(N, \Gamma)$ such that

$$\Phi(F) = f_\xi \quad \text{for all} \quad F \in \xi.$$

It follows that any class $\xi \in \mathcal{L}_0$ is uniquely associated with a function $f_\xi \in B(N, \Gamma)$. Therefore, we can define the mapping

$$\hat{\Phi} : \mathcal{L}_0 \to B(N, \Gamma) : \xi \mapsto \hat{\Phi}(\xi);$$

$$\hat{\Phi}(\xi) = f_\xi = \Phi(F), \quad \text{for any} \quad F \in \xi.$$

PROPOSITION 2.19
  The mapping $\hat{\Phi} : \hat{\mathcal{L}}_0 \to B(N, \Gamma)$ *is bijective and has the following properties:*

  (i) $\hat{\Phi}(\bar{\xi}) = \overline{\hat{\Phi}(\xi)}$;

(ii) $\hat{\Phi}(\xi_1 \vee \xi_2) = \hat{\Phi}(\xi_1) \vee \hat{\Phi}(\xi_2)$;

(iii) $\hat{\Phi}(\xi_{X_j}) = p_j$ *for any* j $(1 \leq j \leq N)$;

(iv) $\hat{\Phi}(\xi_1 \wedge \xi_2) = \hat{\Phi}(\xi_1) \wedge \hat{\Phi}(\xi_2)$;

(v) $\hat{\Phi}(\mathcal{O}) = 0$ *and* $\hat{\Phi}(\Im) = 1$.

PROOF Consider an arbitrary element (function) $f \in B(N, \Gamma)$. From the surjectivity of the mapping Φ it follows that there is some $F \in \mathcal{L}_0$ such that $\Phi(F) = f$. Now taking the class $\xi = \xi_F$ in $\hat{\mathcal{L}}_0$ one obtains

$$\hat{\Phi}(\xi) = \hat{\Phi}(\xi_F) = \Phi(F) = f;$$

this means that the mapping $\hat{\Phi}$ is surjective. It is also injective. Indeed, take $\xi_1, \xi_2 \in \hat{\mathcal{L}}_0$ and suppose $\hat{\Phi}(\xi_1) = \hat{\Phi}(\xi_2)$. There exist formulas $F_1, F_2 \in \mathcal{L}_0$ such that $\xi_1 = \xi_{F_1}$, $\xi_2 = \xi_{F_2}$; hence from equality $\hat{\Phi}(\xi_{F_1}) = \hat{\Phi}(\xi_{F_2})$, we deduce $\Phi(F_1) = \Phi(F_2)$ or $F_1 \approx F_2$. It follows that $\xi_{F_1} = \xi_{F_2}$ or $\xi_1 = \xi_2$ and the injectivity of $\hat{\Phi}$ is therefore proved.

Properties (i) to (v) easily follow from the definition of $\hat{\Phi}$ and the operations $^-, \vee, \wedge$. For instance, taking into account that ξ_{X_j} is the class containing the propositional variable X_j, we get $\hat{\Phi}(\xi_{X_j}) = \Phi(X_j) = p_j$, which proves assertion (iii).

For assertion (i) if we consider an element $\xi \in \hat{\mathcal{L}}_0, \xi = \xi_F$, we obtain successively:

$$\hat{\Phi}(\bar{\xi}) = \hat{\Phi}(\bar{\xi}_F) = \hat{\Phi}(\xi_{\neg F}) = \Phi(\neg F) = \overline{\Phi(F)} = \overline{\hat{\Phi}(\xi_F)} = \overline{\hat{\Phi}(\xi)}.$$

Assertions (ii), (iv) and (v) can be verified in a similar way. ∎

From Proposition 2.19 we can deduce directly the following result:

PROPOSITION 2.20

(a) *The set* $\hat{\mathcal{L}}_0$ *is finite; moreover,* $card\,(\hat{\mathcal{L}}_0) = card(B(N, \Gamma)) = 2^{2^N}$.

(b) $\hat{\Phi}$ *is an isomorphism from* $\hat{\mathcal{L}}_0$ *onto* $B(N, \Gamma)$, *hence* $\hat{\mathcal{L}}_0$ *and* $B(N, \Gamma)$ *are isomorphic Boolean algebras. In other words, the semantic Lindenbaum algebra associated to* $\mathcal{L}_0$ *is isomorphic with the Boolean algebra* $B(N, \Gamma)$.

REMARK 2.11   Using the mappings $\delta : \mathcal{L}_0 \to \hat{\mathcal{L}}_0$ and $\hat{\Phi} : \hat{\mathcal{L}}_0 \to B(N, \Gamma)$ defined above, mapping Φ can be factorized as follows:

$$\Phi = \hat{\Phi} \circ \delta. \quad \blacksquare$$

3

The Truth Structure on $\mathcal{L}_0$ in The Syntactic Version

The purpose of the present chapter is to introduce and to study the syntactic (axiomatic, formal) version of the truth structure on $\mathcal{L}_0$. Roughly speaking, the syntactic version of truth structure is based on concepts such as axioms, inference rules, (syntactic) deductions, (syntactic) proofs and theorems. Theorems play in the syntactic approach the role played by tautologies in the semantic version. In the literature on propositional logic one often considers two inference rules (MP and Substitution) and systems of axioms involving only propositional variables. In our approach we shall use a unique inference rule (MP), but instead of axioms we shall have finite systems of schemes of axioms (involving arbitrary formulas in $\mathcal{L}_0$). There are numerous equivalent systems of (schemes of) axioms; some of them will be presented and commented in Chapter 5. In this chapter and in Chapter 4 we shall study exclusively the system adopted in the book of Hilbert and Ackermann [19], which will be called **the system of Hilbert** and denoted by **H**.

3.1 The System of Hilbert H: Axioms, Inference, Theorems

As was already mentioned, in the system **H** only one unique inference rule is used, namely "Modus Ponens" or, for short, MP. The mechanism of MP will be made clear below.

Let us now present the axioms of **H**.

DEFINITION 3.1   *For any $F, G, H \in \mathcal{L}_0$ the formulas*

(H1) $F \vee F \Rightarrow F$;

(H2) $F \Rightarrow F \vee G$;

(H3) $F \vee G \Rightarrow G \vee F$;

(H4) $(F \Rightarrow G) \Rightarrow ((H \vee F) \Rightarrow (H \vee G))$

are **axioms of H**.

REMARK 3.1   The formulas **(H1)** to **(H4)** are actually axiom schemes because F, G, H are arbitrary formulas in $\mathcal{L}_0$ so that each formula **(H1)** to **(H4)** serves as a pattern for an infinite set of concrete axioms.   ∎

In the following we shall use the following convention: if a formula K is an axiom, we shall write $K \in Ax(\mathbf{H})$.

DEFINITION 3.2   *Let $\mathcal{H}$ be an arbitrary (possibly empty) subset of $\mathcal{L}_0$ called a* **set of hypotheses**. *A finite sequence*

$$(\Delta): \quad F_1, F_2, ...F_r \quad (r \geq 1)$$

of formulas in $\mathcal{L}_0$ is called a **syntactic deduction** *(or simply a deduction) from $\mathcal{H}$ if for every index i $(1 \leq i \leq r)$ the formula F_i satisfies one of the conditions:*

(a) $F_i \in Ax$;

(b) $F_i \in \mathcal{H}$;

(c) there are indices $h, k < i$ $(1 \leq h, k \leq r)$ such that

$$F_k \equiv (F_h) \Rightarrow (F_i).$$

Each formula F which can be inserted in a deduction (Δ) from $\mathcal{H}$ will be called **deducible from $\mathcal{H}$**; *in this case, we shall write*

$$\mathcal{H} \vdash F.$$

In particular, $\mathcal{H} \vdash F_r$ and the sequence (Δ) will be called a **deduction from $\mathcal{H}$ of F_r**.

If the set $\mathcal{H}$ is empty, then instead of $\emptyset \vdash F$ we shall write $\vdash F$; in this case, the formula F will be called a **theorem** *(of $\mathbf{H}$) and the sequence (Δ) will be called* **a proof** *(of its last term $F_r \equiv F$).*

REMARK 3.2

(1) The meaning of the inference rule MP becomes clear from Definition 3.2. Namely, by condition (c) a formula $F_i \equiv B$ can be inserted in a deduction (Δ) from $\mathcal{H}$ if $i > 1$ and there are in (Δ) formulas $F_h \equiv A, F_k \equiv A \Rightarrow B$ with $h, k < i$. In other words, from A and $A \Rightarrow B$ we can deduce B.

(2) All terms of a deduction from $\mathcal{H}$ are deducible from $\mathcal{H}$. In particular, if $\mathcal{H} = \emptyset$, all terms of (Δ) are theorems.

(3) If $\mathcal{H}_1, \mathcal{H}_2$ are two sets of hypotheses and $\mathcal{H}_1 \subseteq \mathcal{H}_2$, then obviously

$$\mathcal{H}_1 \vdash F \quad \text{implies} \quad \mathcal{H}_2 \vdash F.$$

In particular, if $\vdash F$, then $\mathcal{H} \vdash F$ for any $\mathcal{H} \subseteq \mathcal{L}_0$. In other words, if F is a theorem, then F is deducible from any set of hypotheses.

(4) If $F_1, F_2, ..., F_r \equiv F$ is a proof, then the first term F_1 is an axiom.

(5) All axioms are trivial theorems.

(6) If

$$(\Delta_1): \quad A_1, A_2, ..., A_p \quad (p \geq 1)$$

$$(\Delta_2): \quad B_1, B_2, ..., B_q \quad (q \geq 1)$$

are deductions from $\mathcal{H}$, then one can obtain new deductions from $\mathcal{H}$ by "mixing" the terms of (Δ_1) and (Δ_2) provided the mutual order of terms in (Δ_1) and (Δ_2) be unchanged.

(7) We shall use the following conventions: if $\mathcal{H} \cup \mathcal{K} \vdash F$, then we shall often write:

$$\mathcal{H}, \mathcal{K} \vdash F ;$$

if $\mathcal{H} = \{H_1, H_2, ... H_p\}$, $p \geq 1$, then instead of $\mathcal{H} \vdash F$ we shall also write:

$$H_1, H_2, ..., H_p \vdash F .$$

In particular, if $\mathcal{H}$ consists of a single element H, i.e., $\mathcal{H} = \{H\}$, then instead of $\mathcal{H} \vdash F$ we shall write $H \vdash F$. Similarly, if $\mathcal{K} = \{K\}$, then $\mathcal{H}, K \vdash F$ will stand for $\mathcal{H}, \mathcal{K} \vdash F$.

∎

3.2   Metatheorems

The aim of this section is to establish the deducibility and the provability of certain formulas and to give some general properties of these concepts. In order to avoid any confusion in the use of formal concepts like theorems and deductions, all assertions concerning formal theorems and formal deductions will be called **Metatheorems**. It is clear that the "proofs" of metatheorems do not coincide with formal (syntactic) proofs; metatheorems are proved by means of usual (elementary) mathematical arguments (and essentially consist of producing **formal** deductions or proofs) but, for the sake of simplicity, we shall use the word **proof** in both situations, the sense of each following from the context. Metatheorems will be shortly designated as **MTh.**

METATHEOREM 3.1 (Metatheorem of MP)

$$If \ \vdash F \ and \ \vdash F \Rightarrow G, \ then \ \vdash G.$$

PROOF   Consider the following (formal) proof:

$$
\left.
\begin{array}{l}
(1) \\
\quad\vdots \\
(p) \quad F
\end{array}
\right\} \text{a proof of } F;
$$

$$
\left.
\begin{array}{l}
(p+1) \\
\quad\vdots \\
(p+q) \quad F \Rightarrow G
\end{array}
\right\} \text{a proof of } F \Rightarrow G;
$$

$$(p+q+1) \quad G; \quad MP: \ (p), \ (p+q).$$

The sequence (1) - $(p+q+1)$ of formulas being a proof of G, it follows that $\vdash G$.   ∎

METATHEOREM 3.2 (Metatheorem of syllogism)

$$If \ \vdash F \Rightarrow G \ and \ \vdash G \Rightarrow H, \ then \ \vdash F \Rightarrow H.$$

PROOF   Consider the following (formal) proof:

$$
\left.\begin{array}{l}
(1) \\
\quad\vdots \\
(p)\ F \Rightarrow G
\end{array}\right\} \text{a proof of } F \Rightarrow G;
$$

$$
\left.\begin{array}{l}
(p+1) \\
\quad\vdots \\
(p+q)\ G \Rightarrow H
\end{array}\right\} \text{a proof of } G \Rightarrow H;
$$

$(p+q+1)\quad (G \Rightarrow H) \Rightarrow ((\neg F \vee G) \Rightarrow (\neg F \vee H));$ axiom **H4**;

$(p+q+2)\quad (\neg F \vee G) \Rightarrow (\neg F \vee H); MP : (p+q), (p+q+1);$

$(p+q+3)\quad (F \Rightarrow G) \Rightarrow (F \Rightarrow H)$; transcription of $(p+q+2);$

$(p+q+4)\quad F \Rightarrow H; MP : (p), (p+q+3).$

The sequence (1) - $(p+q+4)$ being a proof it follows that $\vdash F \Rightarrow H.$ ▌

Metatheorems 3.1 and 3.2 will be used in proving other metatheorems. They will be denoted by (MP) and (Syl), respectively.

METATHEOREM 3.3

$$
\vdash F \Rightarrow (G \vee F).
$$

PROOF   By axiom (**H2**) we have

$$
\vdash F \Rightarrow (F \vee G)
$$

and by axiom (**H3**) we have also

$$
\vdash (F \vee G) \Rightarrow (G \vee F).
$$

Using (Syl) we get $\vdash F \Rightarrow (G \vee F).$ ▌

METATHEOREM 3.4

$$
\vdash F \Rightarrow F.
$$

PROOF   By (**H4**) we have:

$$
\vdash ((F \vee F) \Rightarrow F) \Rightarrow ((\neg F \vee (F \vee F)) \Rightarrow (\neg F \vee F))
$$

and by (**H1**) we have:

$$
\vdash (F \vee F) \Rightarrow F.
$$

Using (MP) we obtain

$$
\vdash ((\neg F \vee (F \vee F)) \Rightarrow (\neg F \vee F)
$$

or

$$\vdash (F \Rightarrow (F \vee F)) \Rightarrow (F \Rightarrow F).$$

By **(H2)** we have

$$\vdash F \Rightarrow (F \vee F)$$

and using (MP) we get $\vdash F \Rightarrow F.$ ∎

METATHEOREM 3.5

If $\mathcal{H} \vdash G$, then $\mathcal{H} \vdash F \Rightarrow G$ for any F.

PROOF   By MTh. 3.3,

$$\vdash G \Rightarrow (\neg F \vee G) \quad \text{or} \quad \vdash G \Rightarrow (F \Rightarrow G).$$

By assumption, we have also $\mathcal{H} \vdash G$; therefore, using (MP) we obtain $\mathcal{H} \vdash F \Rightarrow G.$ ∎

METATHEOREM 3.6 (Metatheorem of excluded middle: "Tertium non datur")

$$\vdash F \vee \neg F.$$

PROOF   From MTh. 3.4, we have

$$\vdash F \Rightarrow F \quad \text{or} \quad \vdash \neg F \vee F.$$

By axiom **(H2)** we also have

$$\vdash (\neg F \vee F) \Rightarrow (F \vee \neg F).$$

Using (MP) we obtain $\vdash F \vee \neg F.$ ∎

METATHEOREM 3.7 (Direct implication of the double negation law)

$$\vdash F \Rightarrow \neg\neg F.$$

PROOF   From MTh. 3.6, we have $\vdash \neg F \vee \neg\neg F$; hence $\vdash F \Rightarrow \neg\neg F.$
∎

METATHEOREM 3.8 (Metatheorem of contraposition, direct implication)

$$\vdash (F \Rightarrow G) \Rightarrow (\neg G \Rightarrow \neg F).$$

PROOF   By axiom (**H4**) we have

$$\vdash (G \Rightarrow \neg\neg G) \Rightarrow ((\neg F \vee G) \Rightarrow (\neg F \vee \neg\neg G)).$$

Taking into account MTh. 3.7 we have $\vdash G \Rightarrow \neg\neg G$. Using (MP) we obtain

$$\vdash (\neg F \vee G) \Rightarrow (\neg F \vee \neg\neg G).$$

By axiom (**H3**) we also have $\vdash (\neg F \vee \neg\neg G) \Rightarrow (\neg\neg G \vee \neg F)$, hence using (Syl) we get:

$$\vdash (\neg F \vee G) \Rightarrow (\neg\neg G \vee \neg F)$$

or

$$\vdash (F \Rightarrow G) \Rightarrow (\neg G \Rightarrow \neg F). \quad \blacksquare$$

METATHEOREM 3.9

If $\mathcal{H} \vdash F \Rightarrow G$, *then* $\mathcal{H}, F \vdash G$.

PROOF   Consider the following deduction from $\mathcal{H}, F$:

$$\left.\begin{array}{l} (1) \\ \quad\vdots \\ (p) \quad F \Rightarrow G \end{array}\right\} \text{a deduction of } F \Rightarrow G \text{ from } \mathcal{H} \text{ (hence from } \mathcal{H}, F);$$

$(p+1)$ F;   hypothesis belonging to $\mathcal{H} \cup \{F\}$;
$(p+2)$ G;   (MP) : (p), $(p+1)$.

It follows that $\mathcal{H}, F \vdash G$. $\blacksquare$

The following important result represents the reciprocal of MTh. 3.9 and is due to J. Herbrand.

METATHEOREM 3.10 (Metatheorem of (syntactic) deduction)

If $\mathcal{H}, F \vdash G$,

then $\mathcal{H} \vdash F \Rightarrow G$.

PROOF   Let

$$(\Delta) : K_1, K_2, ..., K_r \quad (r \geq 1), \quad K_r \equiv G$$

be a deduction of G from $\mathcal{H}, F$. We shall prove by induction that for all i $(1 \leq i \leq r)$ we have $\mathcal{H} \vdash F \Rightarrow K_i$.
For $i = 1$ the formula K_1 in (Δ) satisfies one of the conditions:

(1) $K_1 \in Ax$;

(2) $K_1 \in \mathcal{H}$;

(3) $K_1 \equiv F$.

In cases (1) and (2) we have $\mathcal{H} \vdash K_1$ hence by MTh. 3.5, $\mathcal{H} \vdash F \Rightarrow K_1$. In case (3), $F \Rightarrow K_1 \equiv F \Rightarrow F$, therefore, by MTh. 3.4, $\vdash F \Rightarrow K_1$, hence $\mathcal{H} \vdash F \Rightarrow K_1$. So the case where $i = 1$ is settled.

Now suppose $i > 1$ and assume that

$$\mathcal{H} \vdash F \Rightarrow K_1, \mathcal{H} \vdash F \Rightarrow K_2, ..., \mathcal{H} \vdash F \Rightarrow K_{i-1}.$$

The formula K_i in (Δ) satisfies one of the conditions: (1), (2), (3) above or is obtained by (MP) from two preceding formulas in (Δ). In cases (1), (2), (3) we get $\mathcal{H} \vdash F \Rightarrow K_i$ as above. It remains to prove that if K_i is obtained in (Δ) by (MP), then also $\mathcal{H} \vdash F \Rightarrow K_i$. Consequently, suppose that there are indices $p, q < i$ such that $K_q \equiv K_p \Rightarrow K_i$. By the inductive assumption we have

$$\mathcal{H} \vdash F \Rightarrow K_p \quad \text{and} \quad \mathcal{H} \vdash F \Rightarrow (K_p \Rightarrow K_i).$$

Consider now the following deduction from $\mathcal{H}$:

(1) $F \Rightarrow K_p$; inductive assumption;

(2) $F \Rightarrow (K_p \Rightarrow K_i)$; inductive assumption;

(3) $(F \Rightarrow K_p) \Rightarrow (\neg K_p \Rightarrow \neg F)$;   MTh. 3.8;

(4) $\neg K_p \Rightarrow \neg F$;   (MP) : (1), (3);

(5) $(\neg K_p \Rightarrow \neg F) \Rightarrow ((K_i \vee \neg K_p) \Rightarrow (K_i \vee \neg F))$;   axiom (**H4**);

(6) $(K_i \vee \neg K_p) \Rightarrow (K_i \vee \neg F)$; (MP); (4), (5);

(7) $(\neg K_p \vee K_i) \Rightarrow (K_i \vee \neg K_p)$; axiom (**H3**);

(8) $(\neg K_p \vee K_i) \Rightarrow (K_i \vee \neg F)$;   (Syl): (7), (6);

(9) $(K_i \vee \neg F) \Rightarrow (\neg F \vee K_i)$;   axiom (**H3**);

(10) $(\neg K_p \vee K_i) \Rightarrow (\neg F \vee K_i)$; (Syl) : (8), (9);

(11) $(K_p \Rightarrow K_i) \Rightarrow (F \Rightarrow K_i)$; transcription of (10);

(12) $F \Rightarrow (F \Rightarrow K_i)$; (Syl): (2), (11);

(13) $\neg F \vee (F \Rightarrow K_i)$; transcription of (12);

(14) $(\neg F \vee (F \Rightarrow K_i)) \Rightarrow ((F \Rightarrow K_i) \vee \neg F)$; axiom (**H3**);

(15) $(F \Rightarrow K_i) \vee \neg F$; (MP): (13), (14);

(16) $\neg F \Rightarrow (\neg F \vee K_i)$; axiom (**H2**);

(17) $(\neg F \Rightarrow (\neg F \vee K_i)) \Rightarrow (((F \Rightarrow K_i) \vee \neg F) \Rightarrow ((F \Rightarrow K_i) \vee (\neg F \vee K_i)))$; axiom (**H4**);

(18) $((F \Rightarrow K_i) \vee \neg F) \Rightarrow ((F \Rightarrow K_i) \vee (\neg F \vee K_i))$; (MP): (16), (17);

(19) $(F \Rightarrow K_i) \vee (\neg F \vee K_i)$; (MP): (15), (18);

(20) $(F \Rightarrow K_i) \vee (F \Rightarrow K_i)$; transcription of (19);

(21) $((F \Rightarrow K_i) \vee (F \Rightarrow K_i)) \Rightarrow (F \Rightarrow K_i)$; axiom (**H1**);

(22) $F \Rightarrow K_i$; (MP): (20), (21).

We have obtained the deduction $\mathcal{H} \vdash F \Rightarrow K_i$. By the principle of mathematical induction it follows that

$$\mathcal{H} \vdash F \Rightarrow K_i \quad \text{for all indices} \quad i \ \ (1 \le i \le r).$$

In particular we have $\mathcal{H} \vdash F \Rightarrow K_r$, hence $\mathcal{H} \vdash F \Rightarrow G$, which concludes the proof. ∎

Using the metatheorem of (syntactic) deduction one can establish many other metatheorems.

METATHEOREM 3.11 (Metatheorem of permutation of antecedents)

$$\vdash (F \Rightarrow (G \Rightarrow H)) \Rightarrow (G \Rightarrow (F \Rightarrow H)).$$

PROOF   Consider the following deduction:

(1) $F \Rightarrow (G \Rightarrow H)$ $\Big\}$
(2) F ; hypotheses;
(3) G
(4) $G \Rightarrow H$; (MP): (1), (2);
(5) H; (MP): (3), (4).

Therefore, $F \Rightarrow (G \Rightarrow H), G, F \vdash H$. Applying three times successively the metatheorem of Herbrand we obtain: $F \Rightarrow (G \Rightarrow H), G \vdash F \Rightarrow H$; $F \Rightarrow (G \Rightarrow H) \vdash G \Rightarrow (F \Rightarrow H)$ and $\vdash (F \Rightarrow (G \Rightarrow H)) \Rightarrow (G \Rightarrow (F \Rightarrow H))$. ∎

REMARK 3.3   From the above metatheorem we immediately get the following assertion:

$$\text{If} \ \ \vdash F \Rightarrow (G \Rightarrow H), \ \ \text{then} \ \ \vdash G \Rightarrow (F \Rightarrow H)$$

and so we can see that in the above two theorems the antecedents F, G are simply permuted. ∎

METATHEOREM 3.12

$$\vdash F \Rightarrow (G \Rightarrow (F \wedge G)).$$

PROOF   By (MP) we have the following deduction:

(1) $F, F \Rightarrow G \vdash G$.
Applying successively the metatheorem of deduction, we obtain:

(2) $\vdash F \Rightarrow ((F \Rightarrow G) \Rightarrow G)$.
By MTh. 3.8, we have:

(3) $\vdash ((F \Rightarrow G) \Rightarrow G) \Rightarrow (\neg G \Rightarrow \neg (F \Rightarrow G))$.
From (2), (3), by (Syl), it follows:

(4) $\vdash F \Rightarrow (\neg G \Rightarrow \neg (F \Rightarrow G))$.
Replacing in (4) G by $\neg G$ we have:

(5) $\vdash F \Rightarrow (\neg \neg G \Rightarrow \neg (F \Rightarrow \neg G))$.
Applying in (5) MTh. 3.11 we get:

(6) $\vdash \neg \neg G \Rightarrow (F \Rightarrow \neg (F \Rightarrow \neg G))$.
From MTh. 3.7 and then from (Syl) we obtain successively:

(7) $\vdash G \Rightarrow \neg \neg G$;

(8) $\vdash G \Rightarrow (F \Rightarrow \neg (F \Rightarrow \neg G))$;

(9) $\vdash G \Rightarrow (F \Rightarrow (F \wedge G))$ (transcription of (8)).
By MTh. 3.11 we have finally:

(10) $\vdash F \Rightarrow (G \Rightarrow (F \wedge G))$,

which completes the proof. ∎

METATHEOREM 3.13

$$If \quad \vdash F \quad and \quad \vdash G, \quad then \quad \vdash F \wedge G.$$

PROOF   We have the following theorems:

(1) $\vdash F$;

(2) $\vdash G$;

(3) $\vdash F \Rightarrow (G \Rightarrow (F \wedge G))$; MTh. 3.12;

(4) $\vdash G \Rightarrow (F \wedge G)$; (MP): (1), (3);

(5) $\vdash F \wedge G$; (MP): (2), (4). ∎

METATHEOREM 3.14

$$\vdash (F \Rightarrow G) \Rightarrow ((F \vee H) \Rightarrow (G \vee H)).$$

PROOF We can obviously write the following deductions:

(1) $\vdash (F \Rightarrow G) \Rightarrow ((H \vee F) \Rightarrow (H \vee G))$; axiom (**H4**);

(2) $F \Rightarrow G \vdash (H \vee F) \Rightarrow (H \vee G)$; MTh. 3.9;

(3) $F \Rightarrow G \vdash (F \vee H) \Rightarrow (H \vee F)$; axiom (**H3**);

(4) $F \Rightarrow G \vdash (F \vee H) \Rightarrow (H \vee G)$; (Syl): (3), (2);

(5) $F \Rightarrow G \vdash (H \vee G) \Rightarrow (G \vee H)$; axiom (**H3**);

(6) $F \Rightarrow G \vdash (F \vee H) \Rightarrow (G \vee H)$ (Syl): (4), (5);

(7) $\vdash (F \Rightarrow G) \Rightarrow ((F \vee H) \Rightarrow (G \vee H))$; MTh. 10 for (6). ∎

METATHEOREM 3.15

If $\vdash F \Rightarrow G$ *and* $\vdash H \Rightarrow K$, *then* $\vdash (F \vee H) \Rightarrow (G \vee K)$.

PROOF Consider the following deductions:

(1) $\vdash F \Rightarrow G$;

(2) $\vdash H \Rightarrow K$;

(3) $\vdash (F \Rightarrow G) \Rightarrow ((F \vee H) \Rightarrow (G \vee H))$; MTh. 3.14;

(4) $\vdash (F \vee H) \Rightarrow (G \vee H)$; (MP) :(1), (3);

(5) $F \vee H \vdash G \vee H$; MTh. 3.9 applied for (4);

(6) $\vdash (H \Rightarrow K) \Rightarrow ((G \vee H) \Rightarrow (G \vee K))$; axiom (**H4**);

(7) $\vdash (G \vee H) \Rightarrow (G \vee K)$; (MP): (2), (6);

(8) $F \vee H \vdash (G \vee H) \Rightarrow (G \vee K)$; from (7);

(9) $F \vee H \vdash G \vee K$; (MP): (5), (8);

now, applying MTh. 3.10, we obtain from (9) that:

$$\vdash (F \vee H) \Rightarrow (G \vee K)$$

and the proof is finished. ∎

METATHEOREM 3.16

$$\vdash (F \vee G) \Rightarrow (\neg\neg F \vee \neg\neg G).$$

PROOF   By MTh. 3.7 we can write the following two theorems:

(1) $\vdash F \Rightarrow \neg\neg F$;

(2) $\vdash G \Rightarrow \neg\neg G$.

From these theorems, applying MTh. 3.15, we obtain
$$\vdash (F \vee G) \Rightarrow (\neg\neg F \vee \neg\neg G). \quad ∎$$

METATHEOREM 3.17 (Converse implication of the first De Morgan law)
$$\vdash (\neg F \wedge \neg G) \Rightarrow \neg(F \vee G).$$

PROOF   By the metatheorem of contraposition (MTh. 3.8) we have:

(1) $\vdash ((F \vee G) \Rightarrow (\neg\neg F \vee \neg\neg G)) \Rightarrow ((\neg(\neg\neg F \vee \neg\neg G)) \Rightarrow \neg(F \vee G))$.
Now, let us write MTh. 3.16:

(2) $\vdash (F \vee G) \Rightarrow (\neg\neg F \vee \neg\neg G)$.
By (MP) applied for (1) and (2) we obtain:

(3) $\vdash \neg(\neg\neg F \vee \neg\neg G) \Rightarrow \neg(F \vee G)$.
By transcription of theorem (3) we get:

$$\vdash (\neg F \wedge \neg G) \Rightarrow \neg(F \vee G). \quad ∎$$

METATHEOREM 3.18 (Converse implication of the double negation law)
$$\vdash \neg\neg F \Rightarrow F.$$

PROOF   By the direct implication of the double negation law (MTh. 3.7) we have:

(1) $\vdash \neg F \Rightarrow \neg\neg\neg F$.
From MTh. 3.4 we have also:

(2) $\vdash F \Rightarrow F$.
From theorems (1) and (2), applying MTh. 3.15, we get:

(3) $\vdash F \vee \neg F \Rightarrow F \vee \neg\neg\neg F$.
But from MTh. 3.6,

(4) $\vdash F \vee \neg F$.
With (MP) from (3) and (4) we obtain:

(5) $\vdash F \vee \neg\neg\neg F$.
By axiom (**H3**) we can write:

(6) $\vdash (F \vee \neg\neg\neg F) \Rightarrow (\neg\neg\neg F \vee F)$.
Applying again (MP), from (5) and (6) we get:

(7) $\vdash \neg\neg\neg F \vee F$.
In transcription, (7) becomes:

(8) $\vdash \neg\neg F \Rightarrow F$. ∎

METATHEOREM 3.19 (Converse implication of contraposition)

$$\vdash (\neg G \Rightarrow \neg F) \Rightarrow (F \Rightarrow G).$$

PROOF By metatheorem 3.8 we get:

(1) $\vdash (\neg G \Rightarrow \neg F) \Rightarrow (\neg\neg F \Rightarrow \neg\neg G)$.
Now using MTh. 3.9 we can obtain from (1) that

(2) $\neg G \Rightarrow \neg F \ \vdash \neg\neg F \Rightarrow \neg\neg G$.
By MTh. 3.7 and MTh. 3.18, respectively, we have:

(3) $\vdash F \Rightarrow \neg\neg F$, hence (3)' $\neg G \Rightarrow \neg F \ \vdash F \Rightarrow \neg\neg F$;

(4) $\vdash \neg\neg G \Rightarrow G$, hence (4)' $\neg G \Rightarrow \neg F \ \vdash \neg\neg G \Rightarrow G$.
Applying (Syl) for (3)' and (2) we get:

(5) $\neg G \Rightarrow \neg F \ \vdash F \Rightarrow \neg\neg G$.
Applying now (Syl) for (5) and (4)' we obtain:

(6) $\neg G \Rightarrow \neg F \ \vdash F \Rightarrow G$.
By the metatheorem of Herbrand we get from (6) that

(7) $\vdash (\neg G \Rightarrow \neg F) \Rightarrow (F \Rightarrow G)$.   ∎

METATHEOREM 3.20 (Direct implication of the first De Morgan law)
$$\vdash \neg(F \vee G) \Rightarrow (\neg F \wedge \neg G).$$

PROOF   By MTh. 3.18 we have:

(1) $\vdash \neg\neg F \Rightarrow F$;

(2) $\vdash \neg\neg G \Rightarrow G$.
Applying now MTh. 3.15 we obtain:

(3) $\vdash (\neg\neg F \vee \neg\neg G) \Rightarrow (F \vee G)$.
From MTh. 3.8 we get:

(4) $\vdash ((\neg\neg F \vee \neg\neg G) \Rightarrow (F \vee G)) \Rightarrow ((\neg(F \vee G)) \Rightarrow (\neg(\neg\neg F \vee \neg\neg G)))$.
By (MP), from (3) and (4) we obtain:

(5) $\vdash \neg(F \vee G) \Rightarrow \neg(\neg\neg F \vee \neg\neg G)$,
which in transcription is the same as

(6) $\vdash \neg(F \vee G) \Rightarrow (\neg F \wedge \neg G)$;   and the proof is finished.   ∎

METATHEOREM 3.21 (Direct implication of the second De Morgan law)

$$\vdash (\neg(F \wedge G)) \Rightarrow (\neg F \vee \neg G)$$

PROOF   By the converse implication of the double negation law we have
$$\vdash \neg\neg(\neg F \vee \neg G) \Rightarrow (\neg F \vee \neg G),$$
which in transcription is the same as
$$\vdash (\neg(F \wedge G)) \Rightarrow (\neg F \vee \neg G). \quad ∎$$

METATHEOREM 3.22 (Converse implication of the second De Morgan law)
$$\vdash (\neg F \vee \neg G) \Rightarrow (\neg(F \wedge G)).$$

PROOF   Applying the direct implication of the double negation law we have
$$\vdash (\neg F \vee \neg G) \Rightarrow (\neg\neg(\neg F \vee \neg G)).$$
which in transcription becomes $\vdash (\neg F \vee \neg G) \Rightarrow (\neg(F \wedge G))$.   ∎

METATHEOREM 3.23

    If $\vdash F \Rightarrow G$ and $\vdash H \Rightarrow K$,   then $\vdash (F \wedge H) \Rightarrow (G \wedge K)$.

PROOF   By contraposition we can write:

(1) $\vdash (F \Rightarrow G) \Rightarrow (\neg G \Rightarrow \neg F)$;

(2) $\vdash (H \Rightarrow K) \Rightarrow (\neg K \Rightarrow \neg H)$.
    By (MP) from $\vdash F \Rightarrow G$ and (1), from $\vdash H \Rightarrow K$ and (2), respectively, we obtain:

(3) $\vdash \neg G \Rightarrow \neg F$;

(4) $\vdash \neg K \Rightarrow \neg H$.
    By MTh. 3.15 we get:

(5) $\vdash (\neg G \vee \neg K) \Rightarrow (\neg F \vee \neg H)$.
    By MTh. 3.21 and MTh. 3.22 we have:

(6) $\vdash (\neg(G \wedge K)) \Rightarrow (\neg G \vee \neg K)$;
    and

(7) $\vdash (\neg F \vee \neg H) \Rightarrow (\neg(F \wedge H))$.
    Applying two times (Syl) from (5), (6), (7) there results:

(8) $\vdash (\neg(G \wedge K)) \Rightarrow (\neg(F \wedge H))$.
    By the converse implication of contraposition we obtain:

(9) $\vdash ((\neg(G \wedge K)) \Rightarrow (\neg(F \wedge H))) \Rightarrow ((F \wedge H) \Rightarrow (G \wedge K))$.
    By (MP), from (8) and (9) we finally get:

(10) $\vdash (F \wedge H)) \Rightarrow (G \wedge K)$.   ∎

METATHEOREM 3.24

$$\vdash (F \Rightarrow G) \Rightarrow ((\neg F \Rightarrow G) \Rightarrow G).$$

PROOF   We can write the following deductions from $\{F \Rightarrow G, \neg F \Rightarrow G\}$:

(1) $F \Rightarrow G$; hypothesis;

(2) $\neg F \vee G$; transcription of (1).

(3) $(\neg F \vee G) \Rightarrow (G \vee \neg F)$; axiom (**H3**);

(4) $G \vee \neg F$; (MP): (2), (3);

(5) $(\neg F \Rightarrow G) \Rightarrow ((G \vee \neg F) \Rightarrow (G \vee G))$; axiom (**H4**);

(6) $\neg F \Rightarrow G$; hypothesis;

(7) $(G \vee \neg F) \Rightarrow (G \vee G)$; (MP): (5), (6);

(8) $G \vee G$; (MP): (4), (7);

(9) $(G \vee G) \Rightarrow G$; axiom (**H1**);

(10) G; (MP): (8), (9).

We have obtained $F \Rightarrow G, \neg F \Rightarrow G \vdash G$. Applying the Metatheorem of Herbrand two times successively we get:

$$F \Rightarrow G \vdash (\neg F \Rightarrow G) \Rightarrow G;$$

$$\vdash (F \Rightarrow G) \Rightarrow ((\neg F) \Rightarrow G) \Rightarrow G),$$

which completes the proof. ∎

METATHEOREM 3.25

$$\vdash (\neg G \Rightarrow \neg F) \Rightarrow ((\neg G \Rightarrow F) \Rightarrow G).$$

PROOF   We can write the following deduction from $\{\neg G \Rightarrow \neg F, \neg G \Rightarrow F\}$:

(1) $\neg G \Rightarrow \neg F$; hypothesis;

(2) $\neg G \Rightarrow F$; hypothesis;

(3) $(\neg G \Rightarrow \neg F) \Rightarrow (F \Rightarrow G)$; MTh. 3.19;

(4) $F \Rightarrow G$; (MP): (1), (3);

(5) $(\neg G \Rightarrow F) \Rightarrow (\neg F \Rightarrow \neg\neg G)$. MTh. 3.8;

(6) $\neg F \Rightarrow \neg\neg G$; (MP): (2), (5);

(7) $\neg\neg G \Rightarrow G$; MTh. 3.18;

(8) $\neg F \Rightarrow G$; (Syl): (6), (7);

(9) $(F \Rightarrow G) \Rightarrow ((\neg F \Rightarrow G) \Rightarrow G)$; MTh. 3.24;

(10) $(\neg F \Rightarrow G) \Rightarrow G$; (MP): (4), (9);

(11) G; (MP): (8), (10).

So we obtained the deduction:

$$\neg G \Rightarrow \neg F, \ \neg G \Rightarrow F \vdash G.$$

Applying the metatheorem of Herbrand two times successively we have:

$$\neg G \Rightarrow \neg F \vdash (\neg G \Rightarrow F) \Rightarrow G.$$

$$\vdash (\neg G \Rightarrow \neg F) \Rightarrow ((\neg G \Rightarrow F) \Rightarrow G). \quad \blacksquare$$

The next metatheorems follow from the already proved ones. We shall list these metatheorems without proof; it will be a good exercise for the reader to prove them.

METATHEOREM 3.26

(a) $\vdash (F \wedge G) \Rightarrow F$;

(b) $\vdash (F \wedge G) \Rightarrow G$.

Or, if $\vdash F \wedge G$, then $\vdash F$ and $\vdash G$.

METATHEOREM 3.27 (commutativity law for "$\wedge$")

$$\vdash (F \wedge G) \Rightarrow (G \wedge F).$$

REMARK 3.4 The theorem $\vdash (F \vee G) \Rightarrow (G \vee F)$ coincides with the axiom (**H3**). $\quad \blacksquare$

METATHEOREM 3.28 (associativity laws)

(a) $\begin{cases} \vdash ((F \vee G) \vee H) \Rightarrow (F \vee (G \vee H)); \\ \vdash (F \vee (G \vee H)) \Rightarrow ((F \vee G) \vee H). \end{cases}$

(b) $\begin{cases} \vdash ((F \wedge G) \wedge H) \Rightarrow (F \wedge (G \wedge H)); \\ \vdash (F \wedge (G \wedge H)) \Rightarrow ((F \wedge G) \wedge H). \end{cases}$

METATHEOREM 3.29 (distributivity laws)

(a) $\begin{cases} \vdash (F \vee (G \wedge H)) \Rightarrow ((F \vee G) \wedge (F \vee H)); \\ \vdash ((F \vee G) \wedge (F \vee H)) \Rightarrow (F \vee (G \wedge H)). \end{cases}$

(b) $\begin{cases} \vdash ((F \wedge (G \vee H)) \Rightarrow ((F \wedge G) \vee (F \wedge H)); \\ \vdash ((F \wedge G) \vee (F \wedge H) \Rightarrow (F \wedge (G \vee H)). \end{cases}$

METATHEOREM 3.30 (absorption laws)

(a) $\begin{cases} \vdash (F \vee (F \wedge G)) \Rightarrow F; \\ \vdash F \Rightarrow (F \vee (F \wedge G)). \end{cases}$

(b) $\begin{cases} \vdash (F \wedge (F \vee G)) \Rightarrow F; \\ \vdash F \Rightarrow (F \wedge (F \vee G)). \end{cases}$

METATHEOREM 3.31

$$\vdash F \Rightarrow (G \Rightarrow F).$$

METATHEOREM 3.32

If $\vdash G$, then

(a) $\vdash F \Rightarrow (F \wedge G)$;

(b) $\vdash (\neg G) \Rightarrow (F \wedge (\neg G))$;

(c) $\vdash (F \vee (\neg G)) \Rightarrow F$.

In the previous chapter we have introduced the set $C^{sem}(\mathcal{H})$ of all semantic consequences from $\mathcal{H}$. Similarly we shall introduce the set $C^{syn}(\mathcal{H})$ of all syntactic consequences from $\mathcal{H}$, that is the set of all formulas $F \in \mathcal{L}_0$ such that $\mathcal{H} \vdash F$; in other words,

$$C^{syn}(\mathcal{H}) = \{F \in \mathcal{L}_0 \mid \mathcal{H} \vdash F\}.$$

By analogy with the characterization of $C^{sem}(\mathcal{H})$, the set $C^{syn}(\mathcal{H})$ can also be described in terms of intersections and characteristic functions. For this purpose, let us denote by $\mathcal{U}_{\mathcal{H}}$ the set of all subsets U in $\mathcal{L}_0$ having the following three properties:

(j) U is MP-closed;

(jj) $U \supseteq Ax$;

(jjj) $U \supseteq \mathcal{H}$.

METATHEOREM 3.33

For any set $\mathcal{H}$ of hypotheses

$$C^{syn}(\mathcal{H}) = \bigcap_{U \in \mathcal{U}_{\mathcal{H}}} U.$$

In terms of characteristic functions, this equality can obviously be expressed by

$$\chi_{\mathcal{C}^{syn}(\mathcal{H})} = \bigwedge_{U \in \mathcal{U}_{\mathcal{H}}} \chi_U.$$

PROOF Let F be a formula in $\mathcal{C}^{syn}(\mathcal{H})$. This means that there is some syntactic deduction from $\mathcal{H}$

$$(\Delta): \quad F_1, F_2, ..., F_r \quad (r \geq 1), \quad F_r \equiv F.$$

Consider now an arbitrary set $U \in \mathcal{U}_{\mathcal{H}}$. We shall show by induction that for all i $(1 \leq i \leq r)$ $F_i \in U$. For $i = 1$, the formula F_1 is either an axiom or an element of $\mathcal{H}$, hence $F_1 \in U$ (because of (jj) and (jjj)) and the case $i = 1$ is settled. For $i > 1$ assume (by the induction hypothesis) that

$$F_1 \in U, \ F_2 \in U, ..., \ F_{i-1} \in U.$$

For the formula F_i in (Δ) there are three possibilities: (1) $F_i \in Ax$; (2) $F_i \in \mathcal{H}$; (3) F_i is obtained by (MP) from two formulas F_p, F_q in (Δ) with $p, q < i$. In cases (1), (2) we have $F_i \in U$ as above. In case (3), denote $F_i \equiv B, F_p \equiv A, F_q \equiv A \Rightarrow B$. Taking into account the inductive assumption, it follows that $A \in U$ and $A \Rightarrow B \in U$, hence $B \in U$ (U being MP-closed). From the principle of mathematical induction we deduce that $F_i \in U$ for all i $(1 \leq i \leq r)$. In particular, $F \equiv F_r \in U$. Therefore, we proved that if $F \in \mathcal{C}^{syn}(\mathcal{H})$, then $F \in U$ for all $U \in \mathcal{U}_{\mathcal{H}}$ or if $F \in \mathcal{C}^{syn}(\mathcal{H})$, then $F \in \bigcap_{U \in \mathcal{U}_{\mathcal{H}}} U$.

So the inclusion $\mathcal{C}^{syn}(\mathcal{H}) \subseteq \bigcap_{U \in \mathcal{U}_{\mathcal{H}}} U$ is established.

On the other hand, it is clear that $\mathcal{C}^{syn}(\mathcal{H}) \supseteq Ax$ and $\mathcal{C}^{syn}(\mathcal{H}) \supseteq \mathcal{H}$, hence $\mathcal{C}^{syn}(\mathcal{H})$ fulfills conditions (jj) and (jjj) from the definition of $\mathcal{U}_{\mathcal{H}}$. It is easy to see that $\mathcal{C}^{syn}(\mathcal{H})$ fulfills also condition (j). Therefore, $\mathcal{C}^{syn}(\mathcal{H}) \in \mathcal{U}_{\mathcal{H}}$; it follows:

$$\mathcal{C}^{syn}(\mathcal{H}) \supseteq \bigcap_{U \in \mathcal{U}_{\mathcal{H}}} U.$$

From the two inclusions above we get the requested equality. ∎

From the equality $\mathcal{C}^{syn}(\mathcal{H}) = \bigcap_{U \in \mathcal{U}_{\mathcal{H}}} U$ it follows that the formula F is deducible from $\mathcal{H}$ $(\mathcal{H} \vdash F)$ if and only if the formula F belongs to all subsets U such that $U \in \mathcal{U}_{\mathcal{H}}$. In terms of characteristic functions this means that

$$\chi_{\mathcal{C}^{syn}(\mathcal{H})}(F) = 1 \text{ if and only if } \chi_U(F) = 1 \text{ for all } U \in \mathcal{U}_{\mathcal{H}}.$$

REMARK 3.5   The set $C^{syn}(\mathcal{H})$ is the smallest subset of $\mathcal{L}_0$ which is MP-closed and contains the sets Ax and $\mathcal{H}$.   ∎

3.3   The Syntactic Lindenbaum Algebra of $\mathcal{L}_0$. Normal Formulas

In Section 2.3, we have introduced the equivalence relation "$\approx$" in $\mathcal{L}_0$ by means of which we have defined the semantic Lindenbaum algebra of $\mathcal{L}_0$ and we have proved that $\hat{\mathcal{L}}_0$ is isomorphic with the Boolean algebra $B(N, \Gamma)$.

In this section another Boolean algebra $\check{\mathcal{L}}_0$ associated to $\mathcal{L}_0$ will be defined exclusively by syntactic means (i.e., using axioms and syntactic deducibility).

We begin by introducing a new binary relation "$\sim$" on $\mathcal{L}_0$.

DEFINITION 3.3   *Let F, G be formulas in $\mathcal{L}_0$. We say that F, G are (**syntactically**) **equivalent** and we denote $F \sim G$ if $F \Rightarrow G$ and $G \Rightarrow F$ are theorems; in other words, $F \sim G$ if and only if $\vdash F \Rightarrow G$ and $\vdash G \Rightarrow F$.*

PROPOSITION 3.1
If $F, G, H \in \mathcal{L}_0$, then the following assertions hold:

(a) $F \sim F$;

(b) if $F \sim G$, then $G \sim F$;

(c) if $F \sim G$ and $G \sim H$, then $F \sim H$.

From (a), (b), (c) it follows that "$\sim$" is an equivalence relation in $\mathcal{L}_0$, so that $\mathcal{L}_0$ decomposes in nonempty, disjoint classes of syntactic equivalent formulas. The set $\mathcal{L}_0 / \sim$ of all these classes will be denoted by $\check{\mathcal{L}}_0$.

PROOF   The assertions are immediate. Indeed, (a) follows from the theorem $\vdash F \Rightarrow F$; (b) follows from the definition of "$\sim$" and (c) is a consequence of the metatheorem of syllogism.   ∎

Taking into account that "$\sim$" is an equivalence relation on $\mathcal{L}_0$, for any formula $F \in \mathcal{L}_0$ there is a unique class $\xi_F \in \check{\mathcal{L}}_0$ such that $\xi_F \ni F$. The mapping
$$F \mapsto \xi_F : \mathcal{L}_0 \to \check{\mathcal{L}}_0$$

is obviously surjective and $\xi_F = \xi_G$ if and only if $F \sim G$.

PROPOSITION 3.2

If F', F'', G', G'' are formulas in $\mathcal{L}_0$, then the following assertions hold:

(a) if $F' \sim F''$, then $\neg F' \sim \neg F''$;

(b) if $F' \sim F''$ and $G' \sim G''$, then $F' \vee G' \sim F'' \vee G''$;

(c) if $F' \sim F''$ and $G' \sim G''$, then $F' \wedge G' \sim F'' \wedge G''$;

PROOF   (a) If $F' \sim F''$, then by definition,

$$(*) \qquad\qquad\qquad \vdash F' \Rightarrow F''$$

and

$$(**) \qquad\qquad\qquad \vdash F'' \Rightarrow F'.$$

From $(**)$ and the metatheorem of contraposition (MTh. 3.8), using (MP), one obtains

$$(*)' \qquad\qquad\qquad \vdash \neg F' \Rightarrow \neg F''.$$

In the same way, from $(*)$ one gets

$$(**)' \qquad\qquad\qquad \vdash \neg F'' \Rightarrow \neg F'.$$

Theorems $(*)'$ and $(**)'$ imply $\neg F' \sim \neg F''$.
(b) Suppose that $F' \sim F''$ and $G' \sim G''$. It follows that

$$\vdash F' \Rightarrow F'' \quad \text{and} \quad \vdash F'' \Rightarrow F';$$

$$\vdash G' \Rightarrow G'' \quad \text{and} \quad \vdash G'' \Rightarrow G'.$$

By MTh. 3.15 one obtains

$$\vdash (F' \vee G') \Rightarrow (F'' \vee G'') \quad \text{and} \quad \vdash (F'' \vee G'') \Rightarrow (F' \vee G').$$

This means that $F' \vee G' \sim F'' \vee G''$.
   The assertion (c) follows in a similar way using MTh. 3.23.   ∎

   The set $\check{\mathcal{L}}_0$ can be organized as a Boolean algebra in a way similar to the set $\hat{\mathcal{L}}_0$ in Section 2.3. For this purpose let us define in $\check{\mathcal{L}}_0$ the following three internal operations:

If ξ_F, ξ_G are arbitrary classes in $\check{\mathcal{L}}_0$, then by definition:

$$\bar{\xi}_F = \xi_{\neg F};$$
$$\xi_F \vee \xi_G = \xi_{F \vee G};$$
$$\xi_F \wedge \xi_G = \xi_{F \wedge G}.$$

By Proposition 3.2 these definitions are correct.

From the metatheorems 3.27 to 3.30 the result below follows directly.

PROPOSITION 3.3

If F, G, H are arbitrary formulas in $\mathcal{L}_0$, then the following assertions hold:

(a) $F \wedge G \sim G \wedge F$;

(b) $F \vee G \sim G \vee F$;

(c) $((F \vee G) \vee H) \sim (F \vee (G \vee H))$;

(d) $((F \wedge G) \wedge H) \sim (F \wedge (G \wedge H))$;

(e) $(F \vee (G \wedge H)) \sim ((F \vee G) \wedge (F \vee H))$;

(f) $(F \wedge (G \vee H)) \sim ((F \wedge G) \vee (F \wedge H))$;

(g) $F \sim F \vee (F \wedge G)$ and $F \sim F \wedge (F \vee G)$;

(h) if $\vdash G$, then

$$F \sim F \vee G;$$
$$F \sim (F \wedge G);$$
$$F \sim F \vee (\neg G);$$
$$\neg G \sim F \wedge (\neg G).$$

From Proposition 3.3 and the definitions above one gets the following equalities:

$$\xi_F \vee \xi_G = \xi_G \vee \xi_F;$$
$$\xi_F \wedge \xi_G = \xi_G \wedge \xi_F;$$
$$\xi_F \vee (\xi_G \wedge \xi_H) = (\xi_F \vee \xi_G) \wedge (\xi_F \vee \xi_H);$$
$$\xi_F \wedge (\xi_G \vee \xi_H) = (\xi_F \wedge \xi_G) \vee (\xi_F \wedge \xi_H);$$
$$\xi_F = \xi_F \vee (\xi_F \wedge \xi_G) = \xi_F \wedge (\xi_F \vee \xi_G).$$

REMARK 3.6   In Proposition 3.3, (c), (d) the following equivalencies were established:

$$(F_1 \vee F_2) \vee F_3 \sim F_1 \vee (F_2 \vee F_3);$$
$$(F_1 \wedge F_2) \wedge F_3 \sim F_1 \wedge (F_2 \wedge F_3).$$

Using an induction procedure, by routine arguments one can extend these equivalencies to any finite number $F_1, F_2, ... F_n$ of formulas. More precisely, let us denote $(...(F_1 \vee F_2) \vee F_3) \vee ...) \vee F_n$ by $F_1 \vee F_2 \vee ... \vee F_n$ and correspondingly, $(...(F_1 \wedge F_2) \wedge F_3) \wedge ...) \wedge F_n$ by $F_1 \wedge F_2 \wedge ... \wedge F_n$.

With these notations, if D and C are a disjunction and a conjunction, respectively, of the same formulas $F_1, F_2, ..., F_n$ having any other distribution of parentheses, then

$$D \sim F_1 \vee F_2 \vee ... \vee F_n,$$
$$C \sim F_1 \wedge F_2 \wedge ... \wedge F_n.$$

In what follows we shall often use the condensed notations:

$$\bigvee_{i=1}^{n} F_i = F_1 \vee F_2 \vee ... \vee F_n;$$
$$\bigwedge_{i=1}^{n} F_i = F_1 \wedge F_2 \wedge ... \wedge F_n.$$

Therefore, in such disjunctions and conjunctions we shall drop parentheses because the distribution of parentheses is immaterial from the point of view of the equivalence relation "$\sim$". ∎

DEFINITION 3.4   *A formula L in $\mathcal{L}$ is called a **literal** if there is a propositional variable X_j $(1 \leq j \leq N)$ such that $L \equiv X_j$ or $L \equiv \neg X_j$; in the first case L is said a **positive** literal, in the second case L is said a **negative** literal.*

It is clear that in $\mathcal{L}_0$ there are $2N$ literals, namely

$$X_1, X_2, ..., X_N, \neg X_1, \neg X_2, ..., \neg X_N.$$

The proposition below is obvious.

PROPOSITION 3.4
If L is a literal containing the propositional variable X_j, then there is a literal $\tilde{L}$ containing the same propositional variable such that $\neg L \sim \tilde{L}$.

PROPOSITION 3.5

For any finite family $L_1, L_2, ...L_p$ of literals (distinct or not), there is a subfamily $L_{i_1}, L_{i_2}, ...L_{i_k}$ ($\{i_1, i_2, ...i_k\} \subset \{1, 2, ..., p\}$) of distinct literals such that

$$L_1 \vee L_2 \vee ... \vee L_p \sim L_{i_1} \vee L_{i_2} \vee ... \vee L_{i_k}$$

and

$$L_1 \wedge L_2 \wedge ... \wedge L_p \sim L_{i_1} \wedge L_{i_2} \wedge ... \wedge L_{i_k}.$$

The assertion is easily proved by induction on $p \geq 1$.

DEFINITION 3.5   A formula $F \in \mathcal{L}_0$ is called

(a)   a **normal disjunctive formula** if

$$F \equiv \bigvee_{j=1}^{m} C_j,$$

where C_j is a conjunction of literals $(1 \leq j \leq m)$;

(b)   a **normal conjunctive formula** if

$$F \equiv \bigwedge_{j=1}^{n} D_j,$$

where D_j is a disjunction of literals $(1 \leq j \leq n)$.

PROPOSITION 3.6

For any $F \in \mathcal{L}_0$ there are a normal disjunctive formula F^d and a normal conjunctive formula F^c such that

$$F \sim F^d \quad and \quad F \sim F^c.$$

Moreover, one can suppose that

$$F^d \equiv \bigvee_{j=1}^{h} C_j, \quad with \quad C_j \equiv M_1^j \wedge ... \wedge M_{p_j}^j$$

and

$$F^c \equiv \bigwedge_{j=1}^{k} D_j, \quad with \quad D_j \equiv N_1^j \vee ... \vee N_{q_j}^j$$

where $M_1^j, ..., M_{p_j}^j$ and $N_1^j, ..., N_{q_j}^j$, respectively, are distinct literals for each j.

PROOF   Taking into account that $F \in \mathcal{L}_0$ there is a formal construction $F_1, F_2, ..., F_r$ $(r \geq 1)$ such that $F_r \equiv F$. We shall prove by induction on i $(1 \leq i \leq r)$ that for each index i there are formulas F_i^d and F_i^c having the required structure, such that

$$(*) \qquad\qquad F_i \sim F_i^d \quad \text{and} \quad F_i \sim F_i^c.$$

For $i = 1$ the formula F_1 is a propositional variable, say $F \equiv X_l$ $(1 \leq l \leq N)$. In this case F_1 is obviously a literal and, consequently, one can take $F_1^d \equiv F_1^c \equiv X_l$.

For $i > 1$ suppose that the assertion $(*)$ holds for any $j < i$. If F_i coincides with a propositional variable, then the assertion follows as for $i = 1$. Thus we have to analyze two cases :

(a) $F_i \equiv \neg F_j$ with $j < i$;   (b) $F_i \equiv F_h \vee F_k$ with $h, k < i$.

In case (a), by the inductive assumption there are formulas

$$F_j^d \equiv \bigvee_{\alpha=1}^{m} C_\alpha \quad \text{and} \quad F_j^c \equiv \bigwedge_{\beta=1}^{n} D_\beta$$

where C_α is a conjunction of distinct literals $(1 \leq \alpha \leq m)$ and D_β is a disjunction of distinct literals $(1 \leq \beta \leq n)$, such that

$$F_j \sim F_j^d \quad \text{and} \quad F_j \sim F_j^c.$$

It follows that

$$\neg F_j \sim \neg F_j^d \quad \text{and} \quad \neg F_j \sim \neg F_j^c$$

$$F_i \sim \neg F_j^d \quad \text{and} \quad F_i \sim \neg F_j^c.$$

By De Morgan's laws $\neg F_j^d \sim \bigwedge\limits_{\alpha=1}^{m} \neg C_\alpha$ and $F_j^c \sim \bigvee\limits_{\beta=1}^{n} \neg D_\beta$. But $C_\alpha \equiv M_1^\alpha \wedge ... \wedge M_{p_\alpha}^\alpha$ and $D_\beta = N_1^\beta \vee ... \vee N_{q_\beta}^\beta$ hence, (by De Morgan laws) $\neg C_\alpha \sim \neg M_1^\alpha \vee ... \vee \neg M_{p_\alpha}^\alpha$ and $\neg D_\beta \sim \neg N_1^\beta \wedge ... \wedge \neg N_{q_\beta}^\beta$. We get, therefore:

$$F_i \sim \bigwedge_{\alpha=1}^{m} (\neg M_1^\alpha \vee ... \vee \neg M_{p_\alpha}^\alpha) \quad \text{and} \quad F_i \sim \bigvee_{\beta=1}^{n} (\neg N_1^\beta \wedge ... \wedge \neg N_{q_\beta}^\beta).$$

By Proposition 3.4 for each M_s^α $(1 \leq s \leq p_\alpha)$ and each N_t^β $(1 \leq t \leq q_\beta)$ there are literals $\tilde{M}_t^\beta$ containing the same variables as M_s^α and N_t^β respectively such that $\neg M_s^\alpha \sim \tilde{M}_s^\alpha$ and $\neg N_t^p \sim \tilde{N}_t^\beta$. The literals $M_1^\alpha, ..., M_{p_\alpha}^\alpha$ respectively $N_1^\beta, ..., N_{q_\beta}^\beta$ being distinct, it follows that the literals $\tilde{M}_1^\alpha, ..., \tilde{M}_{p_\alpha}^\alpha$ and $\tilde{N}_1^\beta, ..., \tilde{N}_{q_\beta}^\beta$ are also distinct. Finally, we get

$$F_i \sim \bigwedge_{\alpha=1}^{m} (\tilde{M}_1^\alpha \vee ... \vee \tilde{M}_{p_\alpha}^\alpha) \equiv F_i^c$$

and

$$F_i \sim \bigvee_{\beta=1}^{n} (\tilde{N}_1^\alpha \wedge ... \wedge \tilde{N}_{q_\beta}^\beta) \equiv F_i^d.$$

This proves the assertion for F_i; hence from the induction principle the assertion is proved for all F_i $(1 \le i \le r)$, in particular for $F_r \equiv F$. ∎

In what follows, the class of all theorems in the system **H** will be denoted by τ, i.e.,

$$\tau = \{F \in \mathcal{L}_0 | \vdash F\}.$$

PROPOSITION 3.7

The set τ is an equivalence class with respect to "$\sim$", so that $\tau \in \check{\mathcal{L}}_0$.

PROOF   Let F be a theorem in (**H**) (i.e., $F \in \tau$) and take an arbitrary $G \in \tau$. By the definition of τ we have $\vdash F$ and $\vdash G$. From the metatheorem MTh. 3.5 it follows that

$$\vdash G \Rightarrow F \quad \text{and} \quad \vdash F \Rightarrow G;$$

hence $G \sim F$, or $G \in \xi_F$. Therefore, $G \in \tau$ implies $G \in \xi_F$, which means that

$$\tau \subset \xi_F.$$

On the other hand, if $H \in \xi_F$, then $H \sim F$. This means that

$$\vdash H \Rightarrow F \quad \text{and} \quad \vdash F \Rightarrow H.$$

From the second theorem (taking into account that $\vdash F$) by (MP) we obtain $\vdash H$, i.e., $H \in \tau$. We proved that $H \in \xi_F$ implies $H \in \tau$, hence

$$\xi_F \subset \tau.$$

From the two inclusions above it follows that

$$\tau = \xi_F,$$

hence τ is an equivalence class, so that $\tau \in \check{\mathcal{L}}_0$. ∎

The complement $\bar{\tau}$ of the class τ will be denoted by θ. By definition of the operation $\bar{\xi}$ it follows that $\theta \in \check{\mathcal{L}}_0$.

REMARK 3.7   It is clear that the class θ consists of all formulas F for which there is some $G \in \tau$ such that $F \sim \neg G$. ∎

PROPOSITION 3.8

The set $\check{\mathcal{L}}_0$ endowed with the operations $\bar{\xi}$, $\xi_1 \vee \xi_2$, $\xi_1 \wedge \xi_2$ is a Boolean algebra, having θ and τ as unit elements.

PROOF   Commutativity, associativity, distributivity and absorption laws in $\check{\mathcal{L}}_0$ follow from the Proposition 3.3.

The equalities $\xi_F \vee \xi_F = \xi_F$ and $\xi_F \wedge \xi_F = \xi_F$ follow immediately from axioms (**H1**) and (**H2**); for the second equality the assertion results from metatheorems $\vdash F \Rightarrow (F \wedge F)$ and $\vdash (F \wedge F) \Rightarrow F$. Equalities

(a) $\tau \vee \xi = \xi$;   (c) $\theta \vee \xi = \xi$
(b) $\tau \wedge \xi = \xi$;   (d) $\theta \wedge \xi = \theta$

can be verified by showing that classes which appear on the two sides of these equalities contain pairs of equivalent formulas. For instance, consider equality (d) and take $H \in \theta, F \in \xi$; it follows that $H \wedge F \in \theta \wedge \xi$. From the definition of θ, there exists some $G \in \tau$ such that $H \sim \neg G$. Therefore, $H \wedge F \sim \neg G \wedge F$ and by Proposition 3.3, (h), $\neg G \wedge F \sim \neg G$, but $\neg G \sim H$ so that

$$\theta \wedge \xi \ni H \wedge F \sim H \in \theta;$$

it follows $\theta \wedge \xi = \theta$.

In the same way we obtain two more equalities:

(e) $\xi \vee \bar{\xi} = \tau$;   (f) $\xi \wedge \bar{\xi} = \theta$.

The equalities (a) to (f) show that θ and τ are the unit elements in $(\check{\mathcal{L}}_0,^-, \vee, \wedge)$. This means that $\check{\mathcal{L}}_0$ is a Boolean algebra.   ∎

REMARK 3.8   The Boolean algebra $(\check{\mathcal{L}}_0,^-, \vee, \wedge)$ will be called the syntactic Lindenbaum algebra of $\mathcal{L}_0$. In Chapter 4 we shall prove that the system **H** is non-contradictory and from this result it will follow that the syntactic Lindenbaum algebra $\check{\mathcal{L}}_0$ is isomorphic with $B(N, \Gamma)$ and with the semantic Lindenbaum algebra $\hat{\mathcal{L}}_0$.   ∎

$$if \quad \vdash F, \quad then \quad \models F.$$

PROOF   In the "Preliminaries" of the book we give a list of tautologies which contain among other things the formulas (**H1**) to (**H4**) of the system **H** (the first four formulas in list (**C**)); anyway, the fact that axioms (**H1**) to (**H4**) are tautologies can immediately be seen if we use, for instance, the properties of the functional interpretation Φ.

Now consider a formula $F \in \mathcal{L}_0$ and assume that F is a theorem. By definition there is a proof

$$(*) \qquad\qquad F_1, F_2, ..., F_r \quad (r \geq 1)$$

with $F_r \equiv F$. We shall show by induction that

$$\models F_i \quad \text{for all} \ i \ (1 \leq i \leq r).$$

For $i = 1$, the formula F_1 is necessarily an axiom and consequently $\models F_1$. Let be $i > 1$ and assume that

$$\models F_1, \ \models F_2, \ ..., \ \models F_{i-1}.$$

For the formula F_i in $(*)$ there are two possibilities :

   (a) $F_i \in Ax$;

   (b)   there are indices $h, k < i$ such that $F_k \equiv F_h \Rightarrow F_i$.

In case (a) it follows as above that $\models F_i$. In case (b), by the inductive assumption we have $\models F_h$ and $\models F_h \Rightarrow F_i$ or equivalently $\Phi(F_h) = 1$ and $\Phi(F_h \Rightarrow F_i) = 1$.

Taking into account that $\Phi(F_h) = 1$, hence $\Phi(\neg F_h) = \overline{\Phi(F_h)} = 0$, we get:

$$1 = \Phi(\neg F_h \vee F_i) = \Phi(\neg F_h) \vee \Phi(F_i) = \Phi(F_i).$$

By the principle of mathematical induction it follows that all formulas in $(*)$ are tautologies; in particular, the formula $F \equiv F_r$ is a tautology.   ∎

REMARK 4.1   The set of all tautologies was denoted in Chapter 2 by $\mathfrak{S}$. If one denotes by τ the set of all theorems, then the preceding metatheorem can be expressed by the inclusion:

$$\tau \subset \mathfrak{S}.$$

In the second section of this chapter we shall prove the converse inclusion, which is essentially deeper.   ∎

The above metatheorem and all results presented in Chapter 3 have been obtained in the system **H**. Consider now an syntactical version **S** of the truth structure on $\mathcal{L}_0$ which is defined by a set **Ax(S)** of axioms and by some set $\mathcal{R}_\mathbf{S}$ of inference rules. The syntactic truth structure on $\mathcal{L}_0$ defined by **Ax(S)** and $\mathcal{R}_\mathbf{S}$ will be called, for short, a system **S**. A formula $F \in \mathcal{L}_0$ is a theorem in **S** and we denote this fact by

$$\vdash_\mathbf{S} F$$

if there is a sequence of formulas $F_1, F_2, ..., F_r$ $(r \geq 1)$ with $F_r \equiv F$ having the property that for each i $(1 \leq i \leq r)$ the formula F_i satisfies one of the conditions :
(a) $F_i \in$ **Ax(S)**;
(b) the formula F_i is obtained from some formulas $F_{h_i}, F_{h_2}, ..., F_{h_j}$ $(h_1, h_2, ..., h_j < i)$ by means of some inference rule $r \in \mathcal{R}_\mathbf{S}$.

DEFINITION 4.1 *A system* **S** *on* $\mathcal{L}_0$ *is called* **normal** *if for any* $F, G \in \mathcal{L}_0$ *the formula* $F \Rightarrow (F \vee G)$ *is a theorem of* **S** *and the rule (MP) belongs to* $\mathcal{R}_\mathbf{S}$.

REMARK 4.2 The system **H** is obviously normal because the formula $F \Rightarrow (F \vee G)$ is an axiom of **H** (it coincides with the axiom **(H2)**) and $\mathcal{R}_\mathbf{H} = \{MP\}$. ∎

DEFINITION 4.2 *We say that a system* **S** *is* **contradictory** *if there is a formula* $F \in \mathcal{L}_0$ *such that*

$$\vdash_\mathbf{S} F \ and \ \vdash_\mathbf{S} \neg F.$$

METATHEOREM 4.2 (Metatheorem of contradictory systems)
If **S** *is a contradictory normal system, then each formula of* $\mathcal{L}_0$ *is a theorem in* **S**.

PROOF Suppose that **S** is a normal contradictory system and take an arbitrary formula $G \in \mathcal{L}_0$; we shall prove that $\vdash_\mathbf{S} G$. The system **S** being contradictory, there is a formula $F \in \mathcal{L}_0$ such that $\vdash_\mathbf{S} F$ and $\vdash_\mathbf{S} \neg F$. Consider the following proof in **S**:

$$\left.\begin{array}{c} F_1 \\ \vdots \\ F_p \equiv F \end{array}\right\} ; \quad \text{the proof of } F;$$

$$\left.\begin{array}{l} F_{p+1} \\ \vdots \\ F_{p+q} \equiv \neg F \end{array}\right\} \quad ; \quad \text{the proof of } \neg F;$$

$F_{p+q+1} \equiv \neg F \Rightarrow \neg F \vee G; \quad$ theorem in **S**;
$F_{p+q+2} \equiv \neg F \vee G; \quad (MP): \ F_{p+q}, F_{p+q+1};$
$F_{p+q+3} \equiv F \Rightarrow G;$ transcription of $F_{p+q+2};$
$F_{p+q+4} \equiv G; \quad (MP): \ F_p, \ F_{p+q+3}.$

The sequence $F_1, ..., F_{p+q+4}$ being a proof in **S**, it follows that $\vdash_{\mathbf{S}} G$. ∎

REMARK 4.3   The above metatheorem shows that contradictory (normal) systems on $\mathcal{L}_0$ are not "good" for scientific purposes because in such systems there is no difference between provable and nonprovable sentences. ∎

METATHEOREM 4.3 (Metatheorem of consistency or of non-contradiction of **H**)
*The system **H** is non-contradictory.*

PROOF   We have already mentioned that **H** is a normal system. Let us suppose that **H** is contradictory. By the preceding metatheorem we have $\vdash_{\mathbf{H}} G$, for any $G \in \mathcal{L}_0$. From the soundness metatheorem we deduce that all formulas in $\mathcal{L}_0$ are tautologies. But this conclusion is false taking into account that, for instance, the propositional variables X_j ($1 \leq j \leq N$) are not tautologies; indeed, $\Phi(X_j) = p_j$, where $p_j \in B(N, \Gamma)$ is a projector which is a non-constant Boolean function. ∎

4.2 All Tautologies are Theorems (Completeness of Propositional Logic)

The aim of this section is to prove that (in **H**)

$$\models F \quad \text{implies} \quad \vdash F.$$

In other words, if the formula $F \in \mathcal{L}_0$ has the property

$$\varphi(F) = 1 \quad \text{for all} \quad \varphi \in \mathcal{V} \quad (\text{or } \Phi(F) = 1),$$

then there is a proof (in **H**)

$$F_1, F_2, ..., F_r \quad (r \geq 1) \quad \text{with} \quad F_r \equiv F.$$

For this purpose we need some preliminary results. Recall that for any $F \in \mathcal{L}_0$ and any $\sigma \in \Gamma$, the formula F^σ coincides with F if $\sigma = 1$ and with $\neg F$ if $\sigma = 0$.

METATHEOREM 4.4 (Kalmar's Lemma)
If $F \in \mathcal{L}_0$, then for any $\varphi \in \mathcal{V}$

$$\{X_1^{\varphi(X_1)}, X_2^{\varphi(X_2)}, ..., X_N^{\varphi(X_N)}\} \vdash F^{\varphi(F)}.$$

PROOF We shall proceed by induction on $\lambda(F) = m \geq 1$. First step, $m = 1$; taking into account that the only formulas of length 1 are the propositional variables, it follows that $F \equiv X_j$ for some j $(1 \leq j \leq N)$. It is clear that for any $\varphi \in \mathcal{V}$ the formula $F^{\varphi(F)} \equiv X_j^{\varphi(X_j)}$ belongs to the set $\mathcal{H}_\varphi = \{X_1^{\varphi(X_1)}, X_2^{\varphi(X_2)}, ..., X_N^{\varphi(X_N)}\}$; therefore,

$$\mathcal{H}_\varphi \vdash F^{\varphi(F)}.$$

Second step, $m > 1$; take as an inductive assumption that for any $F' \in \mathcal{L}_0$ with $\lambda(F') < m$ and for any $\varphi \in \mathcal{V}$

$$\mathcal{H}_\varphi \vdash F'^{\varphi(F')}.$$

Let now $F \in \mathcal{L}_0$ be a formula with $\lambda(F) = m$. Taking into account that $\lambda(F) = m > 1$, the formula F cannot be a propositional variable; hence for F there are two possibilities:
  (a) $F \equiv \neg(G)$; $G \in \mathcal{L}_0$
  (b) $F \equiv (H) \vee (K)$; $H, K \in \mathcal{L}_0$.
In case (a),

$$F^{\varphi(F)} \equiv F^{\overline{\varphi(G)}} \equiv \begin{cases} F, & \text{if } \varphi(G) = 0 \\ \neg F, & \text{if } \varphi(G) = 1 \end{cases} \equiv \begin{cases} \neg\neg G, & \text{if } \varphi(G) = 1 \\ \neg G, & \text{if } \varphi(G) = 0 \end{cases}$$

On the other hand,

$$G^{\varphi(G)} \equiv \begin{cases} G, & \text{if } \varphi(G) = 1 \\ \neg G, & \text{if } \varphi(G) = 0. \end{cases}$$

Since $\lambda(G) < m$, by the induction assumption we have

$$\mathcal{H}_\varphi \vdash G^{\varphi(G)}.$$

If $\varphi(G) = 0$, then

$$\mathcal{H}_\varphi \vdash \neg G \equiv F^{\varphi(F)} \quad (\varphi(F) = 1).$$

If $\varphi(G) = 1$, then $\mathcal{H}_\varphi \vdash G$ and using the metatheorem $\vdash G \Rightarrow \neg\neg G$, we get by MP

$$\mathcal{H}_\varphi \vdash \neg\neg G \equiv F^{\varphi(F)}.$$

Therefore, in case (a) we have

$$\mathcal{H}_\varphi \vdash F^{\varphi(F)}.$$

In case (b)

$$F^{\varphi(F)} \equiv \begin{cases} (H) \vee (K) & , \text{ if } \varphi(F) = 1 \\ \neg((H) \vee (K)) & , \text{ if } \varphi(F) = 0 \end{cases} \equiv$$

$$\equiv \begin{cases} (H) \vee (K) & , \text{ if } \varphi(H) = 1 \text{ or } \varphi(K) = 1 \\ \neg((H) \vee (K)) & , \text{ if } \varphi(H) = \varphi(K) = 0. \end{cases}$$

We have to analyze two subcases: (b_1) $\varphi(F) = 1$; (b_2) $\varphi(F) = 0$.

In subcase (b_1), $\varphi(H) = 1$ or $\varphi(K) = 1$. Suppose, for instance, that $\varphi(H) = 1$; then $H^{\varphi(H)} \equiv H$ and by the inductive assumption

$$\mathcal{H}_\varphi \vdash H^{\varphi(H)} \equiv H.$$

By axiom (**H2**) we also have $\vdash H \Rightarrow (H) \vee (K)$ and by (MP) we get

$$\mathcal{H}_\varphi \vdash (H) \vee (K) \equiv F \equiv F^{\varphi(F)}.$$

In subcase (b_2), $\varphi(H) = \varphi(K) = 0$ and $F^{\varphi(F)} \equiv \neg((H) \vee (K))$. By the inductive assumption $\mathcal{H}_\varphi \vdash H^{\varphi(H)} \equiv \neg H$ and $\mathcal{H}_\varphi \vdash K^{\varphi(K)} \equiv \neg K$. Using MTh. 3.13, Section 3.2, it follows

$$\mathcal{H}_\varphi \vdash (\neg H) \wedge (\neg K).$$

By MTh. 3.17, Section 3.2, we also have

$$\vdash ((\neg H) \wedge (\neg K)) \Rightarrow (\neg(H \vee K)).$$

Using (MP) we get $\mathcal{H}_\varphi \vdash \neg(H \vee K) \equiv F^{\varphi(F)}$. Therefore, in case (b) we have again

$$\mathcal{H}_\varphi \vdash F^{\varphi(F)},$$

which achieves the second step. Now the assertion of lemma follows from the principle of mathematical induction.   ∎

Another preliminary result is the metatheorem below.

METATHEOREM 4.5 (The reduction metatheorem)
      If $\mathcal{H}, F \vdash G$ and $\mathcal{H}, \neg F \vdash G$,   then $\mathcal{H} \vdash G$.

PROOF   In Section 3.2 we have proved the metatheorem

$$(*) \quad \vdash (F \Rightarrow G) \Rightarrow ((\neg F \Rightarrow G) \Rightarrow G).$$

Taking into account that $\mathcal{H}, F \vdash G$ and $\mathcal{H}, \neg F \vdash G$, we obtain (by Herbrand's metatheorem) $\mathcal{H} \vdash F \Rightarrow G$ and $\mathcal{H} \vdash \neg F \Rightarrow G$. The last two assertions with $(*)$ provide, using rule (MP), the final result $\mathcal{H} \vdash G$.   ∎

METATHEOREM 4.6   (Completeness metatheorem)
$$If \quad \models F, \quad then \quad \vdash F.$$

PROOF   Let $F \in \mathcal{L}_0$ be a tautology, i.e., $\models F$. This means that $\varphi(F) = 1$ for all $\varphi \in \mathcal{V}$. By Kalmar's Lemma

$$(1) \quad\quad \{X_1^{\varphi(X_1)}, X_2^{\varphi(X_2)}, ..., X_N^{\varphi(X_N)}\} \vdash F, \quad \text{for all} \quad \varphi \in \mathcal{V}.$$

It is clear that the values $\varphi(X_1), \varphi(X_2), ..., \varphi(X_N)$ can be chosen arbitrarily in $\Gamma = \{0, 1\}$. Hence from (1) we get:

$$\{X_1^{\sigma_1}, X_2^{\sigma_1}, ... X_{N-1}^{\sigma_{N-1}}, X_N\} \vdash F$$

and

$$\{X_1^{\sigma_1}, X_2^{\sigma_1}, ... X_{N-1}^{\sigma_{N-1}}, \neg X_N\} \vdash F.$$

From the reduction metatheorem we obtain $\{X_1^{\sigma_1}, X_2^{\sigma_1}, ... X_{N-1}^{\sigma_{N-1}}\} \vdash F$, for all $\sigma_1, \sigma_2, ..., \sigma_{N-1} \in \Gamma$. In particular,

$$\{X_1^{\sigma_1}, ..., X_{N-2}^{\sigma_{N-2}}, X_{N-1}\} \vdash F,$$

and

$$\{X_1^{\sigma_1}, ..., X_{N-2}^{\sigma_{N-2}}, \neg X_{N-1}\} \vdash F.$$

By the same reduction metatheorem we obtain

$$\{X_1^{\sigma_1}, ..., X_{N-2}^{\sigma_{N-2}}\} \vdash F, \quad \text{for all} \quad \sigma_1, ..., \sigma_{N-2} \in \Gamma.$$

Continuing in the same way, after N steps, we shall get:

$$\vdash F,$$

which completes the proof.   ∎

REMARK 4.4   By the Soundness Metatheorem, all theorems in **H** are tautologies ($\tau \subseteq \mathfrak{I}$) and by the above metatheorem, all tautologies are theorems in **H** ($\mathfrak{I} \subseteq \tau$). Thus, the set of tautologies coincides with the set of theorems in **H** ($\tau = \mathfrak{I}$). In other words, the system **H** of Hilbert produces all tautologies and only them.   ∎

4.3   Another Proof of the Completeness Metatheorem

In the previous section it was proved that the class of theorems coincides with the class of tautologies. In other words, the syntactic version of the truth structure in $\mathcal{L}_0$ is "good" (adequate), because it allows formal proof of all semantically true formulas (tautologies) and only them. This fundamental result belongs to D. Hilbert. In his book [19], written together with W. Ackermann, Hilbert indicated another formulation of the completeness property, suggesting the new proof below of the Completeness Metatheorem.

PROPOSITION 4.1

Let F be a formula in $\mathcal{L}_0$. If F is not a theorem in $\mathbf{H}$, then the system $\mathbf{H}'$ obtained by adding the formula F as a new scheme of axioms will be inconsistent.

PROOF   In Proposition 3.6, we proved that for any formula, in particular for our formula F, there is a formula G in normal conjunctive form such that $F \sim G$. In other words, there are formulas $D_1, D_2, ..., D_M$ each D_j $(1 \le j \le m)$ being some disjunction of distinct literals such that the formula $G \equiv \bigwedge_{j=1}^{m} D_j$ is equivalent to F, i.e.,

$$\vdash F \Longleftrightarrow G$$

If we take into account that, by assumption F is not a theorem in $\mathbf{H}$, i.e., $\text{not}(\vdash F)$, it follows $\text{not}(\vdash G)$. The formula G being a conjunction of D_j s it follows, furthermore, that among $D_1, D_2, ..., D_m$ there is a disjunction $D_{j_0} \equiv D$ such that $\text{not}(\vdash D)$, hence D necessarily has the form

$$(*) \qquad D \equiv X_{i_1} \vee X_{i_2} \vee ... \vee X_{i_p} \vee \neg X_{j_1} \vee \neg X_{j_2} \vee ... \vee \neg X_{j_q},$$

where the variables $X_{i_1}, X_{i_2}, ..., X_{i_p}, X_{j_1}, ..., X_{j_q}$ are all distinct.

On the other hand, let S be an arbitrary formula in $\mathcal{L}_0$ and let S^* and S^{**} be the formulas obtained respectively from S by replacing the propositional variables $X_1, ..., X_N$ in the following manner:

- for S^*, put X_1 instead of $X_{i_1}, X_{i_2}, ..., X_{i_p}$ and $\neg X_1$ instead of X_{j_1}, $X_{j_2}, ..., X_{j_q}$ (the other variables remaining unchanged);

- for S^{**}, put $\neg X_1$ instead of $X_{i_1}, X_{i_2}, ..., X_{i_p}$ and X_1 instead of X_{j_1}, $X_{j_2}, ..., X_{j_q}$ (the other variables remaining unchanged).

With these conventions, it is easy to show that for any $H, K \in \mathcal{L}_0$ the following implication holds

$$(**) \qquad \text{if} \vdash H \Rightarrow K, \quad \text{then} \begin{cases} \vdash H^* \Rightarrow K^* \\ \text{and} \\ \vdash H^{**} \Rightarrow K^{**} \end{cases}$$

(the proof can be performed by induction along the formal proof of $H \Rightarrow K$).

Consider now the following deduction in $(\mathbf{H}, F)$ taking into account that $\vdash F \Leftrightarrow G$ and $\vdash G \Rightarrow D$:

(1) F; F is a theorem in $(\mathbf{H}, F)$;

(2) $F \Rightarrow G$ (Proposition 3.6);

(3) $F^* \Rightarrow G^*$ (from $(**)$ and (2));

(4) $F^{**} \Rightarrow G^{**}$ (from $(**)$ and (2));

(5) F^* (because F is an axiom scheme);

(6) F^{**} (because F is an axiom scheme);

(7) G^*;   (MP): (3), (5);

(8) G^{**};   (MP): (4), (6);

(9) $G^* \Rightarrow D^*$ (from $\vdash G \Rightarrow D$ and $(**)$);

(10) $G^{**} \Rightarrow D^{**}$ (from $\vdash G \Rightarrow D$ and $(**)$);

(11) D^*;   (MP): (7), (9);

(12) D^{**};   (MP): (8), (10).

But it is easy to see that $\vdash D^* \Rightarrow X_1$ and $\vdash D^{**} \Rightarrow \neg X_1$, hence from (11), (12) by MP one gets

$$\vdash_{(\mathbf{H},F)} X_1 \quad \text{and} \quad \vdash_{(\mathbf{H},F)} \neg X_1.$$

This means that the system $(\mathbf{H}, F)$ is contradictory.   ∎

PROPOSITION 4.2
If the formula F is a tautology, then F is a theorem.

PROOF   Let $F \in \mathcal{L}_0$ be a formula such that $\models F$ and assume that F is not a theorem in $\mathbf{H}$, i.e., non($\vdash F$). From Proposition 4.1 it follows that the system $(\mathbf{H}, F)$ is a contradictory one. This means that there is a formula $T \in \mathcal{L}_0$ such that

$$F \vdash T \qquad \text{and} \qquad F \vdash \neg T .$$

Using Herbrand's Metatheorem we get:

$$\vdash F \Rightarrow T \qquad \text{and} \qquad \vdash F \Rightarrow \neg T .$$

The formulas $F \Rightarrow T$ and $F \Rightarrow \neg T$ being theorems in $\mathbf{H}$ are tautologies, hence T and $\neg T$ are both tautologies, too (because F is a tautology), which is impossible. From this contradiction it follows that any tautology is a theorem in $\mathbf{H}$, i.e., the system $\mathbf{H}$ is complete. ∎

5

Other Syntactic Versions of the Truth Structure on $\mathcal{L}_0$

In Chapters 3 and 4 we presented the syntactic version of the truth structure on $\mathcal{L}_0$ within the framework of Hilbert's system **H**. This system is based on axioms (**H1**) to (**H4**) and on the inference rule MP. In Chapter 4 we have proved that the system **H** is sound, consistent and complete. In the same chapter we have introduced the general notion of an axiomatic system S as being some syntactic version of the truth structure on $\mathcal{L}_0$, with certain axioms and some inference rules being given. In the literature one can find numerous such axiomatic systems, for instance, those introduced by G. Frege, B. Russell, J. Lukasiewicz, J. B. Rosser, A. Tarski, S. C. Kleene, etc.

In this chapter we shall present only three examples of such systems and we shall prove that the first two are equivalent to **H**. The problem of the independence of axioms will also be discussed.

5.1 Systems L and M: Equivalence with System H

Systems **L** and **M** below are based on the unique inference rule MP and on the same notion of theorem; hence the difference between these systems consists in the different choices for the set of axioms.

Axioms of system L (J. Łukasiewicz)

(**L1**) $: F \Rightarrow (G \Rightarrow F)$;

(**L2**) $: (F \Rightarrow (G \Rightarrow H)) \Rightarrow ((F \Rightarrow G) \Rightarrow (F \Rightarrow H))$;

(**L3**) $: (\neg G \Rightarrow \neg F) \Rightarrow (F \Rightarrow G)$.

Axioms of system M (see E. Mendelson, [27])

(M1) : $F \Rightarrow (G \Rightarrow F)$;

(M2) : $(F \Rightarrow (G \Rightarrow H)) \Rightarrow ((F \Rightarrow G) \Rightarrow (F \Rightarrow H))$;

(M3) : $(\neg G \Rightarrow \neg F) \Rightarrow ((\neg G \Rightarrow F) \Rightarrow G)$.

REMARK 5.1   Axioms (L1), (L2), (L3) and (M1), (M2), (M3) are actually axiom schemes (as in the case of system **H**) because F, G, H occurring in these axioms represent arbitrary formulas. Moreover, it is not difficult to see that all these axioms are tautologies.   ∎

It is clear that we can compare two systems S' and S'' if they are defined on the same language. In such a case, systems S' and S'' will be called **strongly equivalent** if the following property is true:

$$\vdash_{S'} F \text{ iff } \vdash_{S''} F.$$

In what concerns the systems **L** and **M** above we can consider that formulas involved in axioms of **L** and **M** belong to the language generated by the alphabet $\{X_1, X_2, ..., X_N, \neg, \Rightarrow, (,)\}$. Therefore, these two systems are comparable; moreover, we shall prove that they are strongly equivalent.

On the other hand, the direct comparison of systems **L**, **M** with system **H** is less immediate because the language of system **H** is generated by the alphabet $\{X_1, X_2, ..., X_N, \neg, \vee, (,)\}$.

This difficulty can be overcome in the following manner. Let us denote temporarily by $\mathcal{L}_D$ the language $\mathcal{L}_0$ (of **H**) and let us call it the disjunctive language generated by the propositional variables $X_1, X_2, ..., X_N$. If in $\mathcal{L}_D$ we introduce the operator $\Rightarrow$ as an abbreviation as we did before (i.e., $F \Rightarrow G$ stands for $\neg F \vee G$), we shall get some subset $\mathcal{L}_I$ of $\mathcal{L}_D$ in which all formulas can be expressed by means of $\neg$ and $\Rightarrow$; $\mathcal{L}_I$ will be called the *implicative sublanguage* of $\mathcal{L}_D$. It is easy to see that $\mathcal{L}_I$ coincides with the set of all formulas in $\mathcal{L}_D$ which can be obtained by means of formal constructions defined in Chapter 1, the only difference consisting of the formal use of $\Rightarrow$ instead of $\vee$. Although the language $\mathcal{L}_D$ and its sublanguage $\mathcal{L}_I$ are different ($\mathcal{L}_I \not\subseteq \mathcal{L}_D$), they are, however, sufficiently close for our purposes.

It is easy to see by inductive arguments (on the length of formulas), that Boolean interpretations and functional interpretation of $\mathcal{L}_I$ coincide with the restrictions to $\mathcal{L}_I$ of the corresponding interpretations of $\mathcal{L}_D$. It follows that any universally true formula F' of $\mathcal{L}_I$ is a universally true formula of $\mathcal{L}_D$. The converse of this assertion is not true in the technical sense, but for any formula $F \in \mathcal{L}_D$ one can prove that there is a formula $F' \in \mathcal{L}_I$ such that the formula $F \Leftrightarrow F'$ is universally true; therefore, if a formula $F \in \mathcal{L}_D$ is universally true, then $F' \in \mathcal{L}_I$ is universally true. In this proof one has to use the fact that $(F \vee G) \Leftrightarrow (\neg \neg F \vee G)$ is a universally true formula. In this section we shall prove that the pair $(\mathbf{L}, \mathbf{H})$, respectively, the pair $(\mathbf{M}, \mathbf{H})$ have the following two properties:

(a) if $\vdash_{\mathbf{L}} F$ (respectively $\vdash_{\mathbf{M}} F$), then $\vdash_{\mathbf{H}} F$ $(F \in \mathcal{L}_I)$;

(b) if $\vdash_{\mathbf{H}} F$ $(F \in \mathcal{L}_D)$, then there is a formula $F' \in \mathcal{L}_I$ such that

$$\models F \Leftrightarrow F' \quad \text{and} \quad \vdash_{\mathbf{L}} F' \quad (\text{respectively } \vdash_{\mathbf{M}} F').$$

For this reason we shall say that the pairs $(\mathbf{L}, \mathbf{H})$ and $(\mathbf{M}, \mathbf{H})$ are **simply equivalent**.

In the following we shall also prove that systems $\mathbf{L}$ and $\mathbf{M}$ are strongly equivalent. First, let us prove the following result (the property (a) above).

PROPOSITION 5.1
Let F be a formula in $\mathcal{L}_I$. If

$$\vdash_{\mathbf{L}} F \quad (\text{respectively } \vdash_{\mathbf{M}} F),$$

then $\vdash_{\mathbf{H}} F$.

PROOF   Taking into account that all axioms in $\mathbf{L}$ (and $\mathbf{M}$) are tautologies we deduce that they are theorems in $\mathbf{H}$. Suppose now that a formula $F \in \mathcal{L}_I$ is a theorem in $\mathbf{L}$, i.e.,

$$\vdash_{\mathbf{L}} F.$$

This means that there is a proof in $\mathbf{L}$ of the formula F; in other words, there is a sequence

$$(*) \quad F_1, F_2, ..., F_r \quad (r \geq 1)$$

with $F_r \equiv F$ such that for each index i $(1 \leq i \leq r)$ the formula F_i is an axiom of $\mathbf{L}$ or can be obtained by MP from some formulas F_h, F_k in $(*)$ $(h, k < i)$. We shall prove that all formulas in $(*)$ are tautologies, i.e., for all i $(1 \leq i \leq r)$

$$(**) \quad \models F_i.$$

We shall proceed by (finite) induction on i. For $i = 1$ the formula F_i is an axiom in $\mathbf{L}$, therefore $\models F_1$. Let be $i > 1$ and suppose as an inductive assumption that

$$\models F_1, \quad \models F_2, \quad ..., \quad \models F_{i-1}.$$

For the formula F_i there are two possibilities:
(1) F_i is an axiom of $\mathbf{L}$; (2) F_i is obtained in $(*)$ by MP.
In case (1) F_i is a tautology, i.e., $\models F_i$. In case (2) there are indices $h, k < i$ such that $F_h \equiv A$, $F_k \equiv A \Rightarrow B$, $F_i \equiv B$. By the inductive assumption, A

and $A \Rightarrow B$ are tautologies. It follows that B is also a tautology, hence F_i is a tautology. By the principle of mathematical induction it follows that F_i is a tautology for all i $(1 \leq i \leq r)$; in particular, $F_r \equiv F$ is a tautology and, by the completeness metatheorem, F is also a theorem in **H**, i.e., $\vdash_\mathbf{H} F$. ∎

We shall proceed now to the comparison of **L** and **M** and we shall begin by showing that any theorem in **M** is a theorem in **L**. It is easy to see that for this purpose it is sufficient to prove that all axioms of **M** are theorems in **L**. Taking into account that the two first axioms in **L** and **M** coincide, it remains to prove that the axiom (**M3**) is a theorem in **L**.

PROPOSITION 5.2

$$\vdash_\mathbf{L} F \Rightarrow F.$$

PROOF   Consider the following (formal) proof:

(1) $F \Rightarrow (F \Rightarrow F)$; axiom (**L1**);

(2) $(F \Rightarrow ((F \Rightarrow F) \Rightarrow F)) \Rightarrow ((F \Rightarrow (F \Rightarrow F)) \Rightarrow (F \Rightarrow F))$; axiom (**L2**);

(3) $F \Rightarrow ((F \Rightarrow F) \Rightarrow F)$; axiom (**L1**);

(4) $(F \Rightarrow (F \Rightarrow F)) \Rightarrow (F \Rightarrow F)$; (MP): (2), (3);

(5) $F \Rightarrow F$; (MP): (1), (4).

Therefore, $\vdash_\mathbf{L} F \Rightarrow F$.   ∎

PROPOSITION 5.3

$$\vdash_\mathbf{L} \neg F \Rightarrow (F \Rightarrow G).$$

PROOF   Consider the following (formal) proof:

(1) $\neg F \Rightarrow (\neg G \Rightarrow \neg F)$; axiom (**L1**);

(2) $(\neg G \Rightarrow \neg F) \Rightarrow (F \Rightarrow G)$; axiom (**L3**);

(3) $((\neg G \Rightarrow \neg F) \Rightarrow (F \Rightarrow G)) \Rightarrow (\neg F \Rightarrow ((\neg G \Rightarrow \neg F) \Rightarrow (F \Rightarrow G)))$; axiom (**L1**);

(4) $\neg F \Rightarrow ((\neg G \Rightarrow \neg F) \Rightarrow (F \Rightarrow G))$; (MP): (2), (3);

(5) $(\neg F \Rightarrow ((\neg G \Rightarrow \neg F) \Rightarrow (F \Rightarrow G))) \Rightarrow ((\neg F \Rightarrow ((\neg G \Rightarrow \neg F)) \Rightarrow (\neg F \Rightarrow (F \Rightarrow G)))$; axiom (**L2**);

(6) $(\neg F \Rightarrow (\neg G \Rightarrow \neg F)) \Rightarrow (\neg F \Rightarrow (F \Rightarrow G))$; (MP): (4), (5);

(7) $\neg F \Rightarrow (F \Rightarrow G)$; (MP): (1), (6).

Therefore, $\vdash_{\mathbf{L}} \neg F \Rightarrow (F \Rightarrow G)$. ∎

PROPOSITION 5.4
(Deduction metatheorem in L)
Let $\mathcal{H}$ be a set of formulas in $\mathcal{L}_0$ and $F, G \in \mathcal{L}_0$.

$$\text{If } \mathcal{H}, F \vdash_{\mathbf{L}} G, \quad \text{then } \mathcal{H} \vdash_{\mathbf{L}} F \Rightarrow G.$$

PROOF Suppose that $\mathcal{H}, F \vdash_{\mathbf{L}} G$ and let

$$(*) \quad K_1, K_2, ..., K_r \quad (r \geq 1),$$

$K_r \equiv G$ be a deduction from $\mathcal{H}, F$ in **L** of the formula G. We shall show by induction on $i \geq 1$ that

$$\mathcal{H} \vdash_{\mathbf{L}} F \Rightarrow K_i$$

for all i $(1 \leq i \leq r)$.
In case $i = 1$, there are three possibilities:

(a) K_1 is an axiom of **L**;

(b) $K_1 \in \mathcal{H}$;

(c) $K_1 \equiv F$.

In the situations (a) or (b) we have the following deduction from $\mathcal{H}$:

(1) K_1; axiom in **L** or $K_1 \in \mathcal{H}$;
(2) $K_1 \Rightarrow (F \Rightarrow K_1)$; axiom (**L1**);
(3) $F \Rightarrow K_1$; (MP): (1), (2).

It follows that $\mathcal{H} \vdash F \Rightarrow K_1$. In the situation (c), $F \Rightarrow K_1 \equiv F \Rightarrow F$; by Proposition 5.2 we have $\vdash_{\mathbf{L}} F \Rightarrow F$, hence $\vdash_{\mathbf{L}} F \Rightarrow K_1$ and obviously $\mathcal{H} \vdash_{\mathbf{L}} F \Rightarrow K_1$.
In case $i > 1$, assume that

$$\mathcal{H} \vdash_{\mathbf{L}} F \Rightarrow K_1, ..., \mathcal{H} \vdash_{\mathbf{L}} F \Rightarrow K_{i-1}.$$

If K_i is an axiom of **L** or $K_i \in \mathcal{H}$ or $K_i \equiv F$, then $\mathcal{H} \vdash_\mathbf{L} F \Rightarrow K_i$ as in case $i = 1$. Therefore, we have to analyze the situation when K_i is obtained in (*) by (MP). Consequently, suppose that there are indices $h, l < i$ such that

$$K_h \equiv A, \quad K_l \equiv A \Rightarrow B \quad \text{and} \quad K_i \equiv B.$$

By the inductive assumption we have

$$\mathcal{H} \vdash_\mathbf{L} F \Rightarrow A \quad \text{and} \quad \mathcal{H} \vdash_\mathbf{L} F \Rightarrow (A \Rightarrow B).$$

Consider now the following deduction from $\mathcal{H}$ in **L**:

$$\left.
\begin{array}{ll}
(1) & \\
\vdots & \\
(p) & F \Rightarrow A
\end{array}
\right\} ; \quad \text{the deduction from } \mathcal{H} \text{ of } F \Rightarrow A;$$

$$\left.
\begin{array}{ll}
(p+1) & \\
\vdots & \\
(p+q) & F \Rightarrow (A \Rightarrow B)
\end{array}
\right\} ; \text{the deduction from } \mathcal{H} \text{ of } F \Rightarrow (A \Rightarrow B);$$

$(p+q+1)$ $(F \Rightarrow (A \Rightarrow B)) \Rightarrow ((F \Rightarrow A) \Rightarrow (F \Rightarrow B))$; axiom (**L2**);
$(p+q+2)$ $(F \Rightarrow A) \Rightarrow (F \Rightarrow B)$; (MP): $(p+q)$, $(p+q+1)$;
$(p+q+3)$ $F \Rightarrow B$; (MP): (p), $(p+q+2)$.

It follows that $\mathcal{H} \vdash_\mathbf{L} F \Rightarrow B$ or $\mathcal{H} \vdash_\mathbf{L} F \Rightarrow K_i$. By the induction principle we get $\mathcal{H} \vdash_\mathbf{L} F \Rightarrow K_i$ for all i $(1 \le i \le r)$. In particular, $\mathcal{H} \vdash_\mathbf{L} F \Rightarrow G$ $(G \equiv K_r)$ and the proposition is proved. ∎

PROPOSITION 5.5

$$\vdash_\mathbf{L} (F \Rightarrow G) \Rightarrow ((G \Rightarrow H) \Rightarrow (F \Rightarrow H)).$$

PROOF   Consider the following deduction in **L** from $\mathcal{H} = \{F, F \Rightarrow G, G \Rightarrow H\}$:

(1) $F \Rightarrow G$; hypothesis;

(2) $G \Rightarrow H$; hypothesis;

(3) F; hypothesis;

(4) G; (MP): (1), (3);

(5) H; (MP): (2), (4).

It follows that

$$\{F, F \Rightarrow G, G \Rightarrow H\} \vdash_\mathbf{L} H.$$

Applying the deduction metatheorem three times successively one gets

$$\vdash_{\mathbf{L}} (F \Rightarrow G) \Rightarrow ((G \Rightarrow H) \Rightarrow (F \Rightarrow H)). \quad \blacksquare$$

PROPOSITION 5.6
(Metatheorem of Syllogism)

If $\vdash_{\mathbf{L}} F \Rightarrow G$ *and* $\vdash_{\mathbf{L}} G \Rightarrow H$, *then* $\vdash_{\mathbf{L}} F \Rightarrow H$.

PROOF   Using the previous proposition $\vdash_{\mathbf{L}} (F \Rightarrow G) \Rightarrow ((G \Rightarrow H) \Rightarrow (F \Rightarrow H))$ and using $\vdash_{\mathbf{L}} F \Rightarrow G$ and $\vdash_{\mathbf{L}} G \Rightarrow H$, by (MP) applied twice one obtains $\vdash_{\mathbf{L}} F \Rightarrow H$. $\blacksquare$

PROPOSITION 5.7

$$\vdash_{\mathbf{L}} \neg\neg F F \Rightarrow F.$$

PROOF   Consider the following deduction:

(1) $\neg\neg F$; hypothesis;

(2) $\neg\neg F \Rightarrow (\neg F \Rightarrow \neg\neg\neg F)$; Proposition 5.3;

(3) $\neg F \Rightarrow \neg\neg\neg F$; (MP): (1), (2);

(4) $(\neg F \Rightarrow \neg\neg\neg F) \Rightarrow (\neg\neg F \Rightarrow F)$; axiom (**L3**);

(5) $\neg\neg F \Rightarrow F$; (MP): (3), (4);

(6) F; (MP):(1), (5).

It follows that
$$\neg\neg F \vdash_{\mathbf{L}} F$$
and by the Deduction Metatheorem one gets $\vdash_{\mathbf{L}} \neg\neg F \Rightarrow F$. $\blacksquare$

PROPOSITION 5.8

$$\vdash_{\mathbf{L}} F \Rightarrow \neg\neg F.$$

PROOF   By Proposition 5.7, $\vdash_{\mathbf{L}} \neg\neg\neg F \Rightarrow \neg F$; we have also (axiom (**L3**)),
$$\vdash_{\mathbf{L}} (\neg\neg\neg F \Rightarrow \neg F) \Rightarrow (F \Rightarrow \neg\neg F).$$

From the two theorems above, by (MP), one gets $\vdash_{\mathbf{L}} F \Rightarrow \neg\neg F$. $\blacksquare$

PROPOSITION 5.9

$$\vdash_{\mathbf{L}} (F \Rightarrow G) \Rightarrow (\neg G \Rightarrow \neg F).$$

PROOF   Consider the following deduction in **L**:

(1) $F \Rightarrow G$; hypothesis;

(2) $\neg\neg F$; hypothesis;

(3) $\neg\neg F \Rightarrow F$; Proposition 5.7;

(4) F; (MP): (2), (3);

(5) G; (MP): (1), (4);

(6) $G \Rightarrow \neg\neg G$; Proposition 5.8;

(7) $\neg\neg G$; (MP): (5), (6).

Therefore, $F \Rightarrow G, \neg\neg F \vdash_{\mathbf{L}} \neg\neg G$, hence $F \Rightarrow G \vdash_{\mathbf{L}} \neg\neg F \Rightarrow \neg\neg G$ and using axiom (**L3**) we have $\vdash_{\mathbf{L}} (\neg\neg F \Rightarrow \neg\neg G) \Rightarrow (\neg G \Rightarrow \neg F)$ and by MP, $F \Rightarrow G \vdash_{\mathbf{L}} \neg G \Rightarrow \neg F$. Applying once again the Deduction Metatheorem we obtain $\vdash_{\mathbf{L}} (F \Rightarrow G) \Rightarrow (\neg G \Rightarrow \neg F)$.   ∎

PROPOSITION 5.10
*Axiom (**M3**) is a theorem in **L**, i.e.,*

$$\vdash_{\mathbf{L}} (\neg G \Rightarrow \neg F) \Rightarrow ((\neg G \Rightarrow F) \Rightarrow G).$$

PROOF   Consider the following deduction in **L**:

(1) $F \Rightarrow G$;   hypothesis;

(2) $\neg F \Rightarrow G$;   hypothesis;

(3) $(F \Rightarrow G) \Rightarrow (\neg G \Rightarrow \neg F)$;   Proposition 5.9;

(4) $\neg G \Rightarrow \neg F$;   (MP): (1), (3);

(5) $\neg G \Rightarrow G$;   (syl): (4), (2);

(6) $(\neg G \Rightarrow (G \Rightarrow \neg(\neg G \Rightarrow G))) \Rightarrow ((\neg G \Rightarrow G) \Rightarrow (\neg G \Rightarrow \neg(\neg G \Rightarrow G)))$; axiom (**L2**);

(7) $\neg G \Rightarrow (G \Rightarrow \neg(\neg G \Rightarrow G))$;   Proposition 5.3;

(8) $(\neg G \Rightarrow G) \Rightarrow (\neg G \Rightarrow \neg(\neg G \Rightarrow G))$;   (MP) : (7), (6);

(9) $\neg G \Rightarrow \neg(\neg G \Rightarrow G)$; (MP): (5), (8);

(10) $(\neg G \Rightarrow \neg(\neg G \Rightarrow G)) \Rightarrow ((\neg G \Rightarrow G) \Rightarrow G)$;   axiom **L3**;

(11) $(\neg G \Rightarrow G) \Rightarrow G$: (MP): (9), (10);

(12) G; (MP): (5), (11).

Therefore,
$$F \Rightarrow G, \neg F \Rightarrow G \vdash_{\mathbf{L}} G$$

and by the deduction metatheorem $\vdash_{\mathbf{L}} (F \Rightarrow G) \Rightarrow ((\neg F \Rightarrow G) \Rightarrow G)$. On the other hand, we have $\vdash_{\mathbf{L}} (\neg G \Rightarrow \neg F) \Rightarrow (F \Rightarrow G)$ and by the rule of syllogism it follows that

$$(*) \quad \vdash_{\mathbf{L}} (\neg G \Rightarrow \neg F) \Rightarrow ((\neg F \Rightarrow G) \Rightarrow G).$$

Consider now the following deduction:

(1) $\neg G \Rightarrow F$; hypothesis;

(2) $\neg G \Rightarrow \neg F$; hypothesis;

(3) $(\neg G \Rightarrow \neg F) \Rightarrow ((\neg F \Rightarrow G) \Rightarrow G)$;   theorem $(*)$ above;

(4) $(\neg F \Rightarrow G) \Rightarrow G$; (MP): (3), (2);

(5) $(\neg G \Rightarrow F) \Rightarrow (\neg F \Rightarrow \neg\neg G)$; Proposition 5.9;

(6) $\neg F \Rightarrow \neg\neg G$; (MP): (1), (5);

(7) $\neg\neg G \Rightarrow G$; Proposition 5.7;

(8) $\neg F \Rightarrow G$; (syl): (6), (7);

(9) G; (MP): (4), (8).

Therefore, $\neg G \Rightarrow F, \neg G \Rightarrow \neg F \vdash_L G$ and applying the Deduction Metatheorem twice we get

$$\vdash_L (\neg G \Rightarrow \neg F) \Rightarrow ((\neg G \Rightarrow F) \Rightarrow G). \quad \blacksquare$$

The Proposition 5.10 completes the proof that all theorems in **M** are theorems in **L**, too.

Next we shall prove that each theorem in **L** is a theorem in **M**. For this purpose it is sufficient to prove that the axiom (**L3**) is a theorem in **M** (because (**L1**) and (**L2**) coincide, respectively, with (**M1**) and (**M2**)).

First, let us remark that in the theorem $\vdash_{\mathbf{L}} F \Rightarrow F$ (Proposition 5.2) we have only made use of axioms (**L1**) and (**L2**) (and the rule MP). It follows that $F \Rightarrow F$ is a theorem in **M** too. The same remark also holds for the Deduction Metatheorem in **L** (Proposition 5.4); hence we get the following two propositions.

PROPOSITION 5.11

$$\vdash_M F \Rightarrow F.$$

PROPOSITION 5.12
(Deduction Metatheorem in M)

$$If \ \mathcal{H}, F \vdash_M G, \quad then \ \mathcal{H} \vdash_M F \Rightarrow G.$$

Using the above Deduction Metatheorem one can easily establish the following three propositions.

PROPOSITION 5.13

$$\vdash_M (F \Rightarrow G) \Rightarrow ((G \Rightarrow H) \Rightarrow (F \Rightarrow H)).$$

As a direct corollary of the above proposition one gets the following result.

PROPOSITION 5.14
(Metatheorem of Syllogism in M)

$$If \ \vdash_M F \Rightarrow G \ and \ \vdash_M G \Rightarrow H, \quad then \ \vdash_M F \Rightarrow H.$$

PROPOSITION 5.15

$$\vdash_M (F \Rightarrow (G \Rightarrow H)) \Rightarrow (G \Rightarrow (F \Rightarrow H)).$$

We are now ready to prove that the axiom (**L3**) is a theorem in **M**.

PROPOSITION 5.16

$$\vdash_M (\neg G \Rightarrow \neg F) \Rightarrow (F \Rightarrow G).$$

PROOF   Consider the following deduction in **M**:

(1) $\neg G \Rightarrow \neg F$; hypothesis;

(2) F; hypothesis;

(3) $(\neg G \Rightarrow \neg F) \Rightarrow ((\neg G \Rightarrow F) \Rightarrow G)$; axiom (**M3**);

(4) $(\neg G \Rightarrow F) \Rightarrow G$; (MP): (1), (3);

(5) $F \Rightarrow (\neg G \Rightarrow F)$; axiom (**M1**);

(6) $\neg G \Rightarrow F$; (MP): (2), (5);

(7) G; (MP): (4), (6).

Therefore,

$$\neg G \Rightarrow \neg F, \quad F \vdash_{\mathbf{M}} G.$$

Applying the deduction metatheorem two times successively one gets $\vdash_{\mathbf{M}}$ $(\neg G \Rightarrow \neg F) \Rightarrow (F \Rightarrow G)$. ∎

The above proposition achieves the proof of the fact that systems **L** and **M** are strongly equivalent.

In the remaining part of this section we shall finish the proof that systems **L** and **M** are simply equivalent with **H**. Taking into account that the assertion (a) in the definition of the simple equivalence is already proved (Proposition 5.1), it remains to prove the assertion (b) of the same definition. In order to do this we shall first show that system **M** (therefore the system **L**) is complete, i.e., we shall prove that any tautology (in $\mathcal{L}_I$) is a theorem in **M**. The fact that any theorem in **M** is a tautology in $\mathcal{L}_I$ follows in the usual way, taking into account that the axioms of **M** are tautologies (in $\mathcal{L}_I$). We need the following five preliminary propositions.

PROPOSITION 5.17

$$\vdash_{\mathbf{M}} H \Rightarrow (\neg K \Rightarrow \neg(H \Rightarrow K)).$$

PROOF   It is clear that $H, H \Rightarrow K \vdash_{\mathbf{M}} K$, hence

$$\vdash_{\mathbf{M}} H \Rightarrow ((H \Rightarrow K) \Rightarrow K).$$

We have also $\vdash_{\mathbf{L}} ((H \Rightarrow K) \Rightarrow K) \Rightarrow (\neg K \Rightarrow \neg(H \Rightarrow K))$, hence

$$\vdash_{\mathbf{M}} ((H \Rightarrow K) \Rightarrow K) \Rightarrow (\neg K \Rightarrow \neg(H \Rightarrow K))$$

(because **M** and **L** are strongly equivalent). By the metatheorem of syllogism (Proposition 5.14) we get

$$\vdash_{\mathbf{M}} H \Rightarrow (\neg K \Rightarrow \neg(H \Rightarrow K)). ∎$$

As a corollary of the above proposition, we get the following result.

PROPOSITION 5.18

If $\vdash_M H$ and $\vdash_M \neg K$, then $\vdash_M \neg(H \Rightarrow K)$.

PROPOSITION 5.19

$$\vdash_M (F \Rightarrow G) \Rightarrow ((\neg F \Rightarrow G) \Rightarrow G).$$

PROOF   Consider the following deduction in **M**:

(1) $F \Rightarrow G$; hypothesis;

(2) $\neg F \Rightarrow G$; hypothesis;

(3) $(F \Rightarrow G) \Rightarrow (\neg G \Rightarrow \neg F)$; theorem in **M** because theorem in **L** (Proposition 5.9);

(4) $\neg G \Rightarrow \neg F$; (MP): (1), (3);

(5) $(\neg F \Rightarrow G) \Rightarrow (\neg G \Rightarrow \neg\neg F)$; same argument as for (3);

(6) $\neg G \Rightarrow \neg\neg F$; (MP): (2), (5);

(7) $(\neg G \Rightarrow \neg\neg F) \Rightarrow ((\neg G \Rightarrow \neg F) \Rightarrow G)$; axiom (**M3**);

(8) $(\neg G \Rightarrow \neg F) \Rightarrow G$; (MP): (6), (7);

(9) G; (MP): (4), (8).

Therefore, $F \Rightarrow G, \neg F \Rightarrow G \vdash_M G$. Applying the deduction metatheorem two times successively we get

$$\vdash_M (F \Rightarrow G) \Rightarrow ((\neg F \Rightarrow G) \Rightarrow G). \quad \blacksquare$$

The following proposition is an immediate corollary of the above result.

PROPOSITION 5.20

If $\mathcal{H} \vdash_M F \Rightarrow G$ and $\mathcal{H} \vdash_M \neg F \Rightarrow G$, then $\mathcal{H} \vdash_M G$.

PROPOSITION 5.21
(**Kalmar's Lemma**)
If φ is an arbitrary interpretation of $\mathcal{L}_I$ and $F \in \mathcal{L}_I$, then

$$\mathcal{H}_\varphi \vdash_M F^{\varphi(F)},$$

where $\mathcal{H}_\varphi = \{X_1^{\varphi(X_1)}, X_2^{\varphi(X_2)}, ..., X_N^{\varphi(X_N)}\}$.

PROOF   We shall proceed by induction on the number $\mu(F)$ of occurrences of symbols $\neg$, $\Rightarrow$ in F. If $\mu(F) = 0$, then there obviously exists j $(1 \leq j \leq N)$ such that $F \equiv X_j$. Therefore, $F^{\varphi(F)} \equiv X_j^{\varphi(X_j)} \in \mathcal{H}_\varphi$, hence $\mathcal{H}_\varphi \vdash_M F^{\varphi(F)}$.

Let $\mu(F) = m > 1$ and assume that, for any formula $F' \in \mathcal{L}_I$ with $\mu(F') < m$,

$$\mathcal{H}_\varphi \vdash_M F'^{\varphi(F')}.$$

We have to analyze two cases:

(a) $F \equiv \neg G$;

(b) $F \equiv H \Rightarrow K$.

In case (a), suppose $\varphi(F) = 1$; it follows $F^{\varphi(F)} \equiv F, \varphi(G) = 0, G^{\varphi(G)} \equiv \neg G$. By the induction assumption we have

$$\mathcal{H}_\varphi \vdash_M G^{\varphi(G)} \equiv \neg G \equiv F \equiv F^{\varphi(F)}.$$

Suppose that $\varphi(F) = 0$, then $F^{\varphi(F)} \equiv \neg F, \varphi(G) = 1$ and $G^{\varphi(G)} \equiv G$. By the induction assumption we have $\mathcal{H}_\varphi \vdash_M G^{\varphi(G)} \equiv G$. We have also (Proposition 5.8) $\vdash_L G \Rightarrow \neg\neg G$, hence (by the strong equivalence of **L** and **M**) $\vdash_M G \Rightarrow \neg\neg G$. Using rule MP we get

$$\vdash_M \neg\neg G \equiv \neg(\neg G) \equiv \neg F \equiv F^{\varphi(F)}.$$

Therefore, in case (a) the assertion for F is proved.

In case (b), suppose $\varphi(F) = 1$; it follows that $F^{\varphi(F)} \equiv F \equiv H \Rightarrow K$ and $\varphi(H) = 0$ or $\varphi(K) = 1$. If $\varphi(H) = 0$, then $H^{\varphi(H)} \equiv \neg H$ and

$$\mathcal{H}_\varphi \vdash_M H^{\varphi(H)} \equiv \neg H.$$

On the other hand we have $\vdash_L \neg H \Rightarrow (H \Rightarrow K)$ (Proposition 5.3), hence

$$\vdash_M \neg H \Rightarrow (H \Rightarrow K).$$

Using MP one gets $\mathcal{H}_\varphi \vdash_M H \Rightarrow K \equiv F \equiv F^{\varphi(F)}$. Suppose that $\varphi(F) = 0$, then $F^{\varphi(F)} \equiv \neg F \equiv \neg(H \Rightarrow K)$, $\varphi(H) = 1$ and $\varphi(K) = 0$. By the induction assumption we have

$$\mathcal{H}_\varphi \vdash_M H^{\varphi(H)} \equiv H \quad \text{and} \quad \mathcal{H}_\varphi \vdash_M K^{\varphi(K)} \equiv \neg K.$$

Using the Proposition 5.18 we obtain finally

$$\mathcal{H}_\varphi \vdash_M \neg(H \Rightarrow K) \equiv \neg F \equiv F^{\varphi(F)}.$$

The proof of the required statement follows now from the principle of mathematical induction.   ∎

PROPOSITION 5.22
(The completeness of M)
Let $F \in \mathcal{L}_I$.

$$\text{If} \ \models F, \quad \text{then} \quad \vdash_{\mathbf{M}} F.$$

PROOF   Taking into account that F is a universally true formula we have $\varphi(F) = 1$ for any Boolean interpretation φ of $\mathcal{L}_I$. By Kalmar's Lemma (Proposition 5.21) it follows that

$$\{X_1^{\alpha_1}, X_2^{\alpha_2}, ..., X_N^{\alpha_N}\} \vdash_{\mathbf{M}} F$$

for all $\alpha_1, \alpha_2, ..., \alpha_N \in \Gamma$. If we take interpretations φ', φ'' with

$$\varphi' : \alpha_1, \alpha_2, ..., \alpha_{N-1}, 1,$$

$$\varphi'' : \alpha_1, \alpha_2, ..., \alpha_{N-1}, 0,$$

then

$$\{X_1^{\alpha_1}, X_2^{\alpha_2}, ..., X_{N-1}^{\alpha_{N-1}}, X_N\} \vdash_{\mathbf{M}} F$$

$$\{X_1^{\alpha_1}, X_2^{\alpha_2}, ..., X_{N-1}^{\alpha_{N-1}}, \neg X_N\} \vdash_{\mathbf{M}} F.$$

By the Deduction Metatheorem we have:

$$\{X_1^{\alpha_1}, X_2^{\alpha_2}, ..., X_{N-1}^{\alpha_{N-1}}\} \vdash_{\mathbf{M}} X_N \Rightarrow F$$

and

$$\{X_1^{\alpha_1}, X_2^{\alpha_2}, ..., X_{N-1}^{\alpha_{N-1}}\} \vdash_{\mathbf{M}} \neg X_N \Rightarrow F,$$

for all $\alpha_1, \alpha_2, ..., \alpha_{N-1} \in \Gamma$. Applying Proposition 5.20 we obtain

$$\{X_1^{\alpha_1}, X_2^{\alpha_2}, ..., X_{N-1}^{\alpha_{N-1}}\} \vdash_{\mathbf{M}} F.$$

In this way, after N steps, we finally arrive at $\vdash_{\mathbf{M}} F$.   ∎

REMARK 5.2   Taking into account that systems **L** and **M** are strongly equivalent, from Proposition 5.22 it follows that system **L** is also complete. ∎

As a consequence of Proposition 5.1 we get almost directly the following result.

PROPOSITION 5.23
The systems **L** and **M** are non-contradictory.

PROOF   First, let us remark that **L** is non-contradictory iff **M** is non-contradictory (because **L** and **M** are strongly equivalent). Suppose, for instance, that **L** is contradictory. This means that there is a formula F in $\mathcal{L}_I$ such that

$$\vdash_{\mathbf{L}} F \quad \text{and} \quad \vdash_{\mathbf{L}} \neg F.$$

From Proposition 5.1 it follows that

$$\vdash_{\mathbf{H}} F \quad \text{and} \quad \vdash_{\mathbf{H}} \neg F.$$

which is impossible, because **H** is non-contradictory.   ∎

Finally, we can prove the assertion (b) in the definition of the simple equivalence of **M** and **H** (respectively, **L** and **H**.)

PROPOSITION 5.24

Consider a formula $F \in \mathcal{L}_0$. If $\vdash_{\mathbf{H}} F$, then there is a formula $F' \in \mathcal{L}_I$ such that $\models F' \Leftrightarrow F$ and $\vdash_{\mathbf{M}} F'$ (hence $\vdash_{\mathbf{L}} F'$).

PROOF   As we have already mentioned, for any formula $F \in \mathcal{L}_D$ there is a formula $F' \in \mathcal{L}_I$ such that

$$(*) \qquad\qquad\qquad \models F \Leftrightarrow F'.$$

If $\vdash_{\mathbf{H}} F$, then $\models F$ and from $(*)$ we have $\models F'$. By the completeness metatheorem (Proposition 5.22) it follows that $\vdash_{\mathbf{M}} F'$. Systems **L** and **M** being strongly equivalent we obtain also $\vdash_{\mathbf{L}} F'$.   ∎

REMARK 5.3   From Proposition 5.1 and Proposition 5.24 it follows that both systems **L** and **M** are simply equivalent to **H**.   ∎

5.2   Some Remarks about the Problem of the Independence of Axioms

Let S be a system representing some syntactic truth structure on a language $\mathcal{L}$. We denote by $Ax(S)$ the set of axioms of S. If $S_0 \in Ax(S)$, then we denote by S' the syntactic truth structure on $\mathcal{L}$ with the same set of inference rules and with $Ax(S) \setminus \{S_0\}$ as a set of axioms. The set $Ax(S)$ is called **dependent** if there is an axiom $S_0 \in Ax(S)$ such that $\vdash_{S'} S_0$; otherwise, the set $Ax(S)$ is called **independent**.

The problem of independence for an arbitrary system S is rather difficult, but for systems $\mathbf{H}, \mathbf{M}$ and $\mathbf{L}$ it can be solved pursuing a subtle idea of Hilbert.

Roughly speaking, the idea of Hilbert consists of finding some property p such that all axioms of S' have this property and if formulas A and $A \Rightarrow B$ have the property p, then B also has the property p. It follows that a formula F which does not have the property p cannot be a theorem of S'. Usually the property p is expressed in terms of some "interpretation" or by some arithmetical (computable) means.

PROPOSITION 5.25

The axioms of $\mathbf{H}$ *are independent.*

PROOF   ([19]) For each axiom of $\mathbf{H}$ we shall indicate a property p such that the given axiom does not have this property, while the other axioms have the property p and rule MP preserves this property.

Independence of (H1). Consider the set of "values" $V = \{0, 1, 2\}$ and define the following two operations $\tilde{x}$ and $x * y$ on V;

x	$\tilde{x}$
0	1
1	0
2	2

x	y	$x * y$
0	0	0
1	0	0
2	0	0
0	1	0
1	1	1
2	1	2
0	2	0
1	2	2
2	2	0

By means of these operations we define the "value" $v(F)$ for each formula F of $\mathcal{L}_0 = \mathcal{L}_D$, using the following definitions :

$$v(\neg F) = \widetilde{v(F)}; \quad v(F \vee G) = v(F) * v(G).$$

We shall say that a formula F has the property p if $v(F) = 0$. Our next purpose is to show that the formula $F \vee F \Rightarrow F$ does not have the property

p, while the formulas **H2, H3, H4** have this property. Computing the value of **H1** we have:

$$v[(F \vee F) \Rightarrow F] = v[\neg(F \vee F) \vee F] = v(\widetilde{F \vee F}) * v(F).$$

Suppose now that $v(F) = 2$; then $v(F \vee F) = v(F) * v(F) = 2 * 2 = 0$, hence $v(\widetilde{F \vee F}) = 1$. It follows that $v(\mathbf{H1}) = 1 * 2 = 2$. Therefore, (**H1**) does not have the property p. On the other hand, from the above definition of operations $\tilde{x}$ and $x * y$ we obtain easily the following equalities:

$$x * y = y * x; \quad 0 * x = 0; \quad 1 * x = x;$$

$$x * \tilde{x} = 0; \quad \tilde{\tilde{x}} = x.$$

Now, denoting $v(F) = f$, $v(G) = g$, $v(H) = h$, we can compute successively the "values" $v(\mathbf{H2})$, $v(\mathbf{H3})$ and $v(\mathbf{H4})$.

$v(\mathbf{H2}) = v(F \Rightarrow F \vee G) = v(\neg F \vee (F \vee G)) = \widetilde{v(F)} * (v(F) * v(G)) = \tilde{f} * (f * g)$. If $f = 0$, then $\tilde{f} = 1$ and $f * g = 0$, hence $v(\mathbf{H2}) = 1 * 0 = 0$. If $f = 1$, then $\tilde{f} = 0$ and $v(\mathbf{H2}) = 0 * (f * g) = 0$. If $f = 2$, then $\tilde{f} = 2$ and $v(\mathbf{H2}) = 2 * (2 * g)$. Taking into account that $2 * (2 * x) = 0$ for all $x \in V$ it follows that $v(\mathbf{H2}) = 0$. We have proved that in all possible cases $v(\mathbf{H2}) = 0$. Compute now the "value" of (**H3**). $v(\mathbf{H3}) = v((F \vee G) \Rightarrow (G \vee F)) = v(\neg(F \vee G) \vee (G \vee F)) = (\widetilde{f * g}) * (g * f) = \widetilde{f * g} * (f * g) = 0$. Therefore, $v(\mathbf{H3}) = 0$.

Finally, for (**H4**) we have $v(\mathbf{H4}) = v((F \Rightarrow G) \Rightarrow (H \vee F \Rightarrow H \vee G)) = v(\neg(\neg F \vee G) \vee (\neg(H \vee F) \vee (H \vee G))) = v(\neg \widetilde{F \vee G}) * (v(\widetilde{H \vee F}) * v(H \vee G)) = \tilde{f} * g * ((\widetilde{h * f}) * (h * g))$. If $h = 0$, then $(\widetilde{h * f}) * (h * g) = (\widetilde{h * f}) * 0 = 0$, hence $v(\mathbf{H4}) = 0$. If $h = 1$, then $v(\mathbf{H4}) = (\tilde{f} * g) * (\tilde{f} * g) = 0$. If $h = 2$ and $g = 0$ or $g = 2$, then $(\widetilde{h * g}) * (h * g) = (\widetilde{2 * f}) * (2 * g) = (\widetilde{2 * f}) * 0 = 0$, hence $v(\mathbf{H4}) = 0$. If $h = 2$ and $g = 1$, then $v(\mathbf{H4}) = f * ((\widetilde{2 * f}) * 2)$; if $f = 0$, then $v(\mathbf{H4}) = 0$; if $f = 1$, then $v(\mathbf{H4}) = (\widetilde{2 * 1}) * 2 = 2 * 2 = 0$; if $f = 2$, then $v(\mathbf{H4}) = 2 * ((\widetilde{2 * 2}) * 2 = 2 * (1 * 2) = 2 * 2 = 0$. Therefore, in all possible cases $v(\mathbf{H4}) = 0$.

Let us consider the MP-rule. Suppose that for the formulas A and $A \Rightarrow B$ we have $v(A) = 0$ and $v(A \Rightarrow B) = 0$. It follows $0 = v(\neg A \vee B) = \widetilde{v(A)} * v(B) = 1 * v(B)$, hence $v(B) = 0$. This means that if A and $A \Rightarrow B$ have property p, then B has the property p, too. From the above computations it follows that axiom (**H1**) is independent of axioms (**H2**), (**H3**), (**H4**).

Independence of (H2). Consider the set $V = \{0, 1, 2, 3\}$ and the operations $\tilde{x}$, $x * y$ defined as follows

x	$\tilde{x}$
0	1
1	0
2	3
3	2

$$x * y = \min\{x, y\}.$$

As above, we define $v(F)$ for each formula $F \in \mathcal{L}_0$ by

$$v(\neg F) = \widetilde{v(F)}; \quad v(F \vee G) = v(F) * v(G).$$

We shall say that a formula F has property p if $v(F) = 0$. As in the case of (**H1**), by similar computations, one gets $v(\textbf{H1}) = v(\textbf{H3}) = v(\textbf{H4}) = 0$ and for $f = 2$, $g = 1$, we obtain $v(\textbf{H2}) = 1$. On the other hand, one can easily verify that the MP-rule preserves the property p. It follows that axiom (**H2**) is independent of (**H1**), (**H3**), (**H4**).

Independence of (H3). Consider the set $V = \{0, 1, 2, 3\}$ and the operations $\tilde{x}$, $x * y$ defined:

x	$\tilde{x}$
0	1
1	0
2	0
3	2

$0 * x = x * 0 = 0$;

$1 * x = x * 1 = x$;

$2 * 3 = 0, \ 3 * 2 = 3, \ 2 * 2 = 2, \ 3 * 3 = 3.$

We shall define $v(\neg F) = \widetilde{v(F)}, v(F \vee G) = v(F) * v(G)$ and we shall say that the formula $F \in \mathcal{L}_0$ has property p if $v(F) = 0$. Using the definitions of $\tilde{x}$ and $x * y$ we obtain $v(\textbf{H1}) = v(\textbf{H2}) = v(\textbf{H4}) = 0$ and for $f = 2, g = 3, v(\textbf{H3}) = 3$. One can also see that property p is preserved by the MP-rule. It follows that axiom (**H3**) is independent of (**H1**), (**H2**), (**H4**).

Independence of (H4). Consider the set $V = \{0, 1, 2, 3\}$ and the operations $\tilde{x}, x * y$ defined by:

x	$\tilde{x}$
0	1
1	0
2	3
3	0

$0 * x = x * 0 = 0;$

$1 * x = x * 1 = x;$

$2 * 3 = 3 * 2 = 0;$

$2 * 2 = 2, 3 * 3 = 3.$

We also define $v(\neg F) = \widetilde{v(F)}$, $v(F \vee G) = v(F) * v(G)$ and we say that the formula F has property p if $v(F) = 0$. In the same manner as for axioms $(\mathbf{H1}), (\mathbf{H2}), (\mathbf{H3})$, one gets $v(\mathbf{H1}) = v(\mathbf{H2}) = v(\mathbf{H3}) = 0$ while for $f = 3, g = 1, h = 2, v(\mathbf{H4}) = 2$. On the other hand, the property p is preserved by the MP-rule. Therefore, the axiom $(\mathbf{H4})$ is independent of axioms $(\mathbf{H1}), (\mathbf{H2}), (\mathbf{H3})$.

In conclusion, the axioms of $(\mathbf{H})$ are independent.   ∎

The proof that axioms of $(\mathbf{L})$ and $(\mathbf{M})$ are independent can be performed in a similar way as in the preceding proposition. We shall prove, for instance, the corresponding independence result for $\mathbf{M}$.

PROPOSITION 5.26
The axioms of $(\mathbf{M})$ *are independent.*

PROOF   ([27]) Let us denote, as in the previous proof, $v(F) = f, v(G) = g, v(H) = h$.

Independence of $(\mathbf{M1})$. Consider the set $V = \{0, 1, 2\}$ and define the operation $\tilde{x}, x * y$ by

x	$\tilde{x}$
0	1
1	1
2	0

x	y	$x * y$
0	0	0
1	0	0
2	0	0
0	1	0
1	1	1
2	1	2
0	2	0
1	2	2
2	2	0

We define $v(\neg F) = v(\widetilde{F}), v(F \Rightarrow G) = v(F) * v(G)$ and we say that the formula $F \in \mathcal{L}_I$ has property p if $v(F) = 0$. As in Proposition 5.25, using the definitions of $\tilde{x}$ and $x * y$ we obtain $v(\mathbf{M2}) = v(\mathbf{M3}) = 0$ while for $f = 1, g = 2, v(\mathbf{M1}) = 2$. One can also see that property p is preserved by the MP-rule. Thus, axiom $(\mathbf{M1})$ is independent of $(\mathbf{M2}), (\mathbf{M3})$.

Independence of $(\mathbf{M2})$. Consider the set $V = \{0, 1, 2\}$ and define operations $\tilde{x}, x * y$ by

x	$\tilde{x}$
0	1
1	0
2	1

x	y	$x * y$
0	0	0
1	0	0
2	0	0
0	1	2
1	1	2
2	1	0
0	2	1
1	2	0
2	2	0

We now define $v(\neg F) = v(\widetilde{F}), v(F \Rightarrow G) = v(F) * v(G)$ and we say that the formula $F \in \mathcal{L}_I$ has property p if $v(F) = 0$. From the definitions of operations $\tilde{x}$ and $x * y$ we get $v(\mathbf{M1}) = v(\mathbf{M3}) = 0$ but, for $f = g = 0, h = 1$, one has $v(\mathbf{M2}) = 2$. On the other hand, the MP-rule preserves property p. It follows that axiom $(\mathbf{M2})$ is independent of $(\mathbf{M1}), (\mathbf{M3})$.

Independence of (M3). Consider the set $V = \{0, 1, 2, 3\}$ and define the operations $\tilde{x}, x * y$ by

x	$\tilde{x}$
0	1
1	0
2	3
3	0

x	y	$x * y$
0	0	0
1	0	0
2	0	0
3	0	0
0	1	1
1	1	0
2	1	3
3	1	0
0	2	2
1	2	0
2	2	0
3	2	0
0	3	3
1	3	0
2	3	3
3	3	0

Also define $v(\neg F) = v(\widetilde{F})$, $v(F \Rightarrow G) = v(F) * v(G)$ and say that the formula $F \in \mathcal{L}_I$ has the property p if $v(F) = 0$. From the definitions of $\tilde{x}, x * y$ one obtains $v(\mathbf{M1}) = v(\mathbf{M2}) = 0$ and for $f = 0, g = 2, v(\mathbf{M3}) = 2$. Moreover, the MP-rule preserves the property p. It follows that axiom (M3) is independent of (M1), (M2).

In conclusion, the axioms of (M) are mutually independent. ∎

5.3 System C of Łukasiewicz and Tarski

We shall present below another system **C** which is not equivalent to **H, L, M**. This system was introduced in 1930 by Łukasiewicz and Tarski [24] in order to describe many-valued logics in a more general situation than the one considered by Post in 1921 [38] (see also [44], [6]). System

C is expressed in the "implicative" language $\mathcal{L}_I$. This language may be considered either as a sublanguage of the "disjunctive" language $\mathcal{L}_0 = \mathcal{L}_D$ (Section (5.1)) or as an independent language. In the last case the alphabet contains the symbols $X_1, X_2, ..., X_N$ of propositional variables, the symbols $\neg$ and $\Rightarrow$ of logical operators and the symbols (,) of parentheses; the concept of formula in $\mathcal{L}_I$ is defined in the same way as the corresponding one in $\mathcal{L}_D$, with the difference that instead of the symbol $\vee$ the symbol $\Rightarrow$ is used. It is clear that the language $\mathcal{L}_I$ defined independently is "isomorphic" to the sublanguage $\mathcal{L}_I$ of $\mathcal{L}_0 = \mathcal{L}_D$. For this reason we conserve the notation $\mathcal{L}_I$ in both cases.

System **C** has MP (Modus Ponens) as a unique inference rule, while the axioms of **C** are those listed below.

(**C1**) : $F \Rightarrow (G \Rightarrow F)$;

(**C2**) : $((F \Rightarrow G) \Rightarrow (G \Rightarrow H)) \Rightarrow (F \Rightarrow H)$;

(**C3**) : $((F \Rightarrow G) \Rightarrow G) \Rightarrow ((G \Rightarrow F) \Rightarrow F)$;

(**C4**) : $(\neg G \Rightarrow \neg F) \Rightarrow (F \Rightarrow G)$.

REMARK 5.4   Taking into account the fact that formulas F, G, H in the above axioms are arbitrary, axioms (**C1**) to (**C4**) are, in fact, schemes of axioms.   ■

We shall say that a formula $F \in \mathcal{L}_I$ is a **theorem** in **C**, briefly

$$\vdash_{\mathbf{C}} F,$$

if there is a finite sequence of formulas

$$(*) \quad F_1, F_2, ..., F_r \quad (r \geq 1), \quad F_r \equiv F$$

having the property that for each index i ($1 \leq i \leq r$) the formula F_i satisfies one of two conditions:

(a) F_i is an axiom of **C** (for short, $F_i \in Ax(\mathbf{C})$);

(b) there are indices $h, k < i$ such that $F_k \equiv F_h \Rightarrow F_i$ (F_i is obtained in $(*)$ by the MP-rule).

The sequence $(*)$ is called **a proof** (or a proof for its last formula $F_r \equiv F$).

In the language $\mathcal{L}_I$ we define a natural equivalence relation between formulas in the same way as in the case of system **H**. Namely, we shall say that the formulas $F, G \in \mathcal{L}_I$ are equivalent and we shall write $F \sim G$ if

$$\vdash_{\mathbf{C}} F \Rightarrow G \quad \text{and} \quad \vdash_{\mathbf{C}} G \Rightarrow F.$$

It is easy to verify that "$\sim$" is a genuine equivalence relation, i.e., it is reflexive, symmetric and transitive. It follows that $\mathcal{L}_I$ decomposes into nonempty disjoint classes of equivalence such that each formula F belongs to some class ξ_F and $\bigcup_{F \in \mathcal{L}_I} \xi_F = \mathcal{L}_I$. The set of all these classes (i.e., the quotient set of $\mathcal{L}_I$ with respect to "$\sim$") will be denoted by $\mathcal{L}_I / \sim \, = \hat{\mathcal{L}}_I$. In $\hat{\mathcal{L}}_I$ we define operations $^-, +, \cdot$ as follows:

$$\bar{\xi}_F = \xi_{\neg F};$$
$$\xi_F + \xi_G = \xi_{\neg F \Rightarrow G};$$
$$\xi_F \cdot \xi_G = \xi_{\neg(F \Rightarrow \neg G)}.$$

Moreover, we write $\xi_F = 1$ iff $\vdash_C F$ and $\xi_F = 0$ iff $\vdash_C \neg F$. By a routine argument it can be proved that the operations above are correctly defined, i.e., the classes $\bar{\xi}_F, \xi_F + \xi_G, \xi_F \cdot \xi_G$ are independent of the representatives. The set $\hat{\mathcal{L}}_I$ endowed with these operations satisfies all axioms of an MV-algebra (see APPENDIX B), having classes 1, 0 as unit elements. The MV-algebra $(\hat{\mathcal{L}}_I, -, +, \cdot)$ is no longer a Boolean algebra, because the set of idempotent elements of $\hat{\mathcal{L}}_I$, i.e., of elements x satisfying $x + x = x$ or $x \cdot x = x$, does not coincide with the whole MV-algebra $\hat{\mathcal{L}}_I$; the same is true for elements satisfying the law of the excluded middle, i.e., $x \vee \bar{x} = 1$ or $x \wedge \bar{x} = 0$. At the same time we mention that elements 1 and 0 are idempotent and satisfy the laws $x \vee \bar{x} = 1, x \wedge \bar{x} = 0$.

The purpose of this section is to prove that system **C** is not equivalent to systems **M, L** and **H**; more specifically, we shall show that there is a formula $H \in \mathcal{L}_I$ such that $\vdash_L H$, but it is not a theorem in **C**.

Consider the set $V = \{0, \frac{1}{2}, 1\}$ and define in V the operations $\bar{x}$ and $x \to y$ as follows:

$$\bar{x} = 1 - x \quad \text{and} \quad x \to y = \min\{1, 1 - x + y\}.$$

In Chapter 2 we have proved the existence of Boolean interpretations of the language $\mathcal{L}_0 = \mathcal{L}_D$, hence also of the language $\mathcal{L}_I$. Pursuing the same line of argument the following proposition can be proved: for any sequence $(\alpha_1, \alpha_2, ..., \alpha_N), \alpha_j \in V \ (1 \leq j \leq N)$, there is a unique function $\varphi : \mathcal{L}_I \to V$ with the following properties:

(i) $\varphi(X_j) = \alpha_j$;

(ii) $\varphi(\neg F) = \overline{\varphi(F)}$;

(iii) $\varphi(F \Rightarrow G) = \varphi(F) \to \varphi(G)$.

Consider now function φ with $\varphi(X_1) = \frac{1}{2}$, $\varphi(X_j) = \alpha_j \ (2 \leq j \leq N)$, where $\alpha_2, ..., \alpha_N$ are arbitrary elements in V.

Assertion 1 For any axiom (Ci) $(1 \leq i \leq 4)$ of system $\mathbf{C}$ one has $\varphi(Ci) = 1$.

PROOF   Indeed, $\varphi(\mathbf{C1}) = \varphi(F \Rightarrow (G \Rightarrow F)) = \varphi(F) \to \varphi(G \Rightarrow F) = \varphi(F) \to (\varphi(G) \to \varphi(F))$. Denoting $\varphi(F) = x$, $\varphi(G) = y$ we have $\varphi(\mathbf{C1}) = x \to (y \to x) = \min\{1, 1 - x + (y \to x)\} = \min\{1, 1 - x + \min\{1, 1 - y + x\}\}$. If $x \geq y$, then $\min\{1, 1 - y + x\} = 1$ and $1 - x + \min\{1, 1 - y + x\} = 2 - x \geq 1$; therefore, $\varphi(\mathbf{C1}) = 1$. If $x \leq y$, then $\min\{1, 1 - y + x\} = 1 - y + x$ and $1 - x + \min\{1, 1 - y + x\} = 1 - x + (1 - y + x) = 2 - y \geq 1$; therefore, $\varphi(\mathbf{C1}) = 1$.

By similar computations we get $\varphi(\mathbf{C2}) = \varphi(\mathbf{C3}) = \varphi(\mathbf{C4}) = 1$.   ∎

Assertion 2 If $F, G \in \mathcal{L}_I$ and $\varphi(F) = 1, \varphi(F \Rightarrow G) = 1$, then $\varphi(G) = 1$; in other words, the MP-rule preserves the value 1.

PROOF   For the proof, suppose that $\varphi(F) = 1$ and $\varphi(F \Rightarrow G) = 1$. It follows that $1 = \varphi(F \Rightarrow G) = \varphi(F) \to (G) = \min\{1, 1 - \varphi(F) + \varphi(G)\}$. Taking into account that $\varphi(F) = 1$, hence $1 - \varphi(F) = 0$, we obtain $1 = \min\{1, 0 + \varphi(G)\} = \varphi(G)$ and the assertion is thus proved.   ∎

From Assertions 1 and 2 it follows immediately that if $\vdash_{\mathbf{C}} F$, then $\varphi(F) = 1$.

We shall prove that the formula $T \equiv (X_1 \Rightarrow \neg X_1) \Rightarrow \neg X_1$ is not a theorem in $\mathbf{C}$. Suppose the contrary, i.e., $\vdash_{\mathbf{C}} T$. By the conclusion above, $\varphi(T) = 1$. On the other hand, taking into account that $\varphi(X_1) = \frac{1}{2}$ we have

$$\varphi(T) = (\varphi(X_1) \to \overline{\varphi(X_1)}) \to \overline{\varphi(X_1)} = (\frac{1}{2} \to \frac{1}{2}) \to \frac{1}{2};$$

but $\frac{1}{2} \to \frac{1}{2} = 1$ and

$$1 \to \frac{1}{2} = \min\{1, 1 - 1 + \frac{1}{2}\} = \frac{1}{2},$$

therefore, $\varphi(T) = \frac{1}{2}$. This contradiction shows that T is not a theorem in $\mathbf{C}$. On the other hand, it is easy to see that T is a tautology, hence it is a theorem in $\mathbf{L}$ (and consequently in $\mathbf{M}$ and $\mathbf{H}$).

REMARK 5.5   System $\mathbf{C}$ was introduced by Łukasiewicz and Tarski for the definition of a many-valued logic. The same system will be used in Chapter 6 to introduce fuzzy logic.   ∎

6

Elements of Fuzzy Propositional Logic with a Finite Set of Truth Values

In this chapter we shall present the fundamental ideas of **fuzzy propositional logic**. More precisely, we shall sketch an appropriate framework for modeling the vagueness existing in the natural language and rigorously construct a corresponding language and a truth structure which could be considered an acceptable system for **fuzzy propositional logic**. After Zadeh's fundamental works ([50], [51]) and after the paper by J. A. Goguen [13] it became clear that one of the main tools to tackle this problem is the concept of a fuzzy set allied to that of many-valued logic. Indeed, since 1979 a group of mathematicians and logicians such as J. Pavelka, V. Novák and P. Hájek began to publish a series of remarkable papers ([17], [18], [33], [34], [35], [36], [37]) with great relevance to fuzzy logic. Today, fuzzy logic is a coherent logical mathematical discipline. We agree with Novák's statements that "fuzzy logic is a well-established formal theory which non-trivially generalizes the classical one" and that actually "fuzzy logic very generally means many-valued logic with special properties." Of course, such a development has been possible on the basis of important contributions in the field of MV-algebras, due to C. C. Chang, D. Mundici, A. Di Nola, as well as those of A. Rose, J. B. Rosser, U. Höhle, S. Gottwald, W. Pedrycz, D. Dubois, H. Prade, B. Meunier-Bouchon, G. Klir, D. Butnariu, C. V. Negoiţă, D. Ralescu, P. Klement, R. Yager, M. Sugeno and many others in related fields. At the 70th International Fuzzy System Association World Congress (IFSA '97), in his speech, L. A. Zadeh emphasized the major role played by the Czech school of logic in formalizing fuzzy logic.

Our approach to the subject mainly pursues the line of ideas of Pavelka, Novák, Hájek; its specific characteristic is the systematic use of fuzzy sets having a finite set $\mathcal{A}$ of membership degrees, where the set $\mathcal{A}$ is organized as an MV-algebra. Our option for finite MV-algebras is based on the general remark that in applications we always deal with fuzzy sets having a finite number of membership degrees. We notice that, from a theoretical point of view, this restriction is not essential and can be dropped.

In this chapter we shall often use terms such as **fuzzy set, fuzzy logic,** etc. in contrast with the corresponding usual (non-fuzzy) terms. For this reason the non-fuzzy sets will be called **usual sets** or **crisp sets**; similarly, the non-fuzzy propositional logic (treated in Chapters 1-6) will be designated by expressions such as *usual*, or *classical*, or *crisp* propositional logic.

6.1 Some Elementary Notions about Fuzzy Sets

In Section 2.1 (see also APPENDIX A) we noticed that the set $\Gamma = \{0,1\}$ endowed with the operations (complement), $\vee$ (maximum or lower upper bound or l.u.b.) and $\wedge$ (minimum or greatest lower bound or g.l.b.), i.e.,

$$B - ops: \quad \begin{cases} \bar{x} = 1 - x; \\ x \vee y = \max\{x, y\}; \\ x \wedge y = \min\{x, y\}; \end{cases}$$

is a Boolean algebra with two elements (the smallest non-degenerate Boolean algebra).

If, instead of Γ, we consider the set $I = [0,1]$ of all real numbers x with $0 \leq x \leq 1$, then I endowed with $(B - ops)$ is not a Boolean algebra any more because of the failure of equalities $x \vee \bar{x} = 1$ and $x \wedge \bar{x} = 0$; this new algebraic structure is called by some authors a De Morgan algebra.

The same set I can be organized as an MV-algebra if one defines on I the following internal operations:

$$MV - ops: \quad \begin{cases} \bar{x} = 1 - x; \\ x \oplus y = \min\{1, x + y\}; \\ x \odot y = \max\{0, x + y - 1\} \end{cases}$$

(see APPENDIX B). Recall that an MV-subalgebra of I is a set $J \subset I$ which is closed under MV-ops and contains the elements 0, 1. In a certain sense, the only finite MV-subalgebras of I are the subsets $\mathcal{A} \subseteq I$ of the form

$$\mathcal{A} = \left\{0, \frac{1}{M-1}, \frac{2}{M-1}, ..., \frac{M-1}{M-1} = 1\right\},$$

where M is some natural number, $M \geq 2$. In any MV-algebra one can define internal operations $\vee, \wedge$ (as derived operations) by $x \vee y = (x \odot \bar{y}) \oplus y$; $x \wedge y = (x \oplus \bar{y}) \odot y$.

It is easy to see that in the case of a MV-algebra I and of any MV-subalgebra of I the two operations above coincide with the corresponding operations in (B-ops).

In I and in any MV-subalgebra of I one can also introduce an important derived internal operation " $\to$ " (called "residuation") defined by

$$x \to y = \bar{x} \oplus y$$

that is,

$$x \to y = \min\{1, 1 - x + y\}.$$

Let us now recall the usual concept of fuzzy set. For this purpose consider a set Y, called the universe. If A is a subset of $Y (A \subseteq Y)$, then the characteristic function χ_A, i.e., the function $\chi_A : Y \to \Gamma = \{0, 1\}$ given by

$$\chi_A(y) = \begin{cases} 1, & \text{if } y \in A; \\ 0, & \text{otherwise} \end{cases}$$

is uniquely determined by the set A; moreover, the correspondence

$$A \longmapsto \chi_A$$

is a bijective mapping from the set $\mathcal{P}(Y)$ of all subsets of Y onto the set $\mathcal{F}(Y, \Gamma)$ of all Γ-valued functions defined on Y. The mapping $A \mapsto \chi_A$ has the following remarkable properties:

(a) $A \subseteq B$ if and only if $\chi_A \leq \chi_B$;

(b) $\chi_{A \cup B} = \max\{\chi_A, \chi_B\}$;

(c) $\chi_{A \cap B} = \min\{\chi_A, \chi_B\}$;

(d) $\chi_{CA} = 1 - \chi_A$, where $CA = Y \backslash A$.

The set $\mathcal{F}(Y, \Gamma)$ endowed with the operations $\bar{\chi} = 1 - \chi$, $\chi_1 \vee \chi_2 = \max\{\chi_1, \chi_2\}$, $\chi_1 \wedge \chi_2 = \min\{\chi_1, \chi_2\}$ is a Boolean algebra and the mapping $A \mapsto \chi_A$ is a natural isomorphism of Boolean algebras between $(\mathcal{P}(Y), C, \cup, \cap)$ and $(\mathcal{F}(Y, \Gamma), ^-, \vee, \wedge)$. It follows that the subsets of Y can be identified with the corresponding characteristic functions in $\mathcal{F}(Y, \Gamma)$, so that the theory of the usual subsets of Y is reducible to the theory of functions in $\mathcal{F}(Y, \Gamma)$. This remark leads to the extension of the concept of set by introducing so-called **fuzzy sets**.

Let us denote by $\mathcal{F}(Y, I)$ the set of all I-valued functions defined on Y; it is clear that $\mathcal{F}(Y, \Gamma) \subseteq \mathcal{F}(Y, I)$, so that each usual subset of Y can be identified with some element of $\mathcal{F}(Y, I)$.

DEFINITION 6.1   *Let Y be given. Each function $B : Y \to I$, i.e., any element of $\mathcal{F}(Y, I)$, will be called an **I-fuzzy set** in Y (or, for short,* *a fuzzy set in Y). The number $B(y) \in I$ will be called* **the membership degree of y in the fuzzy set B**. *In particular, it follows that each usual subset B of Y (identified with its characteristic function χ_B) has only two possible degrees of membership (1 and 0).*

DEFINITION 6.2 *Let Y be given and $\mathcal{A} = \{0, \frac{1}{M-1}, ..., 1\}$, a finite MV-subalgebra of I, $M \geq 2$. Each function $B : Y \to \mathcal{A}$ will be called an* **$\mathcal{A}$-fuzzy set in Y**. *$\mathcal{A}$-fuzzy sets in Y will be called often fuzzy sets with a finite (fixed) number of membership degrees.*

If the fuzzy set B viewed as a real function defined on Y is a constant equal 0, i.e., $B(y) = 0$ for all $y \in Y$, then in terms of fuzzy sets we shall write $B \simeq \emptyset$.

The classical operations $\cup$, $\cap$, C for sets can be easily extended to corresponding operations for fuzzy sets. Classical inclusion can also be extended.

DEFINITION 6.3 *Let Y and an MV-subalgebra $\mathcal{A}$ be given and consider two $\mathcal{A}$-fuzzy sets B_1, B_2 in Y.*

(a) *The* **union** *of the $\mathcal{A}$-fuzzy sets B_1, B_2 is the $\mathcal{A}$-fuzzy set $B_1 \cup B_2$ given by*

$$(B_1 \cup B_2)(y) = B_1(y) \vee B_2(y) = max\{B_1(y), B_2(y)\};$$

(b) *The* **intersection** *of the $\mathcal{A}$-fuzzy sets B_1, B_2 is the $\mathcal{A}$-fuzzy set $B_1 \cap B_2$ defined by*

$$(B_1 \cap B_2)(y) = B_1(y) \wedge B_2(y) = min\{B_1(y), B_2(y)\};$$

(c) *If B is an $\mathcal{A}$-fuzzy set in Y, then the* **complement** *CB of B in Y is defined by*

$$(CB)(y) = \overline{B(y)} = 1 - B(y).$$

DEFINITION 6.4 *Let Y and the MV-subalgebra $\mathcal{A}$ be given and consider two $\mathcal{A}$-fuzzy sets B_1, B_2 in Y. We shall say that the $\mathcal{A}$-fuzzy set B_1 is* **included** *in the $\mathcal{A}$-fuzzy set B_2 and we denote it by*

$$B_1 \sqsubseteq B_2$$

if $B_1(y) \leq B_2(y)$ for all $y \in Y$ or, in other words, considering B_1, B_2 as (real) functions, $B_1 \leq B_2$.

REMARK 6.1 The operations $\cup, \cap$ can obviously be extended for any family $(B_\omega)_{\omega \in \Omega}$ of $\mathcal{A}$-fuzzy sets in Y by the equalities

$$(\bigcup_{\omega \in \Omega} B_\omega)(y) = \bigvee_{\omega \in \Omega} B_\omega(y) = \max\{B_\omega(y) \mid \omega \in \Omega\};$$

$$(\bigcap_{\omega \in \Omega} B_\omega)(y) = \bigwedge_{\omega \in \Omega} B_\omega(y) = \min\{B_\omega(y) \mid \omega \in \Omega\}.$$

It is easy to see that $\mathcal{F}(Y, \mathcal{A})$ with the operations $C, \cup, \cap$ is a De Morgan algebra which is a Boolean one if and only if $\mathcal{A} = \Gamma$. In $\mathcal{F}(Y, \mathcal{A})$ one can also define two other operations (corresponding to the operations $\oplus, \odot$ in $\mathcal{A}$) such that $\mathcal{F}(Y, \mathcal{A})$ becomes an MV-algebra. For general considerations about fuzzy sets the reader is invited to see Appendix C at the end of the book. $\blacksquare$

6.2 The Language of Fuzzy Propositional Logic

The language of fuzzy propositional logic will be essentially the same as the language of classical propositional logic. In the classical language of propositional logic we used the primitive operators $\neg, \vee$ and we introduced by abbreviation the operators $\wedge, \Rightarrow, \Leftrightarrow$. For special reasons in the language of fuzzy propositional logic, denoted also by $\mathcal{L}_0$, we shall take the primitive symbols $\neg$ and $\Rightarrow$ instead of $\neg$ and $\vee$ and we shall introduce (as derived operators) the symbols $\vee, \wedge, \Leftrightarrow$. Therefore, in the construction of the language $\mathcal{L}_0$ of fuzzy propositional logic we shall start with the alphabet

$$\{X_1, X_2, ..., X_N, \neg, \Rightarrow, (,)\}.$$

The formal constructions and the formulas in $\mathcal{L}_0$ will be defined in the same way as in Section 1.1 putting $\Rightarrow$ instead of $\vee$. The set of all such formulas will be called the language $\mathcal{L}_0$ of fuzzy propositional logic. In $\mathcal{L}_0$ we introduce the following natural abbreviations:

$$F \wedge G \equiv \neg((F \Rightarrow G) \Rightarrow \neg F); \text{ (conjunction)};$$
$$F \vee G \equiv (F \Rightarrow G) \Rightarrow G); \text{ (disjunction)};$$
$$F \& G \equiv \neg(F \Rightarrow \neg G); \text{ (Lukasiewicz conjunction)};$$
$$F \triangledown G \equiv \neg(\neg F \& \neg G); \text{ (Lukasiewicz disjunction)};$$
$$F \Leftrightarrow G \equiv (F \Rightarrow G) \wedge (G \Rightarrow F); \text{ (equivalence)}.$$

It is clear that $F \wedge G, F \vee G, F \& G, F \triangledown G, F \Leftrightarrow G$ are formulas in $\mathcal{L}_0$.

6.3   The Semantic Truth Structure of Fuzzy Propositional Logic

As in the case of classical propositional logic the truth structure in the semantic version on $\mathcal{L}_0$ of fuzzy propositional logic of the language $\mathcal{L}_0$ is introduced by an appropriate concept of **interpretation**.

In what follows, dealing with fuzzy sets, our universe Y will be always the language $\mathcal{L}_0$ and the set of membership degrees of a fuzzy set will be a fixed finite MV-algebra

$$A = \left\{ 0, \frac{1}{M-1}, \frac{2}{M-1}, ..., \frac{M-1}{M-1} = 1 \right\}, \quad M \geq 2.$$

Therefore, all fuzzy sets will be $\mathcal{A}$-fuzzy sets in $\mathcal{L}_0$. For this reason instead of $\mathcal{A}$-fuzzy sets in $\mathcal{L}_0$ we shall simply use the term "fuzzy sets". In other words, fuzzy sets will be the elements of $\mathcal{F}(\mathcal{L}_0, \mathcal{A})$.

DEFINITION 6.5   *Any $\mathcal{A}$-valued function φ defined on $\mathcal{L}_0$*

$$\varphi : \mathcal{L}_0 \to \mathcal{A} ,$$

having the properties

(i) $\varphi(\neg F) = \overline{\varphi(F)}$;

(ii) $\varphi(F \Rightarrow G) = \varphi(F) \to \varphi(G)$,

where F, G are arbitrary formulas in $\mathcal{L}_0$ and "$\to$" is the residuation operation in $\mathcal{A}$ is called an **MV-interpretation of $\mathcal{L}_0$**. *The set of all MV - interpretations of $\mathcal{L}_0$ will be denoted by $\mathcal{W}$.*

REMARK 6.2   In classical propositional logic we have used the notion of Boolean interpretation and the set of such Boolean interpretations of $\mathcal{L}_0$ has been denoted by $\mathcal{V}$. It is not difficult to verify that any Boolean interpretation of $\mathcal{L}_0$ is also an MV-interpretation of $\mathcal{L}_0$, so that

$$\mathcal{V} \subseteq \mathcal{W},$$

i.e., the set $\mathcal{W}$ is larger than $\mathcal{V}$. This remark shows that $\mathcal{W} \neq \emptyset$. ∎

Taking into account the definitions of formulas $F \wedge G$, $F \vee G$, $F\&G$, $F \triangledown G$ and the properties of any $\varphi \in \mathcal{W}$, one can verify the following equalities:

$$\varphi(F \wedge G) = \min\{\varphi(F), \varphi(G)\} = \varphi(F) \wedge \varphi(G);$$
$$\varphi(F \vee G) = \max\{\varphi(F), \varphi(G)\} = \varphi(F) \vee \varphi(G);$$
$$\varphi(F\&G) = \varphi(F) \odot \varphi(G);$$
$$\varphi(F \triangledown G) = \varphi(F) \oplus \varphi(G).$$

Let us verify, for instance, the third equality. By the definition of $F\&G$ we have:

$$\varphi(F\&G) = \varphi(\neg(F \Rightarrow \neg G)) = \overline{\varphi(F \Rightarrow \neg G)} = \overline{\varphi(F) \rightarrow \overline{\varphi(G)}};$$

if we denote $\varphi(F) = x$ and $\varphi(G) = y$, then

$$\varphi(F) \rightarrow \overline{\varphi(G)} = x \rightarrow \bar{y} = \min\{1, 1 - x + \bar{y}\} = \min\{1, \bar{x} + \bar{y}\} = \bar{x} \oplus \bar{y}.$$

Taking into account that in any MV-algebra $\bar{x} \oplus \bar{y} = \overline{x \odot y}$ and $\bar{\bar{z}} = z$ we have

$$\overline{x \rightarrow \bar{y}} = \overline{\overline{x \odot y}} = x \odot y,$$

hence

$$\varphi(F\&G) = \varphi(F) \odot \varphi(G).$$

The next proposition establishes that in $\mathcal{W}$ there are elements which do not belong to $\mathcal{V}$; moreover, it states that the number of elements of $\mathcal{W}$ is M^N.

PROPOSITION 6.1

For any given element $\gamma \in \mathcal{A}^N$, $\gamma = (a_1, a_2, ..., a_N)$ there is a unique function $\varphi_\gamma : \mathcal{L}_0 \rightarrow \mathcal{A}$ having the properties:

(i) $\varphi_\gamma(\neg F) = \overline{\varphi_\gamma(F)}$;

(ii) $\varphi_\gamma(F \Rightarrow G) = \varphi_\gamma(F) \rightarrow \varphi_\gamma(G)$;

(iii) $\varphi_\gamma(X_j) = a_j, \quad 1 \leq j \leq N.$

The correspondence $\gamma \mapsto \varphi_\gamma$ is a bijective mapping from $\mathcal{A}^N$ onto $\mathcal{W}$, therefore card $(\mathcal{W}) = M^N$.

The assertion can be proved without difficulty in the same manner we have proved the corresponding proposition concerning Boolean interpretations $\varphi_\alpha \in \mathcal{V}$ in Section 2.1.

REMARK 6.3   From the above proposition it follows that any MV-interpretation $\varphi \in \mathcal{W}$ can be indexed by the elements of $\mathcal{A}^N$; for this reason, in the following these interpretations will be denoted equivalently by φ or by φ_γ depending of our needs. ∎

It is clear that any MV-interpretation of $\mathcal{L}_0$ is at the same time a fuzzy set in $\mathcal{L}_0$ (having two additional properties: (i), (ii)).

It is important to notice that $\mathcal{F}(\mathcal{L}_0, \mathcal{A})$ and $\mathcal{W}$ are usual (crisp) sets: the first is the set of all fuzzy sets, the second is the set of all MV-interpretations of $\mathcal{L}_0$. The writings $f \in \mathcal{F}(\mathcal{L}_0, \mathcal{A})$ and $\varphi \in \mathcal{W}$ are meant in the sense of the usual (crisp) membership relation, which can be expressed by the propositions "f is a fuzzy set" and "φ is an MV-interpretation of $\mathcal{L}_0$", respectively. Taking into account the above remarks, it is clear that $\mathcal{W} \subseteq \mathcal{F}(\mathcal{L}_0, \mathcal{A})$, where the symbol "$\subseteq$" is the usual (crisp) inclusion of sets. That is why (among other reasons) for fuzzy set inclusion the symbol "$\sqsubseteq$" is used.

DEFINITION 6.6 *Given an MV-interpretation $\varphi \in \mathcal{W}$ and a formula $F \in \mathcal{L}_0$, the element $\varphi(F) \in \mathcal{A}$ will be called the **truth value of F in (under) the interpretation** φ.*

In Section 2.2 we have defined the set $C^{sem}(\mathcal{H})$ of all classical semantic consequences from $\mathcal{H}$. For fuzzy propositional logic, this set will be similarly defined. In order to do this we shall use for fuzzy sets either set-theoretical notations or equivalent functional notations, depending on the context. For instance, if f, g are fuzzy sets, then instead of $f \sqsubseteq g$, $f \sqcup g$, $f \sqcap g$, etc., we shall also use the equivalent functional notations $f \leq g$, $f \vee g$, $f \wedge g$, etc.

DEFINITION 6.7 *Let $\mathcal{H}$ be a fuzzy set (of hypotheses). The fuzzy set $C^{sem}(\mathcal{H})$ of all fuzzy semantic consequences of $\mathcal{H}$ is defined by the equalities*

$$C^{sem}(\mathcal{H}) = \bigcap_{\varphi \in \mathcal{W}, \varphi \sqsupseteq \mathcal{H}} \varphi = \bigwedge_{\varphi \in \mathcal{W}, \varphi \geq \mathcal{H}} \varphi$$

therefore,

$$C^{sem}(\mathcal{H})(F) = \bigwedge_{\varphi \in \mathcal{W}, \varphi \geq \mathcal{H}} \varphi(F).$$

For a given formula F we shall write

$$\mathcal{H} \models_a F$$

*if $C^{sem}(\mathcal{H})(F) = a$; in this case, we shall say that F is a **fuzzy semantic consequence from $\mathcal{H}$ in degree a**. In particular, if $\mathcal{H} \simeq \emptyset$, then*

$$C^{sem}(\emptyset) = \bigcap_{\varphi \in \mathcal{W}} \varphi = \bigwedge_{\varphi \in \mathcal{W}} \varphi$$

and

$$C^{sem}(\emptyset)(F) = \bigwedge_{\varphi \in W} \varphi(F).$$

*In this case the notation $\models_a F$ means that $C^{sem}(\emptyset)(F) = a$ and F is called an **a-tautology**. If $a = 1$, then instead of $\models_1 F$ we shall write simply $\models F$ and F will be called a **fuzzy tautology**.*

Example 6.1

Let $\mathbf{X_0} = \{X, Y, \neg, \Rightarrow, (,)\}$ be the alphabet of fuzzy propositional logic and $\mathcal{A} = \{0, \frac{1}{3}, \frac{2}{3}, 1\}$. Consider the following fuzzy set $\mathcal{H}$ of hypotheses:

$$\mathcal{H}(F) = \begin{cases} \frac{1}{3}, & \text{if } F \equiv X; \\ \frac{1}{3}, & \text{if } F \equiv Y; \\ 1, & \text{if } F \equiv X \Rightarrow \neg Y; \\ 0, & \text{otherwise.} \end{cases}$$

Let $G \equiv \neg X \Rightarrow Y$ also be a formula of $\mathcal{L}(X_0)$; we intend to compute the value $C^{sem}(\mathcal{H})(G)$. By definition

$$C^{sem}(\mathcal{H})(G) = \bigwedge_{\varphi \in W, \varphi \geq \mathcal{H}} \varphi(G).$$

If we denote $\varphi(X) = x$, $\varphi(Y) = y$, then the condition $\varphi \geq \mathcal{H}$ means

$$x \geq \frac{1}{3}, \quad y \geq \frac{1}{3}, \quad x \to \bar{y} = 1.$$

The third condition is equivalent to the equality $1 = \min\{1, 1-x+(1-y)\} = \min\{1, 2 - x - y\}$, or $x + y \leq 1$. Therefore, the three conditions above can be written

$$(*) \qquad x \geq \frac{1}{3}, \quad y \geq \frac{1}{3}, \quad x + y \leq 1.$$

It is easy to see that system $(*)$ has only three solutions (x, y), namely

$$(\frac{1}{3}, \frac{1}{3}), \quad (\frac{1}{3}, \frac{2}{3}), \quad (\frac{2}{3}, \frac{1}{3}).$$

On the other hand, for each interpretation $\varphi' \in W$, we have

$$\varphi'(G) = \varphi'(\neg X \Rightarrow Y) = \overline{\varphi'(X)} \to \varphi'(Y) = \overline{x'} \to y'$$

where $\varphi'(X) = x'$, $\varphi'(Y) = y'$. If we suppose additionally that $\varphi' \geq \mathcal{H}$ it follows that the set of pairs (x', y') corresponding to all $\varphi' \in W$ with

$\varphi' \geq \mathcal{H}$ coincides with the set $\left\{ \left(\frac{1}{3}, \frac{1}{3} \right), \left(\frac{1}{3}, \frac{2}{3} \right), \left(\frac{2}{3}, \frac{1}{3} \right) \right\} = \mathcal{Z}$ and we obtain

$$
\begin{aligned}
\mathcal{C}^{sem}(\emptyset) &= \bigwedge_{\varphi' \in \mathcal{W}, \varphi' \geq \mathcal{H}} \varphi'(G) = \bigwedge_{(x',y') \in \mathcal{Z}} \overline{x'} \rightarrow y' = \\
&= \bigwedge_{(x',y') \in \mathcal{Z}} (\min\{1, x' + y'\}) = \frac{2}{3}.
\end{aligned}
$$

Therefore, we can write $\mathcal{H} \models_{\frac{2}{3}} G$, which means that G is a semantic consequence from the fuzzy set $\mathcal{H}$ of degree $\frac{2}{3}$. $\square$

PROPOSITION 6.2

If a formula F in $\mathcal{L}_0$ is a fuzzy tautology, then F is a tautology in the classical sense.

PROOF   Let F be a fuzzy tautology. Taking into account that $\mathcal{V} \subseteq \mathcal{W}$ one has: $1 = \mathcal{C}^{sem}(\emptyset)(F) = \bigwedge_{\varphi \in \mathcal{W}} \varphi(F) \leq \bigwedge_{\varphi \in \mathcal{V}} \varphi(F)$. It follows that $\varphi(F) = 1$ for any $\varphi \in \mathcal{V}$. This means that F is a classical tautology. $\blacksquare$

In order to illustrate the notion of fuzzy tautology one can prove that the formulas

$$
\begin{aligned}
C1 &\equiv F \Rightarrow (G \Rightarrow F); \\
C2 &\equiv (F \Rightarrow G) \Rightarrow ((G \Rightarrow H) \Rightarrow (F \Rightarrow H)); \\
C3 &\equiv (\neg G \Rightarrow \neg F) \Rightarrow (F \Rightarrow G); \\
C4 &\equiv ((F \Rightarrow G) \Rightarrow G) \Rightarrow ((G \Rightarrow F) \Rightarrow F);
\end{aligned}
$$

are fuzzy tautologies; therefore, they are also classical tautologies. For $M > 2$, there are formulas which are not fuzzy tautologies but are nevertheless classical tautologies; such a formula is, for instance, $(X \Rightarrow \neg X) \Rightarrow \neg X$, where X is a propositional variable.

Let us prove, for example, that $C2$ is a fuzzy tautology. For this purpose, we shall show that $\varphi(C2) = 1$ for any $\varphi \in \mathcal{W}$. Taking into account the properties of φ and the definition of the residuation "$\rightarrow$" one obtains the following equalities:

$$
\begin{aligned}
\varphi(C2) &= \varphi((F \Rightarrow G) \Rightarrow ((G \Rightarrow H) \Rightarrow (F \Rightarrow H))) = \\
&= \varphi(F \Rightarrow G) \rightarrow \varphi((G \Rightarrow H) \Rightarrow (F \Rightarrow H)) = \\
&= \varphi(F \Rightarrow G) \rightarrow (\varphi(G \Rightarrow H) \rightarrow \varphi(F \Rightarrow H)) = \\
&= (x \rightarrow y) \rightarrow ((y \rightarrow z) \rightarrow (x \rightarrow z)),
\end{aligned}
$$

where $x = \varphi(F)$, $y = \varphi(G)$, $z = \varphi(H)$; that is

$$
\varphi(C2) = \min\{1, 1 - (x \rightarrow y) + ((y \rightarrow z) \rightarrow (x \rightarrow z))\} \text{ or }
$$

$$\varphi(C2) = \min\{1, 1 - \min\{1, 1 - x + y\} + \min\{1, 1 - (y \to z) + (x \to z)\}\},$$

where we have used the formula

$$a \to b = \min\{1, 1 - a + b\}.$$

Consider now two possibilities: (a) $x \le y$; \quad (b) $y \le x$.
In case (a),

$$\varphi(C2) = \min\{1, 0 + \min\{1, 1 - \min\{1, 1 - y + z\} + \min\{1, 1 - x + z\}\}\}, \text{ or}$$

$$\varphi(C2) = \min\{1, \min\{1, 1 - \min\{1, 1 - y + z\} + \min\{1, 1 - x + z\}\}\},$$

Here we have to analyze three subcases :
\quad (a1) $y \le z$; \quad (a2) $x \le z \le y$; \quad (a3) $z \le x$.
In subcase (a1),

$$\varphi(C2) = \min\{1, 1 - 1 + 1\} = 1.$$

In subcase (a2),

$$\varphi(C2) = \min\{1, \min\{1, 1 + y - z\}\} = 1.$$

In subcase (a3),

$$\varphi(C2) = \min\{1, 0 + \min\{1, 1 - (1 - y + z) + (1 - x + z)\}\} =$$
$$= \min\{1, \min\{1, 1 - x + y\}\} = 1.$$

In case (b),

$$\varphi(C2) = \min\{1, x - y + \min\{1, 1 - \min\{1, 1 - y + z) +$$
$$+ \min\{1, 1 - x + z\}\}\} = 1.$$

We have also to analyze three subcases here:
\quad (b1) $x \le z$; \quad (b2) $y \le z \le x$; \quad (b3) $z \le y$.
In subcase (b1),

$$\varphi(C2) = \min\{1, x - y + \min\{1, 0 + 1\}\} = \min\{1, 1 + x - y\} = 1.$$

In subcase (b2),

$$\varphi(C2) = \min\{1, x - y + \min\{1, 0 + 1 - x + z\}\} =$$
$$= \min\{1, x - y + (1 - x + z)\} = \min\{1, 1 + z - y\} = 1.$$

In subcase (b3),

$$\varphi(C2) = \min\{1, x - y + \min\{1, 1 - (1 - y + z) + (1 - x + z)\}\} =$$

$$\min\{1, x - y + \min\{1, 1 - x + y\}\} = \min\{1, x - y + (1 - x + y)\} = 1.$$

In conclusion, for any $\varphi \in \mathcal{W}$, $\varphi(C2) = 1$; therefore, $C2$ is a fuzzy tautology, and therefore, also a classical one. In a similar way we could prove that $C1, C3, C4$ are also fuzzy tautologies.

In what concerns the formula $T \equiv (X \Rightarrow \neg X) \Rightarrow \neg X$ we shall show that one can find an interpretation $\varphi \in \mathcal{W}$ (assuming $M > 2$) such that $\varphi(T) \neq 1$. For an arbitrary $\varphi \in \mathcal{W}$ we have by definition

$$\varphi(T) = \varphi((X \Rightarrow \neg X) \Rightarrow \neg X) = \varphi(X \Rightarrow \neg X) \to \varphi(\neg X) =$$
$$= (\varphi(X) \to \varphi(\neg X)) \to (\varphi(\neg X))$$

or

$$\varphi(T) = (x \to \bar{x}) \to \bar{x},$$

where $x = \varphi(X)$. From this equality we obtain:

$$\varphi(T) = \min\{1, 1 - (x \to \bar{x}) + \bar{x}\} =$$
$$= \min\{1, 1 - \min\{1, 1 - x + (1 - x)\} + (1 - x)\} =$$
$$= \min\{1, 2 - x - \min\{1, 2 - 2x\}\}.$$

Now, choosing an interpretation $\varphi \in \mathcal{W}$ such that

$$1 > \varphi(X) = x \geq \frac{1}{2}$$

(which is possible because $M > 2$) we get:

$$\varphi(T) = \min\{1, 2 - x - (2 - 2x)\} = \min\{1, x\};$$

therefore, $\varphi(T) = x \neq 1$. It follows that T is **not** a fuzzy tautology. On the other hand, it is easy to see that for any interpretation $\varphi \in \mathcal{V}$, $\varphi(T) = 1$, which means that T is a classical tautology.

REMARK 6.4   The last result above shows that the set $\Im$ of all classical tautologies is strictly larger than the set $\Im_f$ of fuzzy tautologies, i.e., $\Im_f \subseteq \Im$ and $\Im_f \neq \Im$.   ∎

From the Definition 7.7 one obtains the following immediate result.

PROPOSITION 6.3
If $\mathcal{H}, \mathcal{K}$ are fuzzy sets (of hypotheses) with $\mathcal{H} \sqsubseteq \mathcal{K}$, then

$$C^{sem}(\mathcal{H}) \sqsubseteq C^{sem}(\mathcal{K})$$

or

$$C^{sem}(\mathcal{H})(F) \leq C^{sem}(\mathcal{K})(F), \quad for \ all \ F \in \mathcal{L}_0.$$

In other words, if $\mathcal{H} \models_a F$, then for every $\mathcal{K}$ such that $\mathcal{H} \subseteq \mathcal{K}$ we have $\mathcal{K} \models_b F$ for some $b \geq a$ $(b = C^{sem}(\mathcal{K})(F))$.

Fuzzy tautologies can be characterized by means of MV-interpretations.

PROPOSITION 6.4
A formula F is a fuzzy tautology if and only if $\varphi(F) = 1$ for all $\varphi \in \mathcal{W}$.

PROOF   By definition $\models F$ if and only if $C^{sem}(\emptyset)(F) = 1$ and taking into account that $C^{sem}(\emptyset)(F) = \bigwedge_{\varphi \in W} \varphi(F)$, it follows that $C^{sem}(\emptyset)(F) = 1$ if and only if $\varphi(F) = 1$ for all $\varphi \in \mathcal{W}$.   ∎

PROPOSITION 6.5
If $\models G$, then $\models F \Rightarrow G$ for any formula F.

PROOF   Let φ be an arbitrary MV-interpretation. By assumption $\varphi(G) = 1$, so we get $\varphi(F \Rightarrow G) = \varphi(F) \rightarrow 1 = 1$ (because, for any $x \in \mathcal{A}$, $x \rightarrow 1 = 1$).   ∎

PROPOSITION 6.6
If $\models F$ and $\models F \Rightarrow G$, then $\models G$.

PROOF   Since $\varphi(F) = 1$ and $\varphi(F \Rightarrow G) = 1$ for any $\varphi \in \mathcal{W}$, we get:

$$1 = \varphi(F \Rightarrow G) = \varphi(F) \rightarrow \varphi(G) = 1 \rightarrow \varphi(G).$$

It follows that $\varphi(G) = 1$, because $1 \rightarrow z = 1$ if and only if $z = 1$.   ∎

REMARK 6.5   In the same way we could establish some other fuzzy tautologies such as $\models \neg\neg F \Leftrightarrow F$ (Double Negation Law), $\models (F \Rightarrow G)$ $\Rightarrow (\neg G \Rightarrow \neg F)$ (Contraposition Law), the Metatheorem of Fuzzy Syllogism, etc.   ∎

In the study of fuzzy propositional logic the natural question of if it is possible to introduce in $\mathcal{L}_0$ an appropriate equivalence relation such that the corresponding set $\hat{\mathcal{L}}_0$ of equivalence classes could be organized as some algebra in a similar manner as we did this for classical propositional logic, where we have constructed the semantic Lindenbaum algebra arises. The answer to this question is affirmative and we shall sketch below how one can construct such a natural algebraic structure on $\hat{\mathcal{L}}_0$ in the case of fuzzy propositional logic. For this purpose we shall denote by $MV(N, \mathcal{A})$ the set of all $\mathcal{A}$-valued functions defined on $\mathcal{A}^N$. This set becomes in a natural way an MV-algebra if we define the operations $\bar{f}, f \oplus g, f \odot g$ as usual by the equalities:

$$\bar{f}(x_1, x_2, ..., x_N) = \overline{f(x_1, x_2, ..., x_N)};$$
$$(f \oplus g)(x_1, x_2, ..., x_n) = f(x_1, x_2, ..., x_N) \oplus g(x_1, x_2, ..., x_N);$$
$$(f \odot g)(x_1, x_2, ..., x_N) = f(x_1, x_2, ..., x_N) \odot g(x_1, x_2, ..., x_N).$$

Let us notice that $MV(N, \mathcal{A})$ is a finite MV-algebra and obviously

$$card[MV(N, \mathcal{A})] = M^{M^N}.$$

Let us now define the algebraic structure in $\hat{\mathcal{L}}_0$. We shall proceed in two steps. First, we shall define an equivalence relation "$\approx$" in $\mathcal{L}_0$ and after we shall define on the set $\mathcal{L}_0/\approx = \hat{\mathcal{L}}_0$ a structure of MV-algebra.

The formulas F, G in $\mathcal{L}_0$ are said to be equivalent and we denote this by $F \approx G$ if $\varphi(F) = \varphi(G)$ for all $\varphi \in \mathcal{W}$. It easy to see that $F \approx G$ if and only if $F \Rightarrow G$ and $G \Rightarrow F$ are fuzzy tautologies or if and only if $F \Leftrightarrow G$ is a fuzzy tautology. The binary relation "$\approx$" is obviously an equivalence relation in $\mathcal{L}_0$, so that the formulas in $\mathcal{L}_0$ are grouped in nonempty, disjoint classes of equivalence. The set $\mathcal{L}_0/\approx$ of all these classes will be denoted by $\hat{\mathcal{L}}_0$. If we denote by ξ_F the class containing the formula F, then obviously $F \approx G$ if and only if $\xi_F = \xi_G$. In $\hat{\mathcal{L}}_0$ one defines the operations $^-, \oplus, \odot$ by $\bar{\xi}_F = \xi_{\neg F}$, $\xi_F \oplus \xi_G = \xi_{F \triangledown G}$, $\xi_F \odot \xi_G = \xi_{F \& G}$. By routine arguments one can show that these definitions are correct, i.e., the result of any such operation is independent of the choice of their representatives. If we denote by $\Im_f$ the set of all fuzzy tautologies, then it is easy to see that $\Im_f$ is an equivalence class, hence $\Im_f \in \hat{\mathcal{L}}_0$. The set $\hat{\mathcal{L}}_0$ endowed with the above three operations will be an MV-algebra in which the elements $\Im_f$ and $\theta_f = \bar{\Im}_f$ are the two unities. In the MV-algebra $\hat{\mathcal{L}}_0$ one also introduces by abbreviation the following derived operations :

$$\xi_F \vee \xi_G = (\xi_F \odot \bar{\xi}_G) \oplus \xi_G,$$
$$\xi_F \wedge \xi_G = (\xi_F \oplus \bar{\xi}_G) \odot \xi_G,$$
$$\xi_F \to \xi_G = \bar{\xi}_F \oplus \xi_G.$$

It is not difficult to verify that

$$\xi_F \vee \xi_G = \xi_{F \vee G}, \quad \xi_F \wedge \xi_G = \xi_{F \wedge G}, \quad \xi_F \to \xi_G = \xi_{F \Rightarrow G}.$$

$\hat{\mathcal{L}}_0$ with respect to the operations $^-, \vee, \wedge$ is a De Morgan algebra in which equalities $\xi \vee \bar{\xi} = \Im_f$ and $\xi \wedge \bar{\xi} = \theta_f$ fail, so that this De Morgan algebra is not a Boolean algebra. It follows that in the initial MV-algebra $(\hat{\mathcal{L}}_0, ^-, \oplus, \odot)$ there exist no **idempotent elements**.

As in the case of classical propositional logic, one can define a mapping Φ which associates any formula F with an element $\Phi(F)$ of $MV(N, \mathcal{A})$. This mapping is defined naturally by the equality $\Phi(F)(\gamma) = \varphi_\gamma(F)$ for all $\varphi \in \mathcal{A}^N$. The mapping Φ is constant on each class $\xi \in \hat{\mathcal{L}}_0$ and has the properties:

(i) $\Phi(\neg F) = \overline{\Phi(F)}$;

(ii) $\Phi(F \Rightarrow G) = \Phi(F) \rightarrow \Phi(G)$;

(iii) $\Phi(X_j) = p_j$ $(1 \leq j \leq N)$ (where the function p_j belongs to $MV(N, \mathcal{A})$ and is defined by $p_j(x_1, x_2, ..., x_N) = x_j$).

Moreover, the mapping Φ is unique and has the following two additional properties:

(iv) $\Phi(F \bigtriangledown G) = \Phi(F) \oplus \Phi(G)$;

(v) $\Phi(F \& G) = \Phi(F) \odot \Phi(G)$.

The mapping $\hat{\Phi} : \hat{\mathcal{L}}_0 \rightarrow MV(N, \mathcal{A})$ induced on $\hat{\mathcal{L}}_0$ by Φ is an isomorphism from $\hat{\mathcal{L}}_0$ onto some MV-subalgebra of $MV(N, \mathcal{A})$. Taking into account that card $[MV(N, \mathcal{A})] = M^{M^N}$ it follows, in particular, that the MV-algebra $\hat{\mathcal{L}}_0$ is finite and can be identified with an MV-algebra of $\mathcal{A}$-valued functions defined on $\mathcal{A}^N$. The image in $MV(N, \mathcal{A})$ of $\hat{\mathcal{L}}_0$ by the mapping $\hat{\Phi}$ is the smallest MV-subalgebra of $MV(N, \mathcal{A})$ containing the projectors $p_1, p_2, ..., p_N$.

6.4 Elements of Fuzzy Propositional Logic in the Syntactic Version

From the important papers by J. Pavelka, V. Novák, P. Hájek and others it follows that fuzzy logic can be essentially regarded as "many-valued logic with special properties" (see Novák [36]). In this section we shall present the syntactic version of the truth structure in fuzzy propositional logic following the main lines of results of Pavelka and Novák results. Let us consider the following four (schemes of) formulas:

C1: $F \Rightarrow (G \Rightarrow F)$;

C2: $(F \Rightarrow G) \Rightarrow ((G \Rightarrow H) \Rightarrow (F \Rightarrow H))$;

C3: $(\neg G \Rightarrow \neg F) \Rightarrow (F \Rightarrow G)$;

C4: $((F \Rightarrow G) \Rightarrow G) \Rightarrow ((G \Rightarrow F) \Rightarrow F)$, where F, G, H are arbitrary formulas in $\mathcal{L}_0$. We shall denote by $\mathbf{C}$ the set of all formulas which can be obtained by the schemes (C1), (C2), (C3), (C4). Each formula $A \in \mathbf{C}$ will be called an **axiom** of fuzzy propositional logic.

Notice that all formulas in $\mathbf{C}$ are fuzzy tautologies (as we proved in 6.3); therefore, they are also tautologies (and theorems) in the classical (crisp) propositional logic.

Let us now consider a fuzzy set of axioms, i.e., a fuzzy set denoted by $\mathbf{ax}$, $\mathbf{ax} : \mathcal{L}_0 \to A$ having the following two properties:

(i) $\mathbf{ax}(A) = 1$ for all $A \in \mathbf{C}$;

(ii) $\mathbf{ax} \leq \varphi$ for all MV- interpretations $\varphi \in \mathcal{W}$.

Such a fuzzy set is, for instance, the characteristic function $\mathbf{Ax}$ of the set $\mathbf{C}$, i.e.,

$$\mathbf{Ax}(F) = \begin{cases} 1, & \text{if } F \in \mathbf{C} \\ 0, & \text{otherwise.} \end{cases}$$

It is clear that, for any $\varphi \in \mathcal{W}$, we have $\mathbf{Ax}(F) \leq \varphi(F)$ for all $F \in \mathcal{L}_0$, hence $\mathbf{Ax} \leq \varphi$ for all $\varphi \in \mathcal{W}$. Therefore,

$$\mathbf{Ax} \leq \mathbf{ax} \leq \bigwedge_{\varphi \in \mathcal{W}} \varphi.$$

DEFINITION 6.8   *A fuzzy set $U : \mathcal{L}_0 \to A$ is said to be **MP-closed** if it satisfies the inequality*

$$U(F) \odot U(F \Rightarrow G) \leq U(G)$$

for all F, G in $\mathcal{L}_0$.

PROPOSITION 6.7
Each interpretation $\varphi \in \mathcal{W}$, viewed as a fuzzy set, is MP-closed.

PROOF   If φ is an MV-interpretation of $\mathcal{L}_0$, then:

$$\varphi(F) \odot \varphi(F \Rightarrow G) = \varphi(F) \odot (\varphi(F) \to \varphi(G)) =$$

$$= \max\{0, \varphi(F) + (\varphi(F) \to \varphi(G)) - 1\} =$$

$$= \max\{0, \varphi(F) - 1 + \min\{1, 1 - \varphi(F) + \varphi(G)\}\}.$$

Denoting $\varphi(F) = x, \varphi(G) = y$ we obtain:

$$\varphi(F) \odot \varphi(F \Rightarrow G) = \max\{0, x - 1 + \min\{1, 1 - x + y\}\}.$$

We have to prove that the right side of this equality is less or equal to y. For this purpose we have to analyze two possibilities:

$$\text{(a)} \quad x \leq y; \quad \text{(b)} \quad y \leq x.$$

In case (a),

$$\max\{0, x - 1 + \min\{1, 1 - x + y\}\} = \max\{0, x - 1 + 1\} = x \leq y.$$

In case (b),

$$\max\{0, x - 1 + \min\{1, 1 - x + y\}\} = \max\{0, x - 1 + (1 - x + y)\} =$$

$$= \max\{0, y\} = y.$$

It follows that $\varphi(F) \odot \varphi(F \Rightarrow G) \leq \varphi(G)$ for all $\varphi \in \mathcal{W}$. Therefore, the fuzzy set φ is MP-closed. ∎

For any fuzzy set $\mathcal{H}$ (of hypotheses), we shall denote by $\mathcal{U}_{\mathcal{H}}$ the (crisp) set of all fuzzy sets U having the following three properties:

(1) U is MP-closed; or   (1) $U(G) \geq U(F) \odot U(F \Rightarrow G)$;

(2) $U \sqsupseteq \mathbf{ax}$;   or   (2) $U(F) \geq \mathbf{ax}(F)$;

(3) $U \sqsupseteq \mathcal{H}$;   or   (3) $U(F) \geq \mathcal{H}(F)$.

DEFINITION 6.9   *The fuzzy set $C^{syn}(\mathcal{H})$ given by the equality*

$$C^{syn}(\mathcal{H}) = \bigcap_{U \in \mathcal{U}_{\mathcal{H}}} U = \bigwedge_{U \in \mathcal{U}_{\mathcal{H}}} U$$

will be called **the fuzzy set of all syntactic consequences from** $\mathcal{H}$. *In other words, for any formula $F \in \mathcal{L}_0$*

$$C^{syn}(\mathcal{H})(F) = \bigwedge_{U \in \mathcal{U}_{\mathcal{H}}} U(F).$$

If $C^{syn}(\mathcal{H})(F) = a$, we shall write equivalently $\mathcal{H} \vdash_a F$ and we shall say that F is **syntactically deducible from** $\mathcal{H}$ **in degree** a.

In the case $\mathcal{H} \simeq \emptyset$, instead of $\emptyset \vdash_a F$ we shall write $\vdash_a F$ and we shall say that F is a **fuzzy theorem in degree** a. *If $a = 1$, then we shall write $\vdash F$ and we shall say that F is a* **fuzzy theorem**.

The next proposition follows immediately from the Definition 6.9.

PROPOSITION 6.8
If $\mathcal{H}, \mathcal{K}$ are fuzzy sets with $\mathcal{H} \sqsubseteq \mathcal{K}$, then:

(a) $\mathcal{U}_{\mathcal{H}} \supseteq \mathcal{U}_{\mathcal{K}}$;

(b) $C^{syn}(\mathcal{H}) \sqsubseteq C^{syn}(\mathcal{K})$.

PROPOSITION 6.9

(a) $C^{syn}(\mathcal{H}) \sqsubseteq C^{sem}(\mathcal{H})$ or (which is the same),

$$C^{syn}(\mathcal{H})(F) \leq C^{sem}(\mathcal{H})(F) \quad \text{for all} \quad F \in \mathcal{L}_0;$$

(b) If $\vdash_x F$ and $\models_y F$, then $x \leq y$;

(c) If F is a fuzzy theorem, then it is a fuzzy tautology.

PROOF By definition $C^{syn}(\mathcal{H}) = \bigcap\limits_{U \in \mathcal{U}_{\mathcal{H}}} U$; in Proposition 6.7 we proved that any $\varphi \in \mathcal{W}$ considered as a fuzzy set is MP-closed; on the other hand by the definition of the fuzzy set **ax** we have $\varphi \sqsupseteq$ **ax**. It follows that any $\varphi \in \mathcal{W}$ satisfying $\varphi \sqsupseteq \mathcal{H}$ belongs to $\mathcal{U}_{\mathcal{H}}$. From these remarks we get:

$$C^{syn}(\mathcal{H}) = \bigcap\limits_{U \in \mathcal{U}_{\mathcal{H}}} U \sqsubseteq \bigcap\limits_{\varphi \in \mathcal{W}, \ \varphi \sqsupseteq \mathcal{H}} \varphi = C^{sem}(\mathcal{H}).$$

In other words, $C^{syn}(\mathcal{H})(F) \leq C^{sem}(\mathcal{H})(F)$ for all $F \in \mathcal{L}_0$. Therefore, assertion (a) is proved.

Assertion (b) is an immediate consequence of (a). From (a), taking $\mathcal{H} \simeq \emptyset$, it follows that if F is a fuzzy theorem, then $1 = C^{syn}(\emptyset)(F) \leq C^{sem}(\emptyset)(F)$, hence $C^{sem}(\emptyset)(F) = 1$, i.e., F is a fuzzy tautology. $\blacksquare$

It is easy to see that any formula $A \in \mathbf{C}$ is a fuzzy theorem. Indeed, taking into account that for each $U \in \mathcal{U}_{\emptyset}$ we have $U \sqsupseteq$ **ax**, hence $U(A) \geq$ **ax**$(A) = 1$, one obtains

$$C^{syn}(\emptyset)(A) = \bigwedge\limits_{U \in \mathcal{U}_{\emptyset}} U(A) = 1,$$

which means that A is a fuzzy theorem.

REMARK 6.6 There are theorems in the classical sense which are not fuzzy theorems; indeed, we have proved in Section 6.3 that the formula

$$T \equiv (X \Rightarrow \neg X) \Rightarrow \Rightarrow \neg X$$

is not a fuzzy tautology, hence it cannot be a fuzzy theorem. $\blacksquare$

In fuzzy propositional logic one can introduce an appropriate concept of syntactic deduction similar to the one defined in the classical (crisp) case. However, this time we have to evaluate each step of the fuzzy deduction by using a certain element $a \in \mathcal{A}$. For this reason fuzzy deductions will be called evaluated deductions.

DEFINITION 6.10 *A finite sequence of pairs*

$$\pi : \quad [F_1; a_1], [F_2; a_2], ..., [F_r; a_r], \quad (r \geq 1)$$

where F_i are formulas and a_i are elements of $\mathcal{A}$ $(1 \leq i \leq r)$ is called an **evaluated deduction from** $\mathcal{H}$ *if, for any index i $(1 \leq i \leq r)$, the element a_i satisfies one of conditions below:*
(1) if there are indices h, k $(h, k < i)$ such that

$$F_k \equiv F_h \Rightarrow F_i,$$

then

$$a_i = a_h \odot a_k \quad or \quad a_i = \mathbf{ax}(F_i) \quad or \quad a_i = \mathcal{H}(F_i);$$

(2) if such indices h, k $(h, k < i)$ do not exist, then

$$a_i = \mathbf{ax}(F_i) \quad or \quad a_i = \mathcal{H}(F_i).$$

The element a_r is called the value of the deduction π and is denoted by $a_r = val_{\mathcal{H}}(\pi)$. If $F_r \equiv F$, then π is called an **evaluated deduction of** F **from** $\mathcal{H}$. *In the case $\mathcal{H} \simeq \emptyset$, the condition $a_i = \mathcal{H}(F_i)$ is dropped and π is called an* **evaluated proof** *(of the formula $F_r \equiv F$).*

REMARK 6.7 (1) It is clear that, for the first pair $[F_1; a_1]$ of an evaluated deduction π from $\mathcal{H}$, we have necessarily

$$a_1 = ax(F_1) \quad or \quad a_1 = \mathcal{H}(F_1).$$

(2) From the Definition 6.10 it follows that if

$$\pi' : \quad [F_1'; a_1'], \ldots, [F_r'; a_r'],$$

$$\pi'' : \quad [F_1''; a_1''], \ldots, [F_s''; a_s'']$$

are evaluated deductions from $\mathcal{H}$, then the interleaving sequences

$$[F_1'; a_1'], \ldots, [F_r'; a_r'], [F_1''; a_1''], \ldots, [F_s''; a_s''],$$

$$[F_1'; a_1'], [F_1''; a_1''], [F_2'; a_2'], [F_2''; a_2''], \quad \text{etc.}$$

are also evaluated proofs from $\mathcal{H}$. ∎

For any fuzzy set $\mathcal{H}$ and for any formula F we shall denote by $\Pi_{\mathcal{H}}(F)$ the (crisp) set of all evaluated deductions π from $\mathcal{H}$ of the formula F. Let us consider now the fuzzy set $\mathcal{Z}_{\mathcal{H}}$ defined by the equality:

$$\mathcal{Z}_{\mathcal{H}}(F) = \bigvee_{\pi \in \Pi_{\mathcal{H}}(F)} val_{\mathcal{H}}(\pi),$$

where $\displaystyle\bigvee_{\pi \in \Pi_{\mathcal{H}}(F)} val_{\mathcal{H}}(\pi)$ is the l.u.b. (lowest upper bound) of all elements $val_{\mathcal{H}}(\pi) \in \mathcal{A}$, when π runs over all $\pi \in \Pi_{\mathcal{H}}(F)$. Taking into account that our MV-algebra is finite and linearly ordered, it is obvious that this l.u.b. (exists and) is an element of $\mathcal{A}$ (this property also holds in the general case of so-called complete infinitely distributive infinite MV-algebras).

PROPOSITION 6.10
For any formula $F \in \mathcal{L}_0$ the following inequality holds:

$$\mathcal{Z}_{\mathcal{H}}(F) \leq \mathcal{C}^{syn}(\mathcal{H})(F).$$

PROOF   Let the sequence

$$\pi : \quad [F_1; a_1], [F_2; a_2], ..., [F_r; a_r],$$

be an evaluated deduction of $F \equiv F_r$ from $\mathcal{H}$. We shall prove by induction that

$(*)$ $\qquad\qquad a_i \leq \mathcal{C}^{syn}(\mathcal{H})(F_i)$, for all i $(1 \leq i \leq r)$.

For $i = 1$, we have $a_1 = \mathbf{ax}(F_1)$ or $a_1 = \mathcal{H}(F_1)$. If U is an arbitrary element of $\mathcal{U}_{\mathcal{H}}$, then $U \sqsupseteq \mathbf{ax}$ and $U \sqsupseteq \mathcal{H}$, hence

$$U(F_1) \geq \mathbf{ax}(F_1) \quad \text{and} \quad U(F_1) \geq \mathcal{H}(F_1).$$

Therefore, $U(F_1) \geq a_1$ for any $U \in \mathcal{U}_{\mathcal{H}}$ and

$$\mathcal{C}^{syn}(\mathcal{H})(F_1) = \bigwedge_{U \in \mathcal{H}} U(F_1) \geq a_1,$$

hence in case $i = 1$ the assertion $(*)$ is verified.
For $i > 1$, assume that the assertion $a_j \leq \mathcal{C}^{syn}(\mathcal{H})(F_j)$ is true for any j $(1 \leq j < i)$. Taking into account Definition 6.10 there are three possibilities:

(1) $a_i = \mathbf{ax}(F_i)$,

(2) $a_i = \mathcal{H}(F_i)$ and

(3) there are indices $p, q < i$ such that

$$F_q \equiv F_p \Rightarrow F_i \quad \text{and} \quad a_i = a_p \odot a_q.$$

Cases (1) and (2) were already discussed above. In case (3), we have $a_i = a_p \odot a_q$ and taking into account the inductive assumption and the inequality $U(F) \geq C^{syn}(\mathcal{H})(F)$, for all $U \in \mathcal{U}_{\mathcal{H}}$ and all $F \in \mathcal{L}_0$, we obtain:

$$a_i = a_p \odot a_q \leq C^{syn}(\mathcal{H})(F_p) \odot C^{syn}(\mathcal{H})(F_q) \leq$$
$$\leq U(F_p) \odot U(F_p \Rightarrow F_i) \leq U(F_i),$$

where the last inequality is due to the MP-closeness of U. The fuzzy set U being an arbitrary element of $\mathcal{U}_{\mathcal{H}}$ it follows that

$$a_i \leq \bigwedge_{U \in \mathcal{U}_{\mathcal{H}}} U(F_i) = C^{syn}(\mathcal{H})(F_i)$$

and assertion (∗) follows from the principle of mathematical induction. In particular,

$$val_{\mathcal{H}}(\pi) = a_r \leq C^{syn}(\mathcal{H})(F_r) = C^{syn}(\mathcal{H})(F),$$

for any deduction π of F from $\mathcal{H}$. Taking into account that $val_{\mathcal{H}}(\pi) \in \mathcal{A}$ (for any deduction π) and that $\mathcal{A}$ is a finite linearly ordered MV-algebra, it follows that

$$\mathcal{Z}_{\mathcal{H}}(F) = \bigvee_{\pi \in \Pi_{\mathcal{H}}} val_{\mathcal{H}}(\pi) \leq C^{syn}(\mathcal{H})(F).$$

∎

REMARK 6.8   One can prove (see Part II, Chapter 13) that the inequality in the above proposition is actually an equality, i.e.,

$$\mathcal{Z}_{\mathcal{H}}(F) = \bigvee_{\pi \in \Pi_{\mathcal{H}}} val_{\mathcal{H}}(\pi) = C^{syn}(\mathcal{H})(F).$$

∎

The next proposition expresses the soundness of syntactic fuzzy propositional logic. This means that we cannot deduce both $\vdash F$ and $\vdash \neg F$.

PROPOSITION 6.11

(Metatheorem of consistency)   *There are no formulas F in $\mathcal{L}_0$ such that and F and $\neg F$ will both be fuzzy theorems.*

PROOF   Suppose that there exists a formula $F \in \mathcal{L}_0$ such that F and $\neg F$ are fuzzy theorems; by Proposition 6.9 (c) it follows that F and $\neg F$ are fuzzy tautologies. But if F and $\neg F$ are fuzzy tautologies, then by definition it follows that $\varphi(F) = 1$ and $1 = \varphi(\neg F) = \overline{\varphi(F)} = 0$ for any $\varphi \in \mathcal{W}$, which is contradictory.   ∎

REMARK 6.9   The above proposition can be proved in another way, too. Indeed, it is not difficult to establish that for any $F, G \in \mathcal{L}_0$, the formula $\neg F \Rightarrow (F \Rightarrow G)$ is a fuzzy theorem (the proof can be performed just as was done in Section 5.2).

Now, supposing that $\vdash F$ and $\vdash \neg F$, then from $\vdash \neg F \Rightarrow (F \Rightarrow G)$ applying (MP) twice one gets $\vdash G$ for any $G \in \mathcal{L}_0$. This is impossible because, for instance, the formula $T \equiv (X \Rightarrow \neg X) \Rightarrow \neg X$ is not a fuzzy theorem (as we proved in the remark following Proposition 6.9).   ∎

Final comments concerning fuzzy propositional logic.

We stop our considerations about syntactic fuzzy logic here. Further details can be found in the papers by J. Pavelka and V. Novák referenced at the end of the book. Here we mention only that in fuzzy propositional logic (as well as in fuzzy predicate logic) one can prove important results such as the Deduction Metatheorem, the Completeness Metatheorem, etc.

7

Applications of Propositional Logic in Computer Science

This chapter contains some applications of propositional calculus. The formalization and study of sentences and connectives in propositional logic have numerous applications in computer science which include Karnaugh maps, switching networks, and logical networks.

7.1   Recall Lindenbaum Algebra of the Language $\mathcal{L}_0$

In Chapter 2 we proved the existence (and uniqueness) of the functional interpretation of $\mathcal{L}_0$, i.e., of the function $\Phi : \mathcal{L}_0 \to B(N, \Gamma)$ having the following properties:

(i) $\Phi(\neg F) = \overline{\Phi(F)}$;

(ii) $\Phi(F \vee G) = \Phi(F) + \Phi(G)$;

(iii) $\Phi(X_j) = p_j$.

$\mathcal{L}_0 = \mathcal{L}(\mathbf{X}_0)$ is the language of propositional logic generated by the propositional variables $X_1, X_2, \ldots, X_N$, while $B(N, \Gamma)$ is the Boolean algebra of all Boolean functions of N variables, i.e.,

$$B(N, \Gamma) = \{f : \Gamma^N \to \Gamma\}.$$

We proved also that the function Φ is surjective, that is, for each $f \in B(N, \Gamma)$ there is some formula $F \in \mathcal{L}_0$ such that $\Phi(F) = f$. In the proof of this assertion we have used the polynomial representation of Boolean functions: each function $f \in B(N, \Gamma)$, $f \neq 0$ can be expressed by the equality

$$f = \sum_{f(\sigma_1, \ldots, \sigma_N) = 1} p_1^{\sigma_1} p_2^{\sigma_2} \ldots p_N^{\sigma_N}$$

equality

$$f = \sum_{f(\sigma_1,\ldots,\sigma_N)=1} p_1^{\sigma_1} p_2^{\sigma_2} \ldots p_N^{\sigma_N}$$

or

$$f(x_1,\ldots,x_N) = \sum_{f(x_1,\ldots,x_N)=1} x_1^{\sigma_1} \ldots x_N^{\sigma_N}, \quad \text{for all} \quad (x_1,\ldots,x_N) \in \Gamma^N.$$

It is easy to see that by the function Φ we have the following correspondence:

$$F \equiv \bigvee_{f(\sigma_1,\ldots,\sigma_N)=1} (X_1^{\sigma_1} \wedge X_2^{\sigma_2} \wedge \ldots \wedge X_N^{\sigma_N})$$
$$\downarrow \qquad\qquad\qquad\qquad \downarrow \quad\; \downarrow \qquad\qquad \downarrow$$
$$f = \sum_{f(\sigma_1,\ldots,\sigma_N)=1} p_1^{\sigma_1} \cdot p_2^{\sigma_2} \;\cdot\ldots\cdot\; p_N^{\sigma_N}$$

which shows that $\Phi(F) = f$.

We mentioned in Part 1, Chapter 2, that although the function Φ is onto it is not one-to-one, i.e., given $f \in B(N,\Gamma)$ there exist many formulas $F', F'', \ldots \in \mathcal{L}_0$ such that $\Phi(F') = \Phi(F'') = \ldots = f$. For this reason we have introduced the equivalence relation "$\approx$" defined by $F' \approx F''$ iff $\Phi(F') = \Phi(F'')$. In such a way the language is decomposed into nonempty equivalence classes of formulas. The set of these classes (the quotient set of $\mathcal{L}_0$ with respect to "$\approx$") was denoted by $\hat{\mathcal{L}}_0$. For each $f \in B(N,\Gamma)$ the set of all formulas F with $\Phi(F) = f$ is an equivalence class $\xi \in \hat{\mathcal{L}}_0$. Moreover, the set $\hat{\mathcal{L}}_0$ can be organized in a natural way as a Boolean algebra with respect to the operations $\bar{\xi}, \xi + \eta, \xi \cdot \eta$ defined as follows: consider arbitrary formulas $F \in \xi, G \in \eta$ and denote by $\bar{\xi}$ the class containing $\neg F$ and by $\xi + \eta, \xi \cdot \eta$ the classes containing $F \vee G, F \wedge G$, respectively. We have already proved that the Boolean algebra $\hat{\mathcal{L}}_0$ is isomorphic with $B(N,\Gamma)$. This isomorphism is given by the correspondence:

$$\xi \mapsto f = \Phi(F), \quad \text{where} \quad F \in \xi.$$

The correspondence $F \mapsto \Phi(F)$ involving implicitly the above correspondence is illustrated by the following diagram:

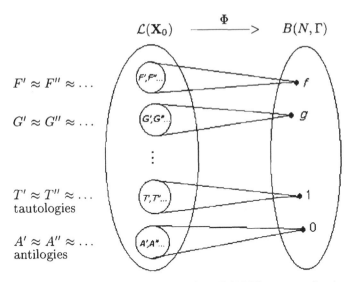

On the other hand, each function $f \in B(N, \Gamma)$ is completely determined by its table of values, often called the truth table of f. For instance, let us consider the function

$$f(x_1, x_2, x_3) = x_1 x_2 x_3 + \bar{x}_1 x_2 x_3 + \bar{x}_1 x_2 \bar{x}_3.$$

It is easy to compute the associate truth table for this function:

x_1	x_2	x_3	f
1	1	1	1
1	1	0	0
1	0	1	0
1	0	0	0
0	1	1	1
0	1	0	1
0	0	1	0
0	0	0	0

Conversely, suppose that this truth table is given and let us define the initial function f. From the truth table we obtain all triples $(\sigma_1, \sigma_2, \sigma_3)$ for which $f(\sigma_1, \sigma_2, \sigma_3) = 1$; they are

$$(\sigma_1, \sigma_2, \sigma_3): \quad (0, 1, 0), \ (0, 1, 1), \ (1, 1, 1).$$

By the theorem of Boolean representation the Boolean function corresponding to this set of triples is

$$\begin{aligned} f'(x_1, x_2, x_3) &= x_1^0 x_2^1 x_3^0 + x_1^0 x_2^1 x_3^1 + x_1^1 x_2^1 x_3^1 \\ &= \bar{x}_1 x_2 \bar{x}_3 + \bar{x}_1 x_2 x_3 + x_1 x_2 x_3 \\ &= f(x_1, x_2, x_3). \end{aligned}$$

Thus, $f' = f$.

The above remark makes clear that instead of working with Boolean functions we can work just as well with truth tables. In our example a formula F having the property $\Phi(F) = f$ can be obtained by the equality

$$F \equiv \bigvee_{f(\sigma_1, \sigma_2, \sigma_3) = 1} (X_1^{\sigma_1} \wedge X_2^{\sigma_2} \wedge X_3^{\sigma_3})$$

that is,

$$F \equiv (\neg X_1 \wedge X_2 \wedge \neg X_3) \vee (\neg X_1 \wedge X_2 \wedge X_3) \vee (X_1 \wedge X_2 \wedge X_3).$$

In practice this approach is used often.

The process of conversion of information can be illustrated by the following diagram:

Truth table $\rightarrow$ Formula F in normal disjunctive form $\rightarrow$ Formula $G \approx F$

but simplified.

7.2   Some Connections of $\mathcal{L}_0$ with Programming Languages

The logical connectives are available in many programming languages. This is shown in the following table:

Logic	English	Pascal	C	Prolog
$\neg$	not	not	!	not
$\vee, +$	or	or	|	;
$\wedge, \cdot$	both...and...	and	&	,
$\Rightarrow, \rightarrow, \supset$	not...without..., if...then			$\rightarrow$, :-
$\Leftrightarrow, \leftrightarrow, \approx$	if and only if			

The truth values of formulas (logical expressions) often determine the direction of the flow control in computer programs. Consider, for example, a conditional branch of a program of the "if-then-else" sort: if the condition is true, then the program will execute one section of code (the one between **then** and **else**); if the value is false, the program will execute different section of code (the one between **else** and **end if**). If the conditional expression is replaced by a simpler but logically equivalent expression, then the truth value of the expression will remain the same. Hence, the flow of control of the program will not be affected. However, the new condition (code) is easier to understand for the reader and may be executed faster by the computer.

Example 7.1
Rewrite the following conditional branching program with a simplified conditional expression:

```
if not(( value1 < value2 ) or odd( number ))
or (not ( value1 < value2 ) and odd( number )) then
  procedure 1
else
  procedure 2;
```

The conditional expression has the form $\neg(P \vee Q) \vee (\neg P \wedge Q)$. Then we get:

$$\neg(P \vee Q) \vee (\neg P \wedge Q) \approx (\neg P \wedge \neg Q) \vee (\neg P \wedge Q) \approx$$
$$\approx \neg P \wedge (\neg Q \vee Q) \approx \neg P$$

Hence, the condition form can be written as:

```
if not( value1 < value2 ) then
  procedure 1
else
  procedure 2;
```

7.3   Karnaugh Maps

Karnaugh maps represent geometrical methods for simplifying formulas of propositional calculus in disjunctive normal form. We shall only treat the case of two, three and four variables; in general Boolean expressions involving at most six variables may be considered by this method. For the simplicity of the notations we shall use the following conventions in this section: X_1' for $\neg X_1$, $X_1 + X_2$ for $X_1 \vee X_2$ and $X_1 X_2$ for $X_1 \wedge X_2$.

DEFINITION 7.1

(i) *A conjunction $X_1 X_2 ... X_n$ is called a* **fundamental conjunction** *if it is composed of a conjunction of literals (propositional variables or complemented propositional variables) in which no two literals involve the same variable.*

(ii) Two fundamental conjunctions are **adjacent** *if they differ in exactly one literal, which must be a complemented variable in one conjunction and uncomplemented in the other.*

(iii) A **Karnaugh map** *is a condensed truth table where the fundamental conjunctions relative to the same variables are represented by squares such that two fundamental conjunctions are adjacent if and only if they have a side in common (are geometrically adjacent).*

Karnaugh maps record the ones (1) of the truth functions in the corresponding squares of arrays, that force products of inputs differing by only one factor to be adjacent. Hence, the sum of two adjacent fundamental conjunctions will be a fundamental conjunction with one less literal. Therefore, we shall obtain a simplified expression.

For example, if $P \equiv X_1X_2X_3$ and $Q \equiv X_1X_2X_3'$, then $P + Q \equiv X_1X_2$. Generally speaking, if we consider two fundamental conjunctions $P \equiv X_1X_2...X_sX_t$ and $Q \equiv X_1X_2...X_sX_t'$, then $P + Q \equiv X_1X_2...X_s$.

The general form for Karnaugh maps for two, three or four variables are as follows:

	X_1	X_1'
X_2		
X_2'		

	X_1X_2	X_1X_2'	$X_1'X_2'$	$X_1'X_2$
X_3				
X_3'				

	X_1X_2	X_1X_2'	$X_1'X_2'$	$X_1'X_2$
X_3X_4				
X_3X_4'				
$X_3'X_4'$				
$X_3'X_4$				

By a **basic rectangle** in the Karnaugh map with four variables we mean a square, two adjacent squares, four squares which form a one-by-four or two-by-two rectangle, or eight squares which form a two-by-four rectangle. These basic rectangles correspond to fundamental conjunctions with four, three, two and one literals, respectively. We shall put loops around the 1s in the maximal basic rectangles, i.e., basic rectangles which are not contained in any larger basic rectangle. In this way we can obtain the simplified formulas, as shown in the following example.

Example 7.2

Use the Karnaugh map to simplify the following formula:

$$X_4' X_2' + X_4' X_1' + X_4 X_1 X_2' + X_3' X_4' X_1 X_2.$$

First, we have to obtain the complete disjunctive normal form (where each conjunction involves all the variables). Starting with the fundamental conjunction $X_2' X_4'$ we can get the following formula:

$X_2' X_4' \approx X_2' X_3 X_4' + X_2' X_3' X_4' \approx X_1 X_2' X_3 X_4' + X_1' X_2' X_3 X_4' + X_1' X_2' X_3' X_4'$
$+ X_1 X_2' X_3' X_4'.$

Then, we mark the corresponding squares by 1s. We shall repeat this procedure for the other conjunctions from the given formula.

	$X_1 X_2$	$X_1 X_2'$	$X_1' X_2'$	$X_1' X_2$
$X_3 X_4$		1		
$X_3 X_4'$		1	1	1
$X_3' X_4'$	1	1	1	1
$X_3' X_4$		1		

From the above table we see that we have three maximal basic rectangles: two rectangles one-by-four and one rectangle two-by-two. Consequently, we get the simplified formula $X_3' X_4' + X_1' X_4' + X_1 X_2'$.

It is worth noting that, because of the idempotence property, any term of a normal disjunctive form can be used as many times as possible.

7.4 Switching Networks

Introductory considerations. In this section we shall describe the structure of some special electrical circuits, called in the following **switching networks** or for short, SN. Roughly speaking, a switching network is a network (a graph) with two terminal points, which is composed by one or more interconnected **binary switches**.

A **binary switch**, abbreviated a switch, is an electrical device which has two possible states, **closed** and **open**, denoted by 1 and 0, respectively. Schematically, a BS can be represented as follows:

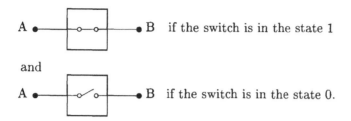

A ———————— B   if the switch is in the state 1

and

A ———————— B   if the switch is in the state 0.

The switches are the simplest switching networks. A remarkable feature of SNs is the possibility of combining them, i.e., of constructing new ones by means of some specific operations such as "parallel connection", "serial connection" and "complementation". Thus, starting from the simplest SNs (i.e., from switches) one can construct successively more and more complex new SNs. The most relevant property of the three mentioned operations consists of the fact that they lead to new circuits also having two terminal points, i.e., to new switching networks and that the states of the resulting networks can be precisely "computed" from the states of already constructed circuits, hence finally from the states of initial switches.

Next we shall informally define the three mentioned operations.

Let C_1 and C_2 be two given switching networks (distinct or not) and suppose that they have the terminal points A_1, B_1 and A_2, B_2, respectively. Let us represent these circuits as follows:

$$A_1 \;\boxed{C_1}\; B_1 \qquad \text{and} \qquad A_2 \;\boxed{C_2}\; B_2$$

Parallel connection. The network obtained from C_1 and C_2 by the identification (the soldering) of A_1 with A_2, of B_1 with B_2 and having as terminal points $A \equiv A_1 \equiv A_2, B \equiv B_1 \equiv B_2$ is a new SN, denoted pC_1C_2 and called the *parallel connection* of C_1, C_2. The switching network pC_1C_2 is represented schematically by the figure below:

$$A \;\left\langle\; \begin{array}{c} \boxed{C_1} \\ \boxed{C_2} \end{array} \;\right\rangle\; B \qquad (A \equiv A_1 \equiv A_2 \,,\, B \equiv B_1 \equiv B_2)$$

It is clear that pC_1C_2 is in the state 1 if and only if at least one of C_1, C_2, is in the state 1 (consequently, pC_1C_2 is in the state 0 if and only if both C_1 and C_2 are in the state 0).

Serial connection. The network having the terminal points A, B which is obtained from C_1 and C_2 by the identification (the soldering) of B_1 with A_2 and by putting $A \equiv A_1, B \equiv B_2$ is a new SN, denoted sC_1C_2 and called the *serial connection* of C_1, C_2. The SN sC_1C_2 is represented schematically by the figure below:

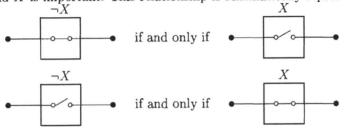

$A \equiv A_1$, $B \equiv B_2$ and $B_1 \equiv A_2$

It follows that sC_1C_2 is in the state 1 if and only if both C_1 and C_2 are in the state 1 (consequently, sC_1C_2 is in the state 0 if and only at least one of C_1, C_2 is in the state 0).

Complementation. For each switching network C one can construct a new SN denoted by $\neg C$ and called the *complement* of C. This construction is based on a recursive procedure which begins with the complementation of switches.

To each switch X one can associate another switch denoted $\neg X$ and called the *complement* of X, which is uniquely defined by the following property: $\neg X$ is in the **closed** state if and only if X is in the **open** state (consequently, $\neg X$ is in the **open** state if and only if X is in the **closed** state). Technically, the construction of $\neg X$ is not difficult and may be performed in various ways, but for us only the relationship between $\neg X$ and X is important. This relationship is schematically represented below:

Thus, the operation $\neg$ is defined for all switches; moreover, if X is a switch, then the complement of $\neg X$ coincides with X, thus we can write $\neg\neg X \equiv X$.

Let us consider now two switching networks C_1, C_2 and suppose that $\neg C_1$ and $\neg C_2$ are already constructed. Then we can construct the complements of pC_1C_2 and of sC_1C_2 putting by definition, $\neg pC_1C_2 \equiv s\neg C_1\neg C_2$, $\neg sC_1C_2 \equiv p\neg C_1\neg C_2$. It is easy to see that if the states of C_1 and C_2 are given, then the states of $\neg pC_1C_2$ and of $\neg sC_1C_2$ are uniquely determined and can be computed.

Based on these preliminaries, the (informal) recursive definition of the SNs can be formulated as follows:

(a) any switch is a SN;

(b) if C_1, C_2 are SNs, then pC_1C_2 is a SN;

(c) if C_1, C_2 are SNs, then sC_1C_2 is a SN;

(d) if C is a SN, then $\neg C$ is a SN;

(e) each SN can be obtained by means of rules (a), (b), (c), (d).

In conclusion, every switching network can be constructed, step by step, starting from some finite set $\{X_1, X_2, ..., X_N\}$ of switches and applying at each step one of the operations $\neg, p, s$. In other words, the construction of SNs is completely similar to the construction of formulas in propositional logic. Therefore, from an abstract point of view, the set of all SNs, which can be constructed using a given finite set $\{X_1, X_2, ..., X_N\}$ of binary switches, coincides with the formal language generated by the alphabet $\{X_1, X_2, ..., X_N, \neg, p, s\}$.

Example 7.3

Let us consider the following two networks

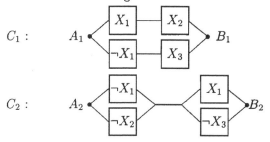

where X_1, X_2, X_3 are switches.

First, it is easily seen that C_1 and C_2 are switching networks. Indeed, they can be obtained step by step by the recursive procedures below:

(1) X_1; initial switch

(2) X_2; initial switch

(3) X_3; initial switch

(4) $\neg X_1$; $\neg$: (1)

(5) sX_1X_2; s: (1), (2);

(6) $s\neg X_1X_3$; s : (4), (3);

(7) $psX_1X_2s\neg X_1X_3 \equiv C_1$; p: (5), (6).

Similarly,

(1) X_1; initial switch

(2) X_2; initial switch

(3) X_3; initial switch

(4) $\neg X_1$; $\neg$: (1)

(5) $\neg X_2$; $\neg$: (2)

(6) $\neg X_3$; $\neg$: (3)

(7) $p\neg X_1 \neg X_2$; p: (4), (5).

(8) $pX_1 \neg X_3$; p: (1), (6).

(9) $sp\neg X_1 \neg X_2 pX_1 \neg X_3 \equiv C_2$; s: (7), (8).

It follows that C_1 and C_2 are switching networks. Moreover, we can see that $\neg C_1 \equiv C_2$. Indeed, using the above definitions one gets successively:

$$\neg C_1 \equiv \neg psX_1 X_2 s\neg X_1 X_3 \equiv s\neg sX_1 X_2 \neg s\neg X_1 X_3 \equiv$$

$$\equiv sp\neg X_1 \neg X_2 pX_1 \neg X_3 \equiv C_2.$$

▯

Furthermore, the set $\Gamma = \{0,1\}$ of (logical) truth values can be interpreted as being the set of all possible states of switches (consequently of SNs) and each Boolean interpretation of the logical language will provide the state of any SN if the states of initial switches are given. In the remaining part of this section we shall develop these ideas in a precise manner.

REMARK 7.1   We have adopted the Polish notation (without parentheses) and we have denoted the complementation, the parallel connection and the serial connection of networks by $\neg C, pC_1 C_2$ and $sC_1 C_2$, respectively. But the formal construction of switching networks can be performed equivalently by means of parentheses (and of the symbol " , ") using, for instance, the notations $\neg(C), p(C_1, C_2), s(C_1, C_2)$. We have preferred the Polish notation because it affords some theoretical advantages.   ∎

Formal definition of switching networks. The construction of switching networks from a formal point of view will be pursued in the same way as the construction of formulas in the formal language of propositional logic. In other words, we shall start with a finite set $X_1, X_2, \ldots, X_N$ of variables (intuitively representing some given set of initial binary switches) and with the symbols $\neg$, p and s (intuitively representing the complementation, the parallel connection and the serial connection). So we have here a formal alphabet and each switching network will be a special sequence of symbols belonging to this alphabet. As we already mentioned the use of parentheses will not be necessary because we have adopted the Polish notation (Section 1.2).

DEFINITION 7.2   *Each finite set*

$$\Delta = \{X_1, X_2, ..., X_N, \neg, p, s\},$$

where $N \geq 1$ is a natural number, is called the **alphabet** *of the (formal) switching networks. The elements of Δ are called* **symbols***. The symbols $X_1, X_2, ..., X_N$ are called* **variables** *and the symbols $\neg, p, s$ are called* **operators** *(of complementation, of parallel connection and of serial connection).*

DEFINITION 7.3   *A* **(formal) word** *in Δ is a finite string*

$$A \equiv \sigma_1 \sigma_2 ... \sigma_m \quad (m \geq 1)$$

of symbols belonging to Δ. The number $\lambda(A) = m$ is called the **length** *of the word A.*

Example 7.4

The strings

$$A_1 \equiv X_3 sp X_1 X_1 \neg,$$

$$A_2 \equiv p X_3 X_1;$$

$$A_3 \equiv \neg s X_2 X_4;$$

$$A_4 \equiv \neg X_2;$$

$$A_5 \equiv X_5 \neg$$

are words in $\Delta = \{X_1, X_2, X_3, X_4, X_5, \neg, p, s\}$ and $\lambda(A_1) = 6$, $\lambda(A_2) = 3$, $\lambda(A_3) = 4$, $\lambda(A_4) = \lambda(A_5) = 2$. It is obvious that the words A_1 and A_5 are meaningless, while A_2, A_3 and A_4 make sense.   ☐

DEFINITION 7.4   *A finite sequence*

$$(\text{f.c.}) \quad A_1, A_2, ..., A_r \quad (r \geq 1)$$

of words in Δ is called a **formal construction** *(in Δ) if, for every index i $(1 \leq i \leq r)$, the word A_i satisfies one of the following conditions:*

(a) A_i is a variable;

(b)   there is j $(1 \leq j < i)$ such that $A_i \equiv \neg A_j$;

(c)   there are h, k $(1 \leq h, k < i)$ such that $A_i \equiv p A_h A_k$ or $A_i \equiv s A_h A_k$.

Each word A which can be inserted in some formal construction in Δ is called an **abstract switching network** *or simply, a* **switching network** *in Δ; in such a case we shall say also that A is an SN. In particular, if the sequence $A_1, A_2, ..., A_r$ is a formal construction, then all terms A_i $(1 \leq i \leq r)$ are SNs and the sequence itself will be said to be the formal construction of its last term A_r.*

REMARK 7.2   If $A_1, A_2, ..., A_r$ is a formal construction, then its first term A_1 is necessarily a variable. Furthermore, suppose that A'_1, $A'_2, ...,$ A'_p and A''_1, $A''_2, ...,$ A''_q are formal constructions and consider an arbitrary *disjoint decomposition* of the set of indices $1, 2, 3, ..., r$, $r = p + q$; $\{1, 2, 3, ..., r\} = \{i_1, i_2, ..., i_p\} \cup \{j_1, j_2, ..., j_q\}$ such that $1 \leq i_1 < i_2 \leq ... < i_p \leq r$, and $1 \leq j_1 < j_2 < ... < j_q \leq r$; then the sequence

$$C_1, C_2, C_3, ..., C_r \quad (r = p + q)$$

where $C_{i_1} \equiv A'_1$, $C_{i_2} \equiv A'_2, ...,$ $C_{i_p} \equiv A'_p$ and $C_{j_1} \equiv A''_1$, $C_{j_2} \equiv A''_2, ...,$ $C_{j_q} \equiv A''_q$ is also a formal construction. For instance, the sequences

$$A'_1, A'_2, ..., A'_p, A''_1, A''_2, ..., A''_q$$

$$A'_1, A''_1, A'_2, A''_2, A'_3, A''_3, ..., \text{etc.}$$

are formal constructions.   ∎

DEFINITION 7.5   *The set $\mathcal{L}(\Delta)$ of all SNs in Δ will be called the* **(formal) language of switching networks** *generated by $X_1, X_2, ..., X_N$.*

The following result is an immediate consequence of the above definitions.

PROPOSITION 7.1
If C, C_1, C_2 are SNs, then $\neg C$, $p C_1 C_2$, $s C_1 C_2$ are also SNs; in other words, if C, C_1, $C_2 \in \mathcal{L}(\Delta)$, then $\neg C$, $p C_1 C_2$, $s C_1 C_2 \in \mathcal{L}(\Delta)$.

REMARK 7.3   The problem of deciding whether a given formal word represents a switching network or not is not trivial. For the simplest (shortest) formal words the problem can be solved by means of Definition 7.4. For instance, taking the formal word $A \equiv \neg p \neg X_1 \neg X_2$, it is easily seen that A is the complement of $B \equiv p \neg X_1 \neg X_2$ which is the parallel connection of $\neg X_1$

and $\neg X_2$. So we get the formal construction $X_1, X_2, \neg X_1, \neg X_2, p\neg X_1\neg X_2$ and $\neg p\neg X_1\neg X_2 \equiv A$ which proves (by Definition 7.4) that A is an SN. However, for long and more complicated formal words Definition 7.4 becomes inefficient. Fortunately, this problem can be solved completely by means of some arithmetic criteria using the notion of a **balanced formal word** which will be introduced below (just as in the case of the language of propositional logic).   ■

DEFINITION 7.6 *For any symbol σ belonging to the alphabet Δ we denote by $\pi(\sigma)$ the number*

$$
\pi(\sigma) = \begin{cases} 0, & \text{if } \sigma \text{ is a variable} \\ 1, & \text{if } \sigma \equiv \neg \\ 2, & \text{if } \sigma \equiv p \text{ or } \sigma \equiv s. \end{cases}
$$

*The number $\pi(\sigma)$ is called the **weight** of σ. If $A \equiv \sigma_1\sigma_2...\sigma_m$ $(m \geq 1)$ is a word, then the number*

$$
\pi(A) = \sum_{i=1}^{m} \pi(\sigma_i) = \pi(\sigma_1) + \pi(\sigma_2) + ... + \pi(\sigma_m)
$$

*is called the **weight** of A.*

Example 7.5
If we consider the following words (in Δ)

$$A_1 \equiv \neg\neg X_1 X_2 X_3 p,$$

$$A_2 \equiv X_3 s\neg,$$

$$A_3 \equiv \neg\neg p X_1 X_2,$$

then $\pi(A_1) = 1 + 1 + 0 + 0 + 0 + 2 = 4$, $\pi(A_2) = 0 + 2 + 1 = 3$, $\pi(A_3) = 1 + 1 + 2 + 0 + 0 = 4$.   ▯

DEFINITION 7.7 *If $A \equiv \sigma_1\sigma_2...\sigma_m$ $(m \geq 2)$ is a word in Δ, then the formal words*

$$S_1 \equiv \sigma_1$$
$$S_2 \equiv \sigma_1\sigma_2$$
$$\vdots$$
$$S_{m-1} \equiv \sigma_1\sigma_2...\sigma_{m-1}$$

*are called the **segments** of A. If the length of the word A is 1, then A has no segments. The fact that some word S is a segment of A ($\lambda(A) > 1$) will be denoted $S \subset A$.*

REMARK 7.4   The symbol $\subset$ which appears in the above definition does not coincide with the inclusion of sets used in set theory. The notation $S \subset A$ in our case means that A and S are words, $A \equiv \sigma_1 \sigma_2 ... \sigma_m$, $\lambda(A) = m > 1$ and there exists i ($1 \leq i < m$) such that $S \equiv \sigma_1 \sigma_2 ... \sigma_i$. It is clear that each word A with $\lambda(A) = m > 1$ has exactly $m - 1$ segments and, for each segment S of A, $1 \leq \lambda(S) < \lambda(A)$; in particular, if $S \subset A$, then $S \not\equiv A$. ∎

DEFINITION 7.8   *A **balanced word** (in Δ) is a word (in Δ) $A \equiv \sigma_1 \sigma_2 ... \sigma_m$ ($m \geq 1$) which satisfies one of the two conditions below:*

(1) $\lambda(A) = 1$ and A coincides with some variable;

(2) $\lambda(A) > 1$ and A has the following properties:

   (a) $\lambda(A) = \pi(A) + 1$;

   (b) if $S \subset A$, then $\lambda(S) \leq \pi(S)$.

The set of all balanced words in Δ will be denoted by $\mathcal{B}(\Delta)$.

REMARK 7.5   All properties of balanced words in the alphabet Δ can be deduced as we did in Section 1.2 for the balanced formal words in the language of propositional logic. For this reason we shall prove (as an example) only two properties of balanced words and shall present the others without proofs. ∎

PROPOSITION 7.2
If $A \in \mathcal{B}(\Delta), \lambda(A) > 1$ and if $S \subset A$, then $S \notin \mathcal{B}(\Delta)$.

PROOF   Let $A \equiv \sigma_1 \sigma_2 ... \sigma_m, \lambda(A) = m > 1$ be a given balanced word in Δ. First let us notice that σ_1 is not a variable. Indeed, let us suppose the contrary: there is some variable X such that $\sigma_1 \equiv X$, then $\pi(\sigma_1) = 0$; but that is impossible because $\sigma_1 \subset A$ and by the property (b) of balanced words gives $1 = \lambda(\sigma_1) \leq \pi(\sigma_1)$.

   Consider now an arbitrary segment S of A. If $\lambda(S) = 1$, then $S \equiv \sigma_1$ and from the above arguments $S \equiv \sigma_1$ does not coincide with a variable, hence $S \notin \mathcal{B}(\Delta)$. If $\lambda(S) > 1$, then supposing that $S \in \mathcal{B}(\Delta)$, we get $\lambda(S) = \pi(S) + 1 > \pi(S)$ (property (a)). On the other hand, taking into account that $A \in \mathcal{B}(\Delta)$ and $S \subset A$ it follows (property (b)) that $\lambda(S) \leq \pi(S)$, which is a contradiction. Therefore, $S \notin \mathcal{B}(\Delta)$. ∎

Let $A \equiv \sigma_1 \sigma_2 ... \sigma_p$ $(p \geq 1)$ and $B \equiv \tau_1 \tau_2 ... \tau_q$ be words. In what follows we shall make use of the notation:

$$AB \equiv \sigma_1 \sigma_2 ... \sigma_p \tau_1 \tau_2 ... \tau_q.$$

It is obvious that AB is a new word and $AB \equiv BA$ if and only if $A \equiv B$. The meaning of the notations $ABC, ABCD$, etc., is also clear, if A, B, C, D, etc., are words.

PROPOSITION 7.3

If $A \in \mathcal{B}(\Delta)$, $\lambda(A) > 1$ and $S \subset A$, then there is a unique $T \in \mathcal{B}(\Delta)$ such that

$$ST \subset A \quad or \quad ST \equiv A.$$

PROOF   Let $A \equiv \sigma_1 \sigma_2 ... \sigma_m$ $(m > 1)$, be a given balanced word and let $S \subset A$, $S \equiv \sigma_1 \sigma_2 ... \sigma_h$ $(1 \leq h < m)$. Let us consider the set:

$$M = \{j \mid (h + 1 \leq< j \leq m) \text{ such that } \lambda(\sigma_{h+1} ... \sigma_j) > \pi(\sigma_{h+1} ... \sigma_j)\}.$$

Notice that $m \in M$, hence the set M is not empty. Indeed, denoting $T' \equiv \sigma_{h+1} ... \sigma_m$ and taking into account that $A \in \mathcal{B}(\Delta)$ and $S \subset A$, we get successively:

$$\lambda(T') = \lambda(\sigma_{h+1} ... \sigma_m) = \lambda(\sigma_1 \sigma_2 ... \sigma_m) - \lambda(\sigma_1 \sigma_2 ... \sigma_h) = \lambda(A) - \lambda(S) =$$

$$= \pi(A) + 1 - \lambda(S) \geq \pi(A) + 1 - \pi(S) = \pi(T') + 1 > \pi(T').$$

It follows that $m \in M$ and M is nonempty. Let k be the smallest element of M, i.e., $k = \min M$.

It is clear that

$$h + 1 \leq k \leq m.$$

We have to analyze two cases:

$$(1) \quad k = h + 1; \quad (2) \quad k > h + 1.$$

In case (1), let us denote $T \equiv \sigma_{h+1}$; taking into account that $k = h + 1 \in M$ we have $\lambda(T) > \pi(T)$, hence

$$1 = \lambda(\sigma_{h+1}) = \lambda(T) > \pi(T) = \pi(\sigma_{h+1});$$

therefore, $\pi(\sigma_{h+1}) = 0$ and consequently, $T \equiv \sigma_{h+1}$ is a variable of Δ. It follows that $T \in \mathcal{B}(\Delta)$ and $ST \subset A$ or $ST \equiv A$.

In case (2), let us denote $T \equiv \sigma_{h+1}...\sigma_k$ and notice that $\lambda(T) > 1$. Taking into account that $k \in M$, it follows that $\lambda(T) > \pi(T)$, or

$$(*) \qquad\qquad \lambda(T) \geq 1 + \pi(T).$$

On the other hand, $\lambda(T) = \lambda(\sigma_{h+1}...\sigma_k) = 1 + \lambda(\sigma_{h+1}...\sigma_{k-1})$. But $k - 1 \notin M$ (because $k = \min M$) and by the definition of the set M we have:

$$\lambda(\sigma_{h+1}...\sigma_{k-1}) \leq \pi(\sigma_{h+1}...\sigma_{k-1}).$$

Therefore, $\lambda(T) \leq 1 + \pi(\sigma_{h+1}...\sigma_{k-1}) \leq 1 + \pi(\sigma_{h+1}...\sigma_k)$; this means that

$$(**) \qquad\qquad \lambda(T) \leq 1 + \pi(T).$$

From the inequalities $(*)$ and $(**)$ one gets $\lambda(T) = 1 + \pi(T)$, hence the word T has the property (a) in the definition of balanced words.

In order to prove that T also has the property (b) let us consider a segment U of T ($U \subset T$), $U \equiv \sigma_{h+1}...\sigma_l$ ($h + 1 \leq l < k$). Taking into account that $l < k$, it follows that $l \notin M$, hence $\lambda(U) \leq \pi(U)$, and T has the property (b), hence $T \in \mathcal{B}(\Delta)$ and $ST \subset A$ or $ST \equiv A$.

So we just proved that, for any $A \in \mathcal{B}(\Delta)$, with $\lambda(A) > 1$ and for any $S \subset A$, there exists $T \in \mathcal{B}(A)$ such that $ST \subset A$ or $ST \equiv A$.

The uniqueness of the balanced word T with the above properties follows almost immediately. Suppose that there are two balanced words T_1 and T_2 such that $ST_1 \subset A$ or $ST \equiv A$ and $ST_2 \subset A$ or $ST_2 \equiv A$. Then necessarily $\lambda(T_1) = \lambda(T_2)$ and $T_1 \equiv T_2$; indeed, if $\lambda(T_1) < \lambda(T_2)$, then $T_1 \subset T_2$ which is impossible because T_1, T_2 are both balanced words.   ∎

Pursuing the same line of ideas as in Section 1.2 we could prove the following result:

THEOREM 7.1

A formal word is a switching network if and only if it is a balanced word, i.e.,

$$\mathcal{L}(\Delta) = \mathcal{B}(\Delta).$$

REMARK 7.6   The above theorem represents a complete and efficient solution to the problem of whether a given formal word is a switching network or not. Indeed, in order to verify that a given formal word is a switching network we have to write the list of its segments, compute the lengths and the weights of all resulting words and check whether the conditions in Definition 7.8 are fulfilled or not.   ∎

The state function of a switching network.
Let $\Delta = \{X_1, X_2, ..., X_N, \neg, p, s\}$ be the alphabet of the language $\mathcal{L}(\Delta)$ of

switching networks, where $N \geq 1$ is a given natural number. Consider also the Boolean algebra $B(N, \Gamma)$ of all Boolean functions

$$f : \Gamma^N \to \Gamma : (x_1, x_2, ..., x_N) \mapsto f(x_1, x_2, ..., x_N)$$

where $\Gamma = \{0, 1\}$, $\Gamma^N = \{(x_1, x_2, ..., x_N) \mid x_i \in \Gamma \ (1 \leq i \leq N)\}$. We recall that in $B(N, \Gamma)$ there are N basic functions $p_1, p_2, ..., p_N$ called projectors and defined by

$$p_j(x_1, x_2, ..., x_j, ..., x_n) = x_j, \quad (1 \leq j \leq N).$$

Using these projectors, every function $f \in B(N, \Gamma)$, $f \neq 0$ can be expressed as a disjunctive polynomial (see APPENDIX A) in the form

$$f = \sum_{f(\sigma_1, ..., \sigma_N) = 1} p_1^{\sigma_1} p_2^{\sigma_2} ... p_N^{\sigma_N}$$

or equivalently in the form

$$f(x_1, ..., x_N) = \sum_{f(\sigma_1, ..., \sigma_N) = 1} x_1^{\sigma_1} x_2^{\sigma_2} ... x_N^{\sigma_N}, \text{ for all } (x_1, x_2, ..., x_N) \in \Gamma^N,$$

where additive and multiplicative notations are used instead of "max" and "min" in the Boolean algebras Γ and $B(N, \Gamma)$.

As in Section 1.2 one can prove the following result.

THEOREM 7.2

There is a unique mapping

$$\Phi : \mathcal{L}(\Delta) \to B(N, \Gamma) : C \mapsto \Phi(C)$$

that has the following properties:

(a) $\Phi(\neg C) = \overline{\Phi(C)}$;

(b) $\Phi(pC_1 C_2) = \Phi(C_1) + (C_2)$;

(c) $\Phi(sC_1 C_2) = \Phi(C_1)\Phi(C_2)$;

(d) $\Phi(X_j) = p_j$, *for all* j $(1 \leq j \leq N)$.

Moreover, for each $f \in B(N, \Gamma)$ *there is some* $C_f \in \mathcal{L}(\Delta)$ *such that* $\Phi(C_f) = f$, *i.e., the function* Φ *is surjective.*

REMARK 7.7   The function Φ is not bijective because it is not injective. Indeed there exist C_1 and C_2 with $C_1 \not\equiv C_2$, such that $\Phi(C_1) = \Phi(C_2)$;

for instance, it is easy to see that $\Phi(\neg s \neg X_1 X_2) = \Phi(p X_1 \neg X_2)$, while $\neg s \neg X_1 X_2 \not\equiv p X_1 \neg X_2$. ∎

DEFINITION 7.9   *The (unique) function $\Phi : \mathcal{L}(\Delta) \to B(N, \Gamma)$ having properties (a)-(d) listed in Theorem 7.2 is called the* **Boolean model** *of the language $\mathcal{L}(\Delta)$. For each switching network $C \in \mathcal{L}(\Delta)$ the Boolean function $\Phi(C) \in B(N, \Gamma)$ is called the* **state function** *of C.*

The state function is the most important mathematical characteristic of switching networks. It allows the analysis and the synthesis of switching networks by showing under which conditions the current flows through the circuit. By using the state function Φ one can also introduce an equivalence relation in the set $\mathcal{L}(\Delta)$ of all SNs. Thus, we shall say (as in the case of $\mathcal{L}_0$) that the switching networks C', C'' are equivalent and we shall write $C' \approx C''$ if they have the same state function ($\Phi(C') = \Phi(C'')$). The set $\widehat{\mathcal{L}}(\Delta)$ of all equivalence classes endowed with the operations $\bar{\xi}$, $p\xi\eta$, $s\xi\eta$ is a Boolean algebra isomorphic with $B(N, \Gamma)$.

If C is a switching network correctly constructed by means of complementation, parallel connection and serial connection from the switches X_1, $X_2, \ldots, X_N$, then the states of C are given by the function $\Phi(C)$; more precisely, if the switches $X_1, X_2, \ldots, X_N$ are respectively in the states $\alpha_1, \alpha_2, \ldots, \alpha_N$ ($\alpha_i \in \Gamma$), then the corresponding state of C is given by the value $\Phi(C)(\alpha_1, \alpha_2, \ldots, \alpha_N) \in \Gamma$. Consider, for instance, the switching network $C \equiv \neg s \neg X_1 X_2$ constructed from the switches X_1, X_2 (i.e., $C \in \mathcal{L}(\Delta)$), where $N = 2$, $\Delta = \{X_1, X_2, \neg, p, s\}$) and suppose that X_1 is in the state $\alpha_1 = 1$ and X_2 is in the state $\alpha_2 = 0$. Taking into account the above properties of the state function Φ, we can compute:

$$\Phi(C) = \Phi(\neg s \neg X_1 X_2) = \overline{\Phi(s \neg X_1 X_2)} = \overline{\Phi(\neg X_1) \cdot \Phi(X_2)} =$$

$$= \overline{\Phi(\neg X_1)} + \overline{\Phi(X_2)} = \Phi(X_1) + \overline{\Phi(X_2)} = p_1 + \bar{p}_2,$$

where $p_1, p_2 \in B(2, \Gamma)$ are the functions (projectors):

$$p_1(x_1, x_2) = x_1 \quad \text{and} \quad p_2(x_1, x_2) = x_2.$$

Therefore, the state of C corresponding to the states $\alpha_1 = 1$, $\alpha_2 = 0$ of X_1, and $\alpha_2 = 0$ of X_2 is given by:

$$\Phi(C)(\alpha_1, \alpha_2) = (p_1 + \bar{p}_2)(\alpha_1, \alpha_2) = p_1(\alpha_1, \alpha_2) + \overline{p_2(\alpha_1, \alpha_2)} =$$

$$= \alpha_1 + \bar{\alpha}_2 = 1 + \bar{0} = 1 + 1 = 1.$$

Thus, if X_1 is in state 1 and X_2 is in state 0, then network C will allow the current to pass. On the other hand, giving an arbitrary function $f \in B(N, \Gamma)$, one can always construct a switching network C for which the state function coincides with f (surjectivity of Φ). The design of C follows from the polynomial representation of f. Furthermore, given the circuit C, it is often possible to transform algebraically the function f in order to obtain a simplified circuit D with $\Phi(D) = f$; thus, the circuit D is simpler and it functions exactly as C does. Let us illustrate these ideas.

Example 7.6

Consider the circuit given in the figure below.

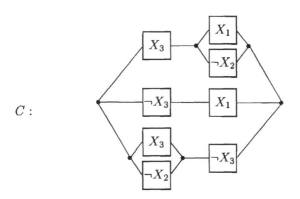

$C:$

We see that C has three branches C_1, C_2, C_3 connected in parallel, i.e.,

$$C \equiv p \, p \, C_1 \, C_2 \, C_3,$$

where

$$C_1 \equiv sX_3pX_1\neg X_2; \; C_2 \equiv s\neg X_3X_1; \; C_3 \equiv spX_3\neg X_2\neg X_3.$$

Let $\Delta = \{X_1, X_2, X_3, \neg, p, s\}$, then $N = 3$ and $C_1, C_2, C_3, C \in \mathcal{L}(\Delta)$. Using properties (a)-(d) in Theorem 7.2, it follows that

$$\Phi(C) = \Phi(ppC_1C_2C_3) = \Phi(pC_1C_2) + \Phi(C_3) = \Phi(C_1) + \Phi(C_2) + \Phi(C_3)$$

and

$$\begin{aligned}
\Phi(C_1) &= \Phi(sX_3pX_1\neg X_2) = \Phi(X_3) \cdot \Phi(pX_1\neg X_2) = \\
&= \Phi(X_3)(\Phi(X_1) + \overline{\Phi(X_2)}) = \\
&= p_3(p_1 + \bar{p}_2); \\
\Phi(C_2) &= \Phi(s\neg X_3X_1) = \Phi(\neg X_3) \cdot \Phi(X_1) = \bar{p}_3p_1; \\
\Phi(C_3) &= \Phi(spX_3\neg X_2\neg X_3) = \Phi(pX_3\neg X_2)\Phi(\neg X_3) = \\
&= (\Phi(X_3) + \overline{\Phi(X_2)})\overline{\Phi(X_3)} = \\
&= (p_3 + \bar{p}_2)\bar{p}_3 = \bar{p}_2\bar{p}_3.
\end{aligned}$$

Hence,

$$\Phi(C) = p_3(p_1 + \bar{p}_2) + \bar{p}_3 p_1 + \bar{p}_2 \bar{p}_3 =$$
$$= p_3 p_1 + p_3 \bar{p}_2 + \bar{p}_3 p_1 + \bar{p}_2 \bar{p}_3 =$$
$$= p_1(p_3 + \bar{p}_3) + \bar{p}_2(p_3 + \bar{p}_3) =$$
$$= p_1 + \bar{p}_2 \quad \text{(because } p_3 + \bar{p}_3 = 1).$$

Thus, by algebraic transformations, we have gotten the equality $\Phi(C) = p_1 + \bar{p}_2$. On the other hand, $p_1 + \bar{p}_2$ is obviously the state function of the switching network

C' :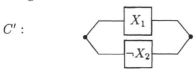

Therefore, the circuits C and C' "behave" identically ($\Phi(C) = \Phi(C')$) but C' is clearly simpler. ⬜

Example 7.7

Consider the problem of turning on/off a bulb by arbitrarily switching one (and only one) of $N \geq 1$ binary switches. Let us denote by $C_N \equiv C_N(X_1, X_2, \ldots, X_N)$ a switching network constructed from N switches $X_1, \ldots, X_N$ and by the given bulb B; consider the diagram

$$\boxed{C_N(X_1, X_2, \ldots, X_N)} \quad\!\!\!\!\! (B)$$

The problem consists of producing a switching network C_N having the property that, if the state of an arbitrary switch changes, then the state of C_N changes too, i.e., if B is turned on, then it turns off and conversely.

For $N = 1$ the problem is trivial: $C_1(X_1) \equiv X_1$.

For $N = 2$, it is easy to verify that the circuit

$$C_2(X_1, X_2) \equiv ps C_1 X_2 s \neg C_1 \neg X_2$$

$C_2(X_1, X_2)$: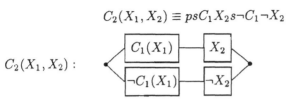

is a solution for the problem.

Consider now a natural number $K \geq 1$ and assume that

$$C_K \equiv C_K(X_1, X_2, \ldots, X_K)$$

is a circuit having the required property. Then it is clear that the switching network

$$C_{K+1} = C_{K+1}(X_1, X_2, \ldots, X_{K+1}) \equiv ps C_K X_{K+1} s \neg C_K \neg X_{K+1}$$

with the diagram:

$$C_{K+1} :$$

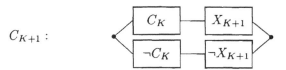

is a solution of the problem. Therefore, for each $N \geq 1$ the problem is solved.

If we denote $\Phi(C_K) = f_K$, then the corresponding state functions f_1, $f_2, \ldots, f_N$ of $C_1, C_2, \ldots, C_N$ are defined recursively as follows:

$$f_1 = p_1 ,$$
$$f_2 = f_1 p_2 + \bar{f}_1 \bar{p}_2 ,$$
$$\vdots$$
$$f_N = f_{N-1} p_N + \bar{f}_{N-1} \bar{p}_N .$$

For instance, if $N = 3$, then

$$f_3(X_1, X_2, X_3) = X_1 X_2 X_3 + \bar{X}_1 \bar{X}_2 \bar{X}_3 + X_1 X_2 \bar{X}_3 + \bar{X}_1 \bar{X}_2 X_3$$

and computing the values of f_3 by means of the table

X_1	X_2	X_3	f_3
0	0	0	1
0	1	0	0
1	0	0	0
1	1	0	1
0	0	1	1
0	1	1	0
1	0	1	0
1	1	1	1

we can see directly that for each passage from one horizontal line (X_1, X_2, X_3) to another of the form $(\bar{X}_1, X_2, X_3)$, $(X_1, \bar{X}_2, \bar{X}_3)$, $(X_1, X_2, \bar{X}_3)$ the corresponding value of the state function changes.   □

7.5 Logical Networks

In 1938 the American mathematician Claude Shannon perceived the parallel between propositional logic and the theory of logical circuits. From then on, he stressed the role played by Boolean algebras in electronics. In the following we do not give details for implementing logical networks. However, it is important to note that technology has progressed from mechanical switches through vacuum tubes and then from transistors to integrated circuits (chips) which may contain millions of transistors.

Logical networks may be viewed as machines which contain one or more inputs and exactly one output. In logical networks, the on/off switches are replaced by logical gates. The logical gates behave like functions. The inputs can have only two possible physical states (high voltage for 1 and low voltage for 0). The best known logical gates are: NOT (inverter), OR and AND. They are symbolized by the following pictures:

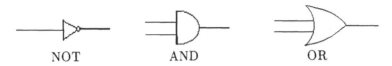

NOT   AND   OR

The truth tables for these gates are the same as for negation, disjunction and conjunction, i.e., the same as for $\neg X_1$, $X_1 \vee X_2$, $X_1 \wedge X_2$. Consequently, the logical networks satisfy the same laws as propositions do; the corresponding set of equivalence classes thus forms a Boolean algebra (two LNs are equivalent if they have the same state function or the same truth table).

For a given logical network L, we can write the associated logical formula, then the corresponding truth table and from it we can see the effect of L for an arbitrary input. Conversely, if a truth table is given, then we can write the disjunctive normal form (DNF) and the associated logical circuit. Therefore, we get the following diagram.

Truth table $\leftrightarrow$ DNF $\leftrightarrow$ Logical network

Logical network are built using integrated circuits technology. An integrated circuit, i.e., an appropiate combination of gates and invertors, is a logical network which represents a certain truth table in order to produce an desired result.

Example 7.8

(1) Let us consider the truth table from Section 7.1. The logical circuit associated to the DNF is:

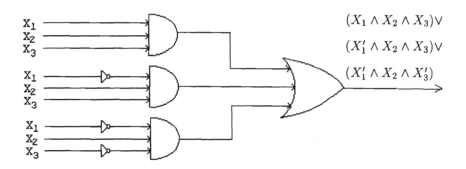

$$(X_1 \wedge X_2 \wedge X_3) \vee$$
$$(X_1' \wedge X_2 \wedge X_3) \vee$$
$$(X_1' \wedge X_2 \wedge X_3')$$

The simplified circuit is:

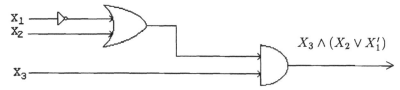

$$X_3 \wedge (X_2 \vee X_1')$$

(2) Let the following circuit be:

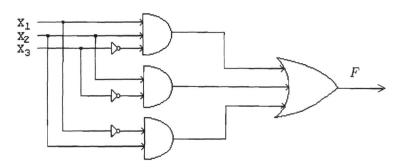

The associated logical formula is:

$$F \equiv (X_1 \wedge X_2 \wedge X_3') \vee (X_2 \wedge X_3') \vee (X_1' \wedge X_2).$$

In order to find the DNF we compute:

$$F \equiv (X_1 \wedge X_2 \wedge X_3') \vee ((X_2 \wedge X_3') \wedge (X_1 \vee X_1')) \vee ((X_1' \wedge X_2) \wedge (X_3 \vee X_3')) \approx$$

$$\approx (X_1 \wedge X_2 \wedge X_3') \vee (X_2 \wedge X_3' \wedge X_1) \vee (X_2 \wedge X_3' \wedge X_1') \vee (X_1' \wedge X_2 \wedge X_3) \vee$$

$$\vee (X_1' \wedge X_2 \wedge X_3') \approx (X_1 \wedge X_2 \wedge X_3') \vee (X_1' \wedge X_2 \wedge X_3') \vee (X_1' \wedge X_2 \wedge X_3).$$

The associated truth table is (by inspection):

X_1	X_2	X_3	F
1	1	1	0
1	1	0	1
1	0	1	0
1	0	0	0
0	1	1	1
0	1	0	1
0	0	1	0
0	0	0	0

Boolean functions model logical networks. To a set of N Boolean values, they associate a new Boolean value. Therefore, the Boolean functions describe logical networks which have N inputs and only one output.

Exercises (Part I)

1. What is the difference between the truth structure on $\mathcal{L}_0$ in the semantic version and the syntactic version of propositional logic?

2. Give the definitions of interpretations of formulas in propositional logic.

3. Write the negation of each of the following sentences as simply as possible:

   (a) He is tall but handsome.

   (b) He has blond hair or blue eyes.

   (c) He is neither rich nor happy.

   (d) He lost his job or he did not go to work today.

   (e) Neither Alain nor Richard is unhappy.

   (f) Nathalie speaks Spanish or French, but not German.

4. Translate each of the following compound sentences into the language of propositional logic using the indicated letters for component sentences:

   (a) Alain will pass the logic course only if he studies. (P, S)

   (b) Alain will not pass the logic course unless he does his homework and studies. (P, H, S)

   (c) Alain will not pass the logic course if he neither does his homework nor studies. (P, H, S)

   (d) If logic is difficult, Alain and Nathalie will pass only if they study. (D, A, N, S)

   (e) If Alain doesn't study and do his homework, then he will not pass the logic course. (S, H, P)

5. Write each of the following sentences in symbolic form by using p and q. Apply the following results $\neg(p \wedge q) \approx \neg p \wedge \neg q$ and $\neg\neg p \approx p$ to simplify the obtained formulas and translate them into natural language.

   (a) It is not true that his mother is English or his father is French.

   (b) It is not true that he studies computer science but not mathematics.

   (c) It is not true that sales are decreasing and prices are rising.

   (d) It is not true that it is cold or it is raining.

6. Verify that

   (a) $((P \Rightarrow Q) \Rightarrow Q) \Rightarrow (\neg(P \Rightarrow Q) \vee Q)$ is a tautology;

   (b) $(P \wedge Q) \wedge \neg(P \vee Q)$ is a contradiction.

7. Among the following formulas indicate those which are universally true and those which are universally false:

   (a) $(\neg P \Rightarrow Q) \vee (P \Rightarrow \neg Q)$

   (b) $(P \wedge Q) \vee (Q \wedge R) \vee (R \wedge P)$

   (c) $(P \wedge S \Rightarrow Q \wedge T) \Rightarrow (P \wedge S \Rightarrow Q \wedge T)$

   (d) $\neg(\neg(P \vee Q \Rightarrow R) \Rightarrow \neg(P \wedge S \Rightarrow R))$.

8. Show that

   (a) $P \Rightarrow Q, \neg Q \models \neg P$;

   (b) $P \Leftrightarrow Q, Q \models P$.

9. Prove that $p \Leftrightarrow q$ logically implies $p \Rightarrow q$.

10. Is it possible to obtain a formula which is in conjunctive normal form and in disjunctive normal form at the same time? If yes, give an example.

11. Transform the formulas below into conjunctive normal form:

   (a) $\neg(\neg p \vee q) \vee (\neg p \wedge q)$;

   (b) $(p \Rightarrow q) \Rightarrow r$.

12. Establish the following deduction

$$P \Rightarrow Q, P \Rightarrow (Q \Rightarrow R) \vdash P \Rightarrow R.$$

13. Let p and q be the sentences "It is raining" and "There are clouds", respectively. Which of the following implications are true in your opinion?

    (a) $p \Rightarrow q$;

    (b) $q \Rightarrow p$;

    (c) $\neg q \Rightarrow \neg p$;

    (d) $\neg p \Rightarrow \neg q$.

14. Verify that the modus ponens tautology $(p \wedge (p \Rightarrow q)) \Rightarrow q$ is logically equivalent to the modus tollens tautology $(\neg q \wedge (p \Rightarrow q)) \Rightarrow \neg p$.

15. Let the following truth table be:

p	q	r	f
1	1	1	0
1	1	0	1
1	0	1	0
1	0	0	1
0	1	1	0
0	1	0	0
0	0	1	0
0	0	0	0

    (a) Find the associated disjunctive normal form;

    (b) Draw the logic circuit which corresponds to this formula;

    (c) Simplify the logic circuit by using both the properties of Boolean algebras and a Karnaugh map. Sketch the new logic circuit.

16. Suppose that we deal with the MV-algebra $A = \{0, \frac{1}{4}, \frac{1}{2}, \frac{3}{4}, 1\}$ and with the alphabet $X_0 = \{X, Y, Z, \neg, \Rightarrow, (,)\}$ and let us consider the following A-fuzzy set of hypotheses

$$\mathcal{H}(F) = \begin{cases} \frac{3}{4}, & \text{if } F \equiv X \Rightarrow Z \\ \frac{1}{4}, & \text{if } F \equiv \neg Y \Rightarrow X \\ \frac{1}{2}, & \text{if } F \equiv (Y \Rightarrow Z) \Rightarrow Z \\ 0, & \text{otherwise.} \end{cases}$$

Compute the value $C^{sem}(\mathcal{H})(G)$, where

$$G \equiv ((\neg X \Rightarrow Y) \Rightarrow Z) \Rightarrow Z.$$

17. Suppose that we deal with the MV-algebra $\mathcal{A} = \{0, \frac{1}{3}, \frac{2}{3}, 1\}$ and with the alphabet $\mathbf{X_0} = \{X, Y, \neg, \Rightarrow, (,)\}$ and let us consider the fuzzy set of axioms $\mathbf{Ax}$ and the following $\mathcal{A}$-fuzzy set of hypotheses

$$\mathcal{H}(F) = \begin{cases} \frac{1}{3}, & \text{if } F \equiv Y \\ 0, & \text{otherwise} \end{cases}$$

Compute the value $\mathcal{C}^{syn}(\mathcal{H})(X \Rightarrow Y)$.

18. Rewrite the following conditional statement with a simplified conditional expression:

```
if (A > B)
    and not ((A > B) and (C < 100))
then
    procedure1
else
    procedure2;
```

Part II

PREDICATE LOGIC

"The triumph of mathematical logic, considered the most theoretical and the most implacable branch of the most abstract mathematics, is the work of today's computers."

(GRIGORE C. MOISIL)

In the second part of the book we shall discuss mathematical logic in a larger framework in which predicators, variables and certain specific operations called quantifiers appear. This more extensive logic is called (first order) predicate logic. The development of topics in this second part will pursue the pattern set forth for propositional logic. Therefore, in what follows, after some informal introductory considerations, we shall present the formal language $\mathcal{L}$ of predicate logic, the semantic and syntactic truth structures on $\mathcal{L}$, some basic elements of fuzzy predicate logic and finally some applications in Computer Science.

8

Introductory Considerations

In Part I, Chapter 1 we discussed some topics concerning sentences and the logico-linguistic structure of sentences; we saw that any sentence has two essential parts: a **subject(s)** and a **predicator**. In propositional logic we have used logical operators $(\neg, \vee, \wedge, \Rightarrow, \Leftrightarrow)$ which modify only the predicator of a sentence, the (fixed, specified) subjects remaining unchanged. In predicate logic we have to penetrate deeper into the structure of sentences also modifying subjects. Therefore, in predicate logic we have to operate with predicates having one or more variable subjects. These (variable) subjects will be called in what follows **individual variables**, intuitively representing the generic names of the **individual objects** we are talking about. Consequently, besides logical operators in predicate logic we have to introduce new operations called **quantifiers** which act on (variable) subjects. Thus, a characteristic feature of predicate logic is the presence of individual variables and quantifiers. It is clear that this new situation demands an enriched logical language $\mathcal{L}$ and a more complex truth structure on $\mathcal{L}$.

As an illustration of the above considerations let us consider the following example.

At the main entrance of a college, a passer-by listens to a piece of conversation between two professors:

- There has been a decision by the Faculty Council, says the first, that all students having the highest average marks will be included in the Dean's Honors List.

- You know, replies the other, that in our college only hard working students have the highest average marks.

The remaining part of the conversation gets lost in the noise of the street. After a while the passer-by makes the following remarks:

- So, if there are some students in this college having the highest average marks, then among hard-working students there are some that will be included in the Dean's Honors List.

Let us analyze the previous conversation and the remark of the passer-by. The sentences Q_1 and Q_2 of the two professors are essentially the following:

Q_1 : All students having the highest average marks are included in the Dean's Honors List

Q_2 : Only hard-working students have the highest average marks.

The remark R of the passer-by is:

R : If there are students with the highest average marks, then among hard-working students there are some included in the Dean's Honors List.

In order to translate the reasoning of the passer-by in a logical language, let us denote by

$$C = \{1, 2, 3, \ldots, k\}$$

the set of the names of all students from this college. Also let us consider the following three sentences with variable subjects (i.e., predicates):

$$A(x) \equiv \text{``}x \text{ has the highest average marks''};$$
$$H(x) \equiv \text{``}x \text{ is a hard-working student''};$$
$$L(x) \equiv \text{``}x \text{ is included in the Dean's Honors List''}.$$

By means of sentences $A(x), H(x), L(x)$ we can construct $3k$ true or false propositions (sentences with fixed subject):

$$A(1), A(2), \ldots, A(k),$$

$$H(1), H(2), \ldots, H(k),$$

$$L(1), L(2), \ldots, L(k).$$

The meaning of the above propositions is clear; for instance, $H(3)$ coincides with the proposition "the student having the name denoted by 3 is hard working" and so on. In general, if $F(x)$ is the sentence "x has the property p", then we get automatically the k propositions

$$F(1), F(2), \ldots, F(k).$$

Therefore, the propositions

$S_1 \equiv$ "all students have property p"
$S_2 \equiv$ "there is some student having property p"

can be expressed, respectively, by the formulas

$$S_1 \equiv F(1) \wedge F(2) \wedge \ldots \wedge F(k) \equiv \bigwedge_{x=1}^{k} F(x)$$

$$S_2 \equiv F(1) \vee F(2) \vee \ldots \vee F(k) \equiv \bigvee_{x=1}^{k} F(x).$$

Returning to Q_1, Q_2 and R we notice first that the sentence $A(x) \Rightarrow L(x)$ means that if the student x has the highest average marks, then x is included in the Dean's Honors List. Taking into account that in Q_1 this implication is true for all students, we can express Q_1 by the formula

$$Q_1 \equiv \bigwedge_{x=1}^{k} (A(x) \Rightarrow L(x)).$$

In what concerns Q_2, we remark that the sentence "Only a hard-working student x has the highest average marks" is equivalent with the sentence "If x has the highest average marks, then x is a hard-working student", i.e., with the implication expressed by the formula:

$$A(x) \Rightarrow H(x).$$

So, $Q_2 \equiv \bigwedge_{x=1}^{k} (A(x) \Rightarrow H(x)).$

In a similar way we remark that the comment of the passer-by coincides with the sentence: $R \equiv$ "If there are students having the highest average marks", then "There are hard-working students who are included in the Dean's Honors List". This sentence can be expressed in formal language by the formula:

$$R \equiv (\bigvee_{x=1}^{k} A(x)) \Rightarrow (\bigvee_{x=1}^{k} (H(x) \wedge L(x))).$$

Using the above notations, the reasoning of the passer-by consists of the intuitive proof of the implication:

$$\text{If } Q_1 \wedge Q_2, \text{ then } R$$

or equivalently:

$$\text{If } [\bigwedge_{x=1}^{k} (A(x) \Rightarrow L(x))] \wedge [\bigwedge_{x=1}^{k} (A(x) \Rightarrow H(x))],$$

$$\text{then } (\bigwedge_{x=1}^{k} A(x)) \Rightarrow (\bigvee_{x=1}^{k} (H(x) \wedge L(x))).$$

Intuitively, by natural reasoning, it is not difficult to see that the deduction of the passer-by is correct. Next, we shall show how to perform syntactically this deduction in the frame of propositional logic. So we have to prove formally the syntactic deduction: $Q_1, Q_2 \vdash R$.

For this purpose let us fix an arbitrary element (student) y in C and consider the following syntactic deduction with the hypotheses Q_1, Q_2 and (temporarily) $A(y)$:

1. $\bigwedge\limits_{x=1}^{k} (A(x) \Rightarrow L(x))$; hypothesis Q_1;

2. $\bigwedge\limits_{x=1}^{k} (A(x) \Rightarrow H(x))$; hypothesis Q_2;

3. $A(y)$; the third hypothesis;

4. $[\bigwedge\limits_{x=1}^{k} (A(x) \Rightarrow L(x))] \Rightarrow (A(y) \Rightarrow L(y))$; the tautology $A_1 \wedge \ldots \wedge A_k \Rightarrow A_i$;

5. $[\bigwedge\limits_{x=1}^{k} (A(x) \Rightarrow H(x))] \Rightarrow (A(y) \Rightarrow H(y))$; same argument;

6. $A(y) \Rightarrow L(y)$; (MP): 1,4;

7. $A(y) \Rightarrow H(y)$; (MP): 2,5;

8. $L(y)$; (MP): 3,6;

9. $H(y)$; (MP): 3,7;

10. $L(y) \wedge H(y)$; metatheorem of conjunction;

11. $(L(y) \wedge H(y)) \Rightarrow \bigvee\limits_{x=1}^{k} (L(x) \wedge H(x))$; tautology resulting from axiom II: $F \Rightarrow F \vee G$ and from the metatheorem $F \Rightarrow G \vee F$;

12. $\bigvee\limits_{x=1}^{k} (L(x) \wedge H(x))$; (MP):10,11.

Therefore, we obtained for each y in $\mathcal{C}$ the following deduction:

$$Q_1, Q_2, A(y) \vdash \bigvee_{x=1}^{k} (L(x) \wedge H(x)).$$

Applying the metatheorem of Herbrand we get

$$Q_1, Q_2 \vdash A(y) \Rightarrow \bigvee_{x=1}^{k} (L(x) \wedge H(x)).$$

Now taking into account that y is an arbitrary element of $\mathcal{C}$ we obtain the following k deductions:

$$Q_1, Q_2 \vdash A(1) \Rightarrow \bigvee_{x=1}^{k} (L(x) \wedge H(x))$$

$$Q_1, Q_2 \vdash A(2) \Rightarrow \bigvee_{x=1}^{k} (L(x) \wedge H(x))$$

$$\vdots$$

$$Q_1, Q_2 \vdash A(k) \Rightarrow \bigvee_{x=1}^{k} (L(x) \wedge H(x)).$$

In Section 3.2, MTh. 3.15 we have proved the following result: if $\vdash F \Rightarrow G$ and $\vdash H \Rightarrow K$, then $\vdash (F \vee H) \Rightarrow (G \vee K)$. It is not difficult to prove the following more general result: if $\mathcal{H} \vdash F_1 \Rightarrow G_1, \mathcal{H} \vdash F_2 \Rightarrow G_2, \ldots, \mathcal{H} \vdash F_k \Rightarrow G_k$, then $\mathcal{H} \vdash (F_1 \vee \ldots \vee F_k) \Rightarrow (G_1 \vee \ldots \vee G_k)$; moreover, if $G_1 \equiv G_2 \equiv \ldots \equiv G_k \equiv G$, then $\mathcal{H} \vdash (F_1 \vee \ldots \vee F_k) \Rightarrow G$.

Using this result, from the above k deductions, we get:

$$Q_1, Q_2 \vdash (\bigvee_{x=1}^{k} A(y)) \Rightarrow (\bigvee_{x=1}^{k} (L(x) \wedge H(x))).$$

Hence the reasoning of the passer-by is proved syntactically in the framework of propositional logic.

As we already mentioned, for a predicate $F(x) \equiv$ "x has the property p", the sentence $S_1 \equiv$ "for all x, x has the property p" $\equiv$ "for all x, $F(x)$" can be expressed by the formula

$$S_1 \equiv \bigwedge_{x=1}^{k} F(x).$$

Similarly, the sentence $S_2 \equiv$ "there are some x such that x has property p" $\equiv$ "for some x, $F(x)$" can be expressed by the formula:

$$S_2 \equiv \bigvee_{x=1}^{k} F(x).$$

The sentence S_1 represents a universal assertion ("for all") concerning the whole set $\mathcal{C}$. The operation of passing from the given predicate $F(x)$ to the (universal) sentence S_1 is called the **universal quantification** of $F(x)$ and the resulting sentence will be denoted by $\forall x(F(x))$. Thus

(∗)
$$S_1 \equiv \bigwedge_{x=1}^{k} F(x) \equiv \forall x(F(x)).$$

On the other hand, the sentence S_2 represents an existential assertion ("there is", "there are", "there exists") also concerning the set $\mathcal{C}$. The operation of passing from the predicate $F(x)$ to the (existential) sentence

S_2 is called the **existential quantification** of the predicate $F(x)$ and will be denoted by $\exists\, x(F(x))$. Thus,

$$(**) \qquad\qquad S_2 \equiv \bigvee_{x=1}^{k} F(x) \equiv \exists\, x(F(x)).$$

Sentences like $\forall\, x(F(x))$ and $\exists\, x(F(x))$ (often written simply $\forall x F(x)$ and $\exists x F(x)$, respectively) are characteristic of predicate logic. In our example, the sentences $\forall\, x(F(x))$ and $\exists\, x(F(x))$ are not essentially new ones, because they are only abbreviated notations for a finite conjunction and finite disjunction of k propositions. So, in our example, the sentences $\forall\, x(F(x))$ and $\exists\, x(F(x))$ belong to the language of propositional logic.

We notice that the sentence $S_1 \equiv \forall\, x(F(x))$ is a proposition (it does not contain individual variables), because the meaning of $\forall x F(x)$ is $F(1) \wedge F(2) \wedge \ldots \wedge F(k)$. For this reason we shall say that in $F(x)$, x is a free variable while, in the sentence $\forall x F(x)$, x is a bound variable, which means that, actually, the formula $\forall x F(x)$ does not contain the variable x; hence, instead of $\forall x F(x)$ we can write equivalently $\forall y F(y)$, $\forall t F(t)$ and so on. It is obvious that same remarks are valid for the formula $S_2 \equiv \exists x F(x) \equiv \exists y F(y) \equiv \exists t F(t)$.

Formulas $(*)$ and $(**)$ show that if the set of the individual objects (in our case, the set $\mathcal{C}$) is finite, then the universal and existential quantifications are reduced to a finite conjunction and a finite disjunction, respectively. But this is a particular situation. In the applications of predicate logic, the sets of individual objects are not supposed to be finite. Therefore, such reductions are not anymore possible. Hence in predicate logic we have to operate directly with quantifiers $\forall$ and $\exists$.

Now, let us analyze the above proof of the deduction $Q_1, Q_2 \vdash R$ from the point of view of predicate logic. For this purpose let us translate the deduction 1-12 in the new notations as follows:

1. $\forall\, x(A(x) \Rightarrow L(x))$; hypothesis Q_1;

2. $\forall\, x(A(x) \Rightarrow H(x))$; hypothesis Q_2;

3. $A(y)$; hypothesis;

4. $\forall\, x(A(x) \Rightarrow L(x))) \Rightarrow (A(y) \Rightarrow L(y))$;

5. $\forall\, x(A(x) \Rightarrow H(x))) \Rightarrow (A(y) \Rightarrow H(y))$;

6. $A(y) \Rightarrow L(y)$; (MP): 1,4;

7. $A(y) \Rightarrow H(y)$; (MP): 2,5;

8. $L(y)$; (MP): 3,6;

9. $H(y)$; (MP): 3,7;

10. $H(y) \wedge L(y)$; metatheorem of conjunction;

11. $(H(y) \wedge L(y)) \Rightarrow (\exists x (H(x) \wedge L(x)))$;

12. $\exists x (H(x) \wedge L(x))$; (MP): 10,11;

At steps 4 and 5 we have used the metatheorem of propositional logic (i.e., the tautology) $E_1 \wedge E_2 \wedge \ldots \wedge E_k \Rightarrow E_i$. In predicate logic we shall adopt the axiom (a_1): $\forall x (F(x)) \Rightarrow F(y)$, so that steps 4 and 5 follow by the application of this axiom. The steps 6, 7, 8, 9 are obtained by the MP-rule. Step 10 also remains valid in predicate logic. Step 11 in the initial proof has been obtained from the tautologies $F \Rightarrow F \vee G$ and $G \Rightarrow F \vee G$. In predicate logic we shall adopt the axiom (a_2): $F(y) \Rightarrow \exists x F(x)$; so that in predicate logic, step 11 is obtained by the application of this axiom. Step 12 has been obtained by the MP-rule. Consequently, in order to perform the deduction 1-12 in predicate logic we have to adopt, besides the axioms of propositional logic, two new axioms, namely (a_1) and (a_2). Therefore, we have within predicate logic the deduction

$$Q_1, Q_2, A(y) \vdash \exists x (H(x) \wedge L(x)).$$

We shall see that in predicate logic one can also prove, under certain conditions (which are fulfilled in our case), a theorem analogous to the one of Herbrand. Therefore, we get the deduction

$(***)$ $\qquad Q_1, Q_2 \vdash A(y) \Rightarrow \exists x (H(x) \wedge L(x)).$

In the initial proof from this deduction we inferred:

$$Q_1, Q_2 \vdash (\exists x A(x)) \Rightarrow \exists x (H(x) \wedge L(x)).$$

This inference has been obtained by using a slight generalization of MTh. 3.15, Section 3.2. In predicate logic, instead of this metatheorem we have to adopt the following new inference rule: if $F(x)$ is a predicate containing the variable x and G is a sentence (proposition or predicate) which does not contain the variable x, then from $F(x) \Rightarrow G$ one can deduce $\exists x (F(x)) \Rightarrow G$. This new inference rule will be denoted by (IR_1). We notice that in predicate logic there is also a second specific inference rule, denoted by (IR_2), which allows to infer $G \Rightarrow \forall x F(x)$ from $G \Rightarrow F(x)$. These inference rules can be applied also in a more general case, namely

from $H_1, H_2 \vdash F(x) \Rightarrow G$, one can infer $H_1, H_2 \vdash \exists x (F(x)) \Rightarrow G$

and

from $H_1, H_2 \vdash G \Rightarrow F(x)$, one can infer $H_1, H_2 \vdash G \Rightarrow \forall x (F(x))$, where sentences H_1, H_2 do not contain the variable x. Applying the first generalized inference rule to the above deduction $(* * *)$ one gets the deduction

$$Q_1, Q_2 \vdash \exists y(A(y)) \Rightarrow \exists x(H(x) \wedge L(x));$$

but as we have already mentioned $\exists y A(y) \equiv \exists x A(x)$ hence we finally obtain the required deduction:

$$Q_1, Q_2 \vdash \exists x(A(x)) \Rightarrow \exists x(H(x) \wedge L(x)).$$

The last deduction represents the syntactic proof of the passer-by's reasoning in the framework of predicate logic. This was possible by adding to the set of axioms and inference rule of propositional logic two natural new axioms $((a_1)$ and $(a_2))$ and the inference rule (IR_1) (we did not used the rule (IR_2)).

REMARK 8.1 An important conclusion drawn from these considerations is that, in the simplest situations (when the universe of individual objects is finite), the framework of propositional logic is quite sufficient for the task of reasoning. However, in general situations, when the universe of individual objects is not supposed to be finite (such situations are typical in mathematics), it is necessary to consider the quantifications as essentially new operations and to postulate explicitly the operating rules with them. Naturally, these rules are suggested by the case of finite universe, but they are maintained in general situations with nonempty domains. ∎

REMARK 8.2 Before ending these preliminary considerations we have to mention another typical element, called **functor**, which plays a specific role in predicate logic. A functor f is intuitively a (mathematical) function which associates to each **object** x another **object** $f(x)$; for instance, if the universe of individual objects is the set $\mathbf{N}$ of all natural numbers, then we can define the functor $s : \mathbf{N} \to \mathbf{N}$, where $s(x) = x + 1$. A functor (as mathematical function) depends on one or more variables; in the formal framework one will say that the functor f has one or more free places. Consequently, each functor f has certain arity $m \geq 0$ (i.e., m free places). For instance, by means of a given binary functor f, with $m = 2$, we can construct new formal words such as $f(x, y)$ where x, y are arbitrary individual variables. In what follows variables and all formal words like $f(x, y)$, $g(x, y, z)$, etc., where f, g are functors and x, y, z, etc. are variables, will be called **terms**. There is some similarity between functors and predicates. Indeed, each predicator P also has certain arity $m \geq 0$ and for each sequence $u_1, u_2, \ldots, u_m$ of variables the formal word $P(u_1, u_2, \ldots, u_n)$ is a sentence (a predicate) with m free places. Intuitively, each predicator P

having, for instance, one free place can be considered as being a (mathematical) function $x \mapsto P(x)$ which associates to each object x the sentence $P(x)$. However, we have to stress the following essential difference existing between functors and predicators: while for the functors f, g, h, etc., the formal words $f(x,y), g(x,y,z), h(x)$, etc. are **terms** (intuitively representing some individual objects), for the predicators F, G, H, etc., the formal words $F(x,y), G(x,y,z), H(x)$, etc. are **formulas** (which intuitively represent some true or false sentences about the objects x, y, z and so on).

9

The Formal Language of Predicate Logic

The construction of the formal language of predicate logic is similar to the one of propositional logic. This means that we shall start with a given (formal) alphabet **X** and we shall define successively (formal) **words**, **terms** and **formulas** in **X**. Of course, in comparison to the language of propositional logic, the language of predicate logic will have some additional specific characteristics. Such additional characteristics will be, for instance, the appearance of individual **variables**, **terms** and **quantifications**.

9.1 The Formal Alphabet of Predicate Logic: Formal Words

DEFINITION 9.1 *The (formal) alphabet of predicate logic is the set* **X** *of all symbols belonging to the three lists below.*

(1) **Literal symbols**
(a) a countable set of individual variables:
$$\mathbf{V} = \{x_1, x_2, x_3, \ldots\};$$
(b) a finite or countable set of individual constants:
$$\mathbf{F}_0 = \{a_1, a_2, a_3, \ldots\};$$
(c) a finite or countable set of functors:
$$\mathbf{F} = \{f_1, f_2, f_3, \ldots\};$$
(d) a finite or countable set of propositional variables:
$$\mathbf{P}_0 = \{p_1, p_2, p_3, \ldots\};$$
(e) a finite or countable set of predicators:
$$\mathbf{P} = \{P_1, P_2, P_3, \ldots\}.$$
*Each functor $f \in \mathbf{F}$ and each predicator $P \in \mathbf{P}$ is provided with a natural number $\alpha(f) \geq 1$ and $\alpha(P) \geq 1$ called, respectively, the **arity of f** and the*

arity of P. *The arities* $\alpha(f)$ *and* $\alpha(P)$ *represent, respectively, the number of free places of* f *and of* P *(i.e., of arguments required by* f *or by* P*). Each constant* $a_i \in \mathbf{F}_0$ *and each propositional variable* $p_j \in \mathbf{P}_0$ *can be considered as being a functor or a predicator without free places, respectively, hence as having the arity zero. It is supposed that the set* $\mathbf{P}$ *is not empty, while the sets* $\mathbf{F}_0, \mathbf{F}, \mathbf{P}_0$ *could be empty.*

(2) **Logical symbols**
  (a) logical operators: $\neg, \vee$*;*
  (b) the quantifier: $\exists$*.*

(3) **Auxiliary symbols**
  (a) parentheses: ( );
  (b) the comma: , ;
  (c) the square: $\square$*;*
  (d) the clip: $\sqcap$ *(the clip is an overline symbol).*

REMARK 9.1   The set $\mathbf{X}_0$ containing only propositional variables, logical operators, parentheses and the comma coincides with the alphabet of propositional logic generated by $p_1, p_2, p_3, \ldots$. Therefore, $\mathbf{X}_0$ is a subalphabet of $\mathbf{X}$ and this will imply that the language of propositional logic is a sublanguage of the language of predicate logic.   ▌

REMARK 9.2   The meaning of literal symbols and logical symbols is clear enough. We notice only that the sets $\mathbf{F}_0$, $\mathbf{F}$ and $\mathbf{P}_0$ can be empty. In what concerns the auxiliary symbols some comments are necessary. These symbols play a technical role, as tools for the correct writing of terms and formulas. Parentheses and comma are introduced for reasons of convenience, to make the formal expressions more readable. The square and the clip are necessary for the rigorous definition of quantifications. In order to be more explicit, let us recall that in formulas like $\exists x(F(x))$ and $\forall x(F(x))$ (the fact that the quantifier $\forall$ does not belong to the alphabet $\mathbf{X}$ is immaterial in this context because it will be defined in what follows by means of $\neg$ and $\exists$), the individual variable x is bound (i.e., not free). Hence we can write

$$\exists x(F(x)) \equiv \exists y(F(y)) \equiv \exists t(F(t)) \text{ etc.}$$
$$\forall x(F(x)) \equiv \forall y(F(y)) \equiv \forall t(F(t)) \text{ etc.}$$

We shall see that the expressions $\exists x(F(x))$ and $\forall x(F(x))$ are only notations (abbreviations) for some well-defined formal words in which the square and the clip symbols play a specific role. The scheme of use of the square and clip symbols is as follows:

$$\exists x(F(x)) \equiv \exists \overline{\square(F(\square))} \ \text{ and } \ \forall x(F(x)) \equiv \forall \overline{\square(F(\square))}.$$

The clip indicates, therefore, textual coreference.

Consider, for instance, the following sentence (predicate) belonging to the language of set theory

$$H(x, y, z) \equiv (x \in y) \Rightarrow (x \in z)$$

which means that if x belongs to (the set) y, then x belongs to (the set) z. The universal quantification with respect to x of the predicate $H(x, y, z)$ leads to the sentence (predicate)

$$K(y, z) \equiv \forall x((x \in y) \Rightarrow (x \in z))$$

which means that every element of y belongs to z, hence coincides with the set-theoretical inclusion, $y \subseteq z$. In the formal language, the predicate $K(y, z) \equiv y \subseteq z$ can be expressed by means of squares and clips as follows:

$$y \subseteq z \equiv \forall \overline{\square((\square \in y)} \Rightarrow (\square \in z)).$$

From this example we can see that the square always replaces variables and the length of a clip depends on the context; moreover, the clips always connect square symbols. ∎

DEFINITION 9.2 *A finite string (sequence) of symbols in* **X**

$$A \equiv \sigma_1 \sigma_2 \ldots \sigma_m \qquad (m \geq 1)$$

provided with clips or without clips, is called a **(formal) word** *in* **X**. *The two ends of each clip are written over two distinct squares as in*

$$\sigma_1 \ldots \overline{\square \ldots \square} \ldots \sigma_m.$$

The number $\lambda(A) = m \geq 1$ is called **the length** *of A. If σ is a symbol in* **X** *(distinct from the clip), then the set of all occurrences (appearances) of the symbol σ in A, i.e., the set of all indices i for which $\sigma_i \equiv \sigma$, is denoted by $\omega(\sigma; A)$; hence*

$$\omega(\sigma; A) = \{i \mid 1 \leq i \leq m, \ \sigma_i \equiv \sigma\}.$$

Obviously, the set $\omega(\sigma; A)$ can be empty for some A and σ.

DEFINITION 9.3

(a) *If $A \equiv \sigma_1 \sigma_2 \ldots \sigma_m$ is a word, then the word $\neg(\sigma_1 \sigma_2 \ldots \sigma_m)$, denoted by $\neg(A)$, is called* **the negation** *of A.*

(b) If $A \equiv \sigma_1 \sigma_2 \ldots \sigma_p$ and $B \equiv \tau_1 \tau_2 \ldots \tau_q$ are words, then the word

$$(\sigma_1 \ldots \sigma_p) \vee (\tau_1 \ldots \tau_q),$$

*denoted by $(A) \vee (B)$, is called **the disjunction** of A, B.*

(c) If $A(x)$ is a word, x is an individual variable and the set $\omega(x; A)$ of all occurrences of x in A is not empty, $\omega(\sigma; A) = \{i_1, i_2, \ldots, i_k\}$, then the word

$$\exists \Box (\sigma_1 \ldots \underset{i_1}{\Box} \ldots \underset{i_k}{\Box} \ldots \sigma_m)$$

*is denoted by $\exists x(A(x))$ and is called **the existential quantification** of $A(x)$ **with respect to** x. The clips of $A(x)$ are maintained in $\exists x(A(x))$.*

REMARK 9.3　　If in the formal word $A(x)$ we replace the variable x by a new variable y (distinct from all other individual variables in $A(x)$), then we obtain a new word $A(y)$ with $\omega(y; A(y)) = \omega(x; A(x))$ and obviously $\exists y(A(y)) \equiv \exists x(A(x))$. That is why in some texts the bound variable x in $\exists x(F(x))$ is called also a **dummy variable**.　∎

As in the case of the language of propositional logic we introduce some derived operations by the following abbreviations:

$$(A) \wedge (B) \equiv \neg((\neg(A)) \vee (\neg(B)))$$
$$(A) \Rightarrow (B) \equiv (\neg(A)) \vee (B)$$
$$(A) \Leftrightarrow (B) \equiv ((A) \Rightarrow (B)) \wedge ((B) \Rightarrow (A))$$
$$\forall x(A(x)) \equiv \neg(\exists x(\neg(A(x)))).$$

The last formula is called **the universal quantification** of $A(x)$ with respect to x. Sometimes, when confusions are excluded, we shall simplify the above notation by dropping some parentheses and we shall write simply:

$$(A) \wedge (B) \equiv \neg(\neg A \vee \neg B),$$
$$(A) \Rightarrow (B) \equiv \neg A \vee B,$$
$$(A) \Leftrightarrow (B) \equiv (A \Rightarrow B) \wedge (B \Rightarrow A),$$
$$\forall x(A(x)) \equiv \neg \exists x(\neg A(x)).$$

However, when we have to compute exactly the length of these formal words we must take into consideration all parentheses.

DEFINITION 9.4 *Let $A \equiv \sigma_1 \sigma_2 \ldots \sigma_p$ $(p \geq 1)$ and $B \equiv \tau_1 \tau_2 \ldots \tau_q$ $(q \geq 1)$ be two formal words, with or without clips. We shall say that words A, B are* **equal** *(coincide) and we shall denote $A \equiv B$ if the following conditions are fulfilled:*

(a) $\lambda(A) = \lambda(B) = m$ $(p = q = m)$;

(b) $\sigma_i \equiv \tau_i$ for all i $(1 \leq i \leq m)$;

(c) A clip connects the squares $\sigma_i \equiv \sigma_j \equiv \square$ $(i \neq j)$ in A if and only if the squares $\tau_i \equiv \tau_j \equiv \square$ are connected in B by a clip.

If at least one of conditions (a), (b), (c) is not fulfilled, then we shall say that A, B are **distinct** *and we shall write $A \not\equiv B$.*

9.2 Terms and Formulas

In this section we shall introduce the essential distinction between general formal words (i.e., meaningless finite strings of symbols in **X**) and the special formal words which make sense (well-formed words) which are of interest for predicate logic.

We shall begin with the definition of terms.

DEFINITION 9.5 *A finite sequence of words in* **X**

$$T_1, T_2, \ldots, T_r \qquad (r \geq 1)$$

is called a **(formal) construction of terms** *if for each index i $(1 \leq i \leq r)$ the word T_i satisfies one of the following two conditions:*

(a) T_i is an individual variable or an individual constant;

*(b) there are indices $i_1, i_2, \ldots, i_n < i$ (distinct or not) and a functor $f \in$ **F** with $\alpha(f) = n \geq 1$ such that $T_i \equiv f(T_{i_1}, T_{i_2}, \ldots, T_{i_n})$. Each word T which can be inserted in a (formal) construction of terms is called a* **term** *in* **X**.

Example 9.1

Let a_1, a_2 be constants and f_5, f_{10} be two functors with $\alpha(f_5) = 3$, $\alpha(f_{10}) = 2$. It is easy to see that the formal word

$$T \equiv f_5(a_2, a_1, f_{10}(x_3, x_4))$$

is a term. Indeed, consider the following formal construction of terms:

(1) $T_1 \equiv x_3$; x_3 is an individual variable;

(2) $T_2 \equiv x_4$; x_4 is an individual variable;

(3) $T_3 \equiv f_{10}(x_3, x_4)$; f_{10} is a functor with $\alpha(f_{10}) = 2$ and $x_3 \equiv T_1$, $x_4 \equiv T_2$;

(4) $T_4 \equiv a_1$; a_1 is a constant;

(5) $T_5 \equiv a_2$; a_2 is a constant;

(6) $T_6 \equiv f_5(a_2, a_1, f_{10}(x_3, x_4))$; f_5 is a functor with $\alpha(f_5) = 3$ and $a_2 \equiv T_5$, $a_1 \equiv T_4$, $f_{10}(x_3, x_4) \equiv T_3$. Therefore, $T_6 \equiv T$, hence T can be inserted in the formal construction of terms $T_1 - T_6$, i.e., T is a term.

⬜

The definition of formulas will be given below. We shall see that it is similar to that of formulas in propositional logic with the difference that the role of sentences will be now played by the so-called **elementary formulas**.

DEFINITION 9.6   *A formal word A is called an **elementary formula** if it satisfies one of the conditions below:*
   (a) $A \equiv p_j \in \mathbf{P_0}$;
   (b) $A \equiv P_j(T_1, T_2, \ldots, T_n)$, where $P_j \in \mathbf{P}$ is a predicator with $n \geq 1$ places $(\alpha(P_j) = n)$ and $T_1, T_2, \ldots, T_n$ are terms.

Example 9.2
If p_1 is a propositional variable (i.e., $p_1 \in \mathbf{P_0}$), then p_1 is an elementary formula. If P_4 is a predicator with 3 places $(\alpha(P_4) = 3)$ and if f_5 is a functor with two places $(\alpha(f_5) = 2)$, then the word

$$P_4(x_3, a_2, f_5(x_4, x_7))$$

is an elementary formula.   ⬜

Now we are ready to define the formulas in $\mathbf{X}$.

DEFINITION 9.7   *A finite sequence of words in $\mathbf{X}$*

$$F_1, F_2, \ldots, F_r \qquad (r \geq 1)$$

*is called a **construction of formulas** if for each index i $(1 \leq i \leq r)$ the word F_i satisfies one of the following four conditions:*

(a) F_i is an elementary formula:

(b) there is an index j $(j < i)$ such that

$$F_i \equiv \neg(F_j);$$

(c) there are indices h, k $(h, k < i)$ distinct or not, such that

$$F_i \equiv (F_h) \vee (F_k);$$

(d) there is an index j $(j < i)$ and an individual variable x contained in $F_j \equiv F_j(x)$ such that
$$F_i \equiv \exists\, x(F_j(x)).$$

Each word which can be inserted in a construction of formulas is called a formula in **X**.

Example 9.3
Consider the words

$$F \equiv \exists\, x_2(P_2(x_2, x_3, a_4)),$$
$$G \equiv \neg((P_7(a_2, a_4)) \vee (\exists\, x_3(P_4(x_3, f_9(x_4))))),$$
$$H \equiv \vee P_7(\neg x_2),$$

where P_2, P_4, P_7 are predicators with $\alpha(P_2) = 3$, $\alpha(P_4) = 2$, $\alpha(P_7) = 2$ and f_9 is a functor with $\alpha(f_9) = 1$. It is easy to see that F, G are formulas, while H is not a formula. []

REMARK 9.4   It is obvious that if A, B, C are formulas in **X**, then $(B) \wedge (C)$, $(B) \Rightarrow (C)$, $(B) \Leftrightarrow (C)$ and $\forall\, x(A(x))$ are also formulas in **X**.

DEFINITION 9.8   The set of all formulas in **X** is denoted by $\mathcal{F} = \mathcal{F}(\mathbf{X})$. The set of all formulas and all terms is called the (**formal**) **language of predicate logic** and is denoted by $\mathcal{L} = \mathcal{L}(\mathbf{X})$. It is clear that $\mathcal{F}(\mathbf{X})$ is a subset of $\mathcal{L}(\mathbf{X})$.

REMARK 9.5   For certain authors, the language of predicate logic coincides with $\mathcal{F}(\mathbf{X})$, the set of terms being considered separately. ▮

In Section 9.1, Remark 9.3, we have discussed the formulas $A(x)$ and $A(y)$. We have mentioned that if x is a free variable in $A(x)$, then replacing the variable x by the variable y in order to obtain the formula $A(y)$ we must make sure that the variable y is distinct from all other individual variables (free or not) of $A(x)$. That is because $A(x)$ is, in general, only a notation of some complex formula containing, for instance, the variable y as a quantified variable. Let us consider, for instance, the formula $A(x) \equiv (P_1(x, z)) \Rightarrow (\exists y(P_2(x, y)))$; if in this formula we replace without care the variable x by y we get the formula $B(y) \equiv (P_1(y, z)) \Rightarrow (\exists y(P_2(y, y)))$.

It is clear that this substitution is not correct; indeed writing $A(x)$ and $B(y)$ correctly in the symbols of the alphabet **X**, we get:

$$A(x) \equiv (P_1(x, z)) \Rightarrow (\exists \overline{\Box(P_2(x, \Box))});$$

$$B(y) \equiv (P_1(y, z)) \Rightarrow (\exists \overline{\Box(P_2(\Box, \Box))}).$$

The correct substitution will be

$$A(y) \equiv (P_1(y, z)) \Rightarrow (\exists \overline{\Box(P_2(y, \Box))}),$$

and it is obvious that $A(y) \not\equiv B(y)$. So, mechanically replacing in a formula $A(x)$ the variable x by y, can produce an incorrect result. Such a situation can also arise in cases when we replace in a formula $A(x)$ the free variable x by an arbitrary term T because a term contains, in general, constants and variables; moreover, the variables contained in terms are always free.

DEFINITION 9.9   *Let $A(x_i) \in \mathcal{L}(\mathbf{X})$ be a formula containing the free variable x_i and T be an arbitrary term. We shall say that the term T is* **free for** x_i *in $A(x_i)$ if each free occurrence of x_i in $A(x_i)$ does not belong to any subformula of $A(x_i)$ which is quantified with respect to a variable x_j of the term T.*

10

The Semantic Truth Structure on the Language $\mathcal{L}$ of Predicate Logic

The truth structure on $\mathcal{L}$ in the semantic version is based, as in the case of propositional logic, on the notion of interpretation. It is easy to understand that the interpretations of $\mathcal{L}$ will be more sophisticated than the ones of $\mathcal{L}_0$. This is because in predicate logic we have to interpret additionally variables, constants, terms, quantifications, etc. Variables will be interpreted as generic elements of some set $\mathcal{D}$ of individual objects called the domain of the interpretation; constants will be interpreted as some special objects in $\mathcal{D}$. Functors and predicators will be interpreted as some functions, etc.

10.1   The Notion of Interpretation of the Language $\mathcal{L}$

The interpretation of $\mathcal{L}$ is defined successively starting from an interpretation of literal symbols in **X**, generating subsequently an interpretation of terms and an interpretation of formulas. We shall begin with the interpretation of literal symbols.

DEFINITION 10.1   *The system*

$$I_0 = (\mathcal{D}, \xi, \eta_0, \eta, \zeta_0, \zeta)$$

is called an **interpretation of literal symbols of X,** *if*

(a) $\mathcal{D}$ is a given nonempty set called the **domain of the interpretation;**

(b) $\xi : x_i \longmapsto \xi(x_i)$ is a function which associates to each individual variable $x_i \in \mathbf{V}$ an element $\xi(x_i) \in \mathcal{D}$ (called the interpretation of x_i);

(c) $\eta_0 : a_i \longmapsto \eta_0(a_i)$ is a function which associates to each individual constant $a_i \in \mathbf{F}_0$ a distinguished element $\eta_0(a_i) \in \mathcal{D}$ (called the interpretation of a_i);

(d) $\eta : f_i \longmapsto \eta(f_i)$ is a function which associates to each functor $f_i \in \mathbf{F}$ with $\alpha(f_i) = n \geq 1$, a function $\eta(f_i) : \mathcal{D}^n \to \mathcal{D}$ (called the interpretation of f_i);

(e) $\zeta_0 : p_i \longmapsto \zeta_0(p_i)$ is a function which associates to each propositional variable $p_i \in \mathbf{P}_0$ an element $\zeta_0(p_i) \in \Gamma, \Gamma = \{0,1\}$ (called the interpretation of p_i);

(f) $\zeta : P_i \longmapsto \zeta(P_i)$ is a function which associates to each predicator $P_i \in \mathbf{P}$ with $\alpha(P_i) = n \geq 1$, a function $\zeta(P_i) : \mathcal{D}^n \to \Gamma, \Gamma = \{0,1\}$ (called the interpretation of P_i).

Example 10.1

Consider the alphabet $\mathbf{X}$ of predicate logic in which the literal symbols are the following:

$$\mathbf{V} = \{x_1, x_2, x_3, \ldots\};$$
$$\mathbf{F}_0 = \{a_1, a_2, a_3\};$$
$$\mathbf{F} = \{f_1, f_2\}, \ \alpha(f_1) = 2, \alpha(f_2) = 2;$$
$$\mathbf{P}_0 \text{ is empty};$$
$$\mathbf{P} = \{P_1, P_2\}, \ \alpha(P_1) = 2, \ \alpha(P_2) = 3.$$

The interpretation $I_0 = (D, \xi, \eta_0, \eta, \zeta_0, \zeta)$ of literal symbols of $\mathbf{X}$ is given (in this example) by:

$\mathcal{D} = \mathbf{N}$ (the set of natural numbers ≥ 0);

$\xi(x_i) = i \in \mathbf{N}$, for each $x_i \in \mathbf{V}$;

$\eta_0(a_1) = 4, \eta_0(a_2) = 0, \eta_0(a_3) = 7$;

$\eta(f_1) : \mathbf{N} \times \mathbf{N} \to \mathbf{N}, \eta(f_1)(i,j) = i + j$;

$\eta(f_2) : \mathbf{N} \times \mathbf{N} \to \mathbf{N}, \eta(f_2)(i,j) = i \cdot j$;

the function ζ_0 is not considered because $\mathbf{P}_0$ is empty;

$$\zeta(P_1) : \mathbf{N} \times \mathbf{N} \to \Gamma, \zeta(P_1)(i,j) = \begin{cases} 1, & \text{if } i \leq j \\ 0, & \text{otherwise}; \end{cases}$$

$$\zeta(P_2) : \mathbf{N} \times \mathbf{N} \times \mathbf{N} \to \Gamma, \zeta(P_2)(i,j,k) = \begin{cases} 1, & \text{if } 2i + j = k \\ 0, & \text{otherwise}. \end{cases} \quad \Box$$

Starting with an interpretation I_0 of literal symbols one can construct the corresponding interpretation of terms $T \longmapsto I(T)$. The construction of the interpretation $T \longmapsto I(T)$ will be performed inductively in a natural way. In order to do this rigorously we need some preliminary notations and results. Sometimes, in the following, the function $\eta(f_i)$ will be denoted by $\hat{f}_i$.

Let us consider a nonempty set $\mathbf{S}$ and a natural number $q \geq 1$. A finite string

$$S \equiv s_1 s_2 \ldots s_q, \text{ with } s_i \in \mathbf{S} \ (1 \leq i \leq q)$$

will be called a finite sequence in $\mathbf{S}$; the number $\lambda(S) = q$ will be called **the length of S**. If the sequence $S = s_1 s_2 \ldots s_q$ is given, then for each element $s \in \mathbf{S}$ we shall denote by $\omega(s; S)$ the set of occurrences of s in S, i.e.,

$$\omega(s; S) = \{i \mid 1 \leq i \leq q \ , \ s_i = s\}.$$

Obviously, the set $\omega(s; S)$ can be empty. If $\omega(s; S)$ is nonempty, then S has the form

$$S \equiv s_1 \ldots s_{i_1} \ldots s_{i_2} \ldots \ldots s_{i_k} \ldots s_q,$$

where $\{i_1, i_2, \ldots i_k\} = \omega(s; S)$ and $s_{i_1} = \ldots = s_{i_k} = s$; in such a case we shall sometimes write $S \equiv S(s)$ and will say that S contains s.

If s is an element of $\mathbf{S}$ and $S \equiv S(s)$ is a finite sequence in $\mathbf{S}$ (containing s), then for each finite sequence in $\mathbf{S}, Y \equiv y_1 y_2 \ldots y_p \ (p \geq 1)$ we shall denote by $S(Y)$, the finite sequence in $\mathbf{S}$ obtained from S by putting Y instead of each occurrence of s in S; hence $S(Y)$ is obtained from $S(s)$ by the operation:

$$S(s) \equiv s_1 \ldots \underset{\underset{\downarrow}{i_1}}{s} \ldots \underset{\underset{\downarrow}{i_2}}{s} \ldots \ldots \underset{\underset{\downarrow}{i_k}}{s} \ldots s_q$$

$$S(Y) \equiv s_1 \ldots \overbrace{y_1 \ldots y_p} \ldots \overbrace{y_1 \ldots y_p} \ldots \overbrace{y_1 \ldots y_p} \ldots s_q$$

It is clear that

$$\lambda(S(Y)) = \lambda(S(s)) + k(\lambda(Y) - 1) = q + k(p - 1).$$

If $\omega(s; S)$ is empty then, by definition, $S(Y) = S$.

Example 10.2

If, for instance, $\mathbf{S}$ is the alphabet $\mathbf{X}$ of predicate logic, then it is clear that each finite sequence in $\mathbf{X}$ is a word in $\mathbf{X}$. Consider, for example, the finite sequence (the word)

$$S \equiv x_1 \neg x_1 a_2, \ \lambda(S) = 4.$$

It is visible that $\omega(x_1; S) = \{1,3\}$, hence S contains x_1 and we can write $S \equiv S(x_1)$. If we now take the finite sequence $Y \equiv)P_2 \vee x_2 x_3 \Box$, then

$$S(Y) \equiv)P_2 \vee x_2 x_3 \Box \neg)P_2 \vee x_2 x_3 \Box a_2$$

and $\lambda(S) = q = 4, \lambda(Y) = p = 6, \lambda(S(Y)) = 4 + 2(6 - 1) = 14.$ $\quad\Box$

If k finite sequences in **S** are given,

$$\begin{aligned}
S_1 &\equiv s_1^1 s_2^1 \ldots s_{l_1}^1, \quad \lambda(S_1) = l_1, \\
S_2 &\equiv s_1^2 s_2^2 \ldots s_{l_2}^2, \quad \lambda(S_2) = l_2, \\
&\vdots \\
S_k &\equiv s_1^k s_2^k \ldots s_{l_k}^k, \quad \lambda(S_k) = l_k,
\end{aligned}$$

then the finite sequence in **S**

$$S \equiv s_1^1 s_2^1 \ldots s_{l_1}^1 s_1^2 s_2^2 \ldots s_{l_2}^2 \ldots s_1^k s_2^k \ldots s_{l_k}^k$$

will be denoted by

$$S \equiv S_1 S_2 \ldots S_k$$

and obviously, $\lambda(S) = \lambda(S_1) + \lambda(S_2) + \ldots + \lambda(S_k)$.

In the following, besides finite sequences in **X** (formal words), we shall also deal with finite sequences in the set of all variables and constants of **X** and with finite sequences in the domain of interpretation.

Let $I_0 = (\mathcal{D}, \xi, \eta_0, \eta, \zeta_0, \zeta)$ be an interpretation of literal symbols (of **X**) and let us denote by $\mathcal{U}$ the set of all variables and constants in **X**, i.e., $\mathcal{U} = \mathbf{V} \cup \mathbf{F}_0$.

For each $u \in \mathcal{U}$ we denote by $\hat{u} \in \mathcal{D}$ the element

$$\hat{u} = \begin{cases} \xi(u), & \text{if } u \in \mathbf{V}, \\ \eta_0(u), & \text{if } u \in \mathbf{F}_0; \end{cases}$$

thus, we get a correspondence $u \longmapsto \hat{u} : \mathcal{U} \to \mathcal{D}$ which associates to each symbol u in $\mathcal{U}$ (i.e., a variable or a constant) its interpretation $\hat{u}$ in $\mathcal{D}$. If $U \equiv u_1 u_2 \ldots u_q$ is a finite sequence in $\mathcal{U}$, then we denote by $\hat{U} \equiv \hat{u}_1 \hat{u}_2 \ldots \hat{u}_q$ the corresponding sequence in $\mathcal{D}$. Therefore, the correspondence $u \longmapsto \hat{u}$ generates in a natural way a correspondence $U \longmapsto \hat{U}$ from sequences in $\mathcal{U}$ to sequences in $\mathcal{D}$. This correspondence obviously has the following two properties:

(a) If $U \equiv u_1 u_2 \ldots u_q$ is a finite sequence in $\mathcal{U}$ which contains the symbol (the variable) $x \in \mathbf{V}, \omega(x; U) = \{i_1, i_2, \ldots, i_k\}$,

$$U \equiv U(x) \equiv u_1 \underset{i_1}{\ldots} x \underset{i_2}{\ldots} x \underset{}{\ldots\ldots} x \underset{i_k}{\ldots} u_q,$$

then
$$\hat{U} \equiv \widehat{U(x)} = \hat{u}_1 \ \underset{i_1}{\ldots} \ \hat{x} \ \underset{i_2}{\ldots} \ \hat{x} \ \ldots\ldots \ \underset{i_k}{\hat{x}} \ \ldots \ \hat{u}_q$$

hence
$$\widehat{U(x)} = \hat{U}(\hat{x});$$

(b) If $U \equiv u_1 u_2 \ldots u_q$ is a finite sequence in $\mathcal{U}$ which contains the variable $x \in \mathbf{V}$, and if $W \equiv w_1 w_2 \ldots w_p$ is also a finite sequence in $\mathcal{U}$, then it is easy to see that
$$\widehat{U(W)} = \hat{U}(\hat{W}).$$

DEFINITION 10.2 *Let T be a term. The finite sequence $U = u_1 u_2 \ldots u_q$ in $\mathcal{U}$ containing all variables and constants of T (possible with repetitions) and written in the order of their occurrence in T will be called the* **sequence of the arguments** *of T; in this case we shall write*
$$T \equiv T(U) \equiv T(u_1 u_2 \ldots u_q).$$

Example 10.3
Let us consider the term
$$T' \equiv f_5(x_1, f_2(a_2, x_1, x_3), a_3, x_1);$$

from the above definition, the sequence of the arguments of T' is $U' \equiv x_1 a_2 x_1 x_3 a_3 x_1$, hence we can write
$$T' \equiv T'(U') \equiv T'(x_1 a_2 x_1 x_3 a_3 x_1).$$

▯

Finally, let us adopt a useful convention concerning the notation of the elements of Cartesian products such as $\mathbf{S}^m$. Let $\mathbf{S}$ be a nonempty set and consider a natural number $m \geq 1$. We know that $\mathbf{S}^m$ is by definition the Cartesian product
$$\underbrace{\mathbf{S} \times \mathbf{S} \times \ldots \times \mathbf{S}}_{\text{m-times}};$$

hence $\mathbf{S}^m$ is the set of all m-tuples (ordered systems) $(s_1, s_2, \ldots, s_m)$, with $s_i \in \mathbf{S}$ $(1 \leq i \leq m)$. It is obvious that any m-tuple $(s_1, s_2, \ldots, s_m)$ is completely determined by the corresponding finite sequence $S \equiv s_1 s_2 \ldots s_m$ in $\mathbf{S}$. For this reason we shall identify in the following the elements of $\mathbf{S}^m$ with the finite sequences S in $\mathbf{S}$ having the property that $\lambda(S) = m$.

Now we shall introduce a natural operation with functions, called **composition**, which will play a basic role in the interpretation of terms.

DEFINITION 10.3   *Let* S, M *be arbitrary nonempty sets. If* $g : S^n \to$ M *is a function defined on the Cartesian product*

$$S^n = \underbrace{S \times \ldots \times S}_{n\text{-times}},$$

then the natural number $n \geq 1$ *is called the* **arity** *of* g *(with respect to* S*) and is denoted* $n = \alpha(g)$.

DEFINITION 10.4   *For each system of functions*

$$g : S^n \to M \text{ and } g_i : S^{l_i} \to M \ (1 \leq i \leq n)$$

we denote by

$$g(g_1, g_2, \ldots, g_n) : S^l \to M, \ l = l_1 + \ldots + l_n$$

the function defined as follows: if $S \in S^l$*, then we represent* S *in the form*

$$S \equiv S_1 S_2 \ldots S_n, \text{ with } S_i \in S^{l_i} \ (1 \leq i \leq n)$$

and put

$$g(g_1, g_2, \ldots, g_n)(S) = g[g_1(S_1), g_2(S_2), \ldots, g_n(S_n)].$$

The function $g(g_1, g_2, \ldots, g_n)$ *is called the* **composition** *of* g *with* $g_1, g_2, \ldots,$ g_n*; obviously,*

$$\alpha(g(g_1, g_2, \ldots, g_n)) = \alpha(g_1) + \alpha(g_2) + \ldots + \alpha(g_n) = l_1 + l_2 + \ldots + l_n = l.$$

Next we shall define a correspondence

$$T \longmapsto g_T$$

which associates to each term T a function g_T having in a certain sense the same structure as T.

PROPOSITION 10.1

If $I_0 = (D, \xi, \eta_0, \eta, \zeta_0, \zeta)$ *is an interpretation of literal symbols (of* X*), then there is a unique correspondence*

$$T \longmapsto g_T$$

which associates to each term T *a function* g_T *such that:*

   (a) *if* $T \equiv x_h$ *or* $T \equiv a_h$ *(i.e.,* T *is a variable or a constant), then* $g_T = \mathbf{1}$, *where* $\mathbf{1}$ *is the identity function on* D *(i.e.,* $\mathbf{1}(d) = d$ *for each* $d \in D$*);*

(b) *if $T \equiv f_h(T_1, T_2, \ldots, T_n)$, where $f_h \in \mathbf{F}, \alpha(f_h) = n$ and $T_1, T_2, \ldots, T_n$ are terms, then*

$$g_T = \eta(f_h)(g_{T_1}, g_{T_2}, \ldots, g_{T_n}).$$

Moreover, the correspondence $T \longmapsto g_T$ has the following additional property:

(c) *if $T \equiv T(U)$ is a term and U is the sequence of the arguments of T, then the arity of g_T coincides with the length of U $(\alpha(g_T) = \lambda(U))$.*

PROOF   The assertion can be established by induction using the recursive definition of terms. Indeed, let $T \equiv T(U)$ be a term; by the definition of terms, there is a sequence of formal words

$$(*) \qquad\qquad t_1(U_1), t_2(U_2), \ldots, t_r(U_r) \quad (r \leq 1)$$

with $t_r(U_r) \equiv T(U)$ such that, for each index i $(1 \leq i \leq r)$, $t_i(U_i)$ satisfies one of the conditions:

(1) $t_i(U_i) \equiv x_h$ or $t_i(U_i) \equiv a_h$;

(2) there is some functor $f_h \in \mathbf{F}$, with $\alpha(f_h) = k$ and there are k terms $t'_1, t'_2, \ldots, t'_k$, preceding t_i in $(*)$, such that

$$t_i \equiv f_h(t'_1, t'_2, \ldots, t'_k).$$

Now we shall define the functions g_{t_i} inductively. For $i = 1$, the term $t_1(U_1)$ is necessarily a variable or a constant. In this case we put

$$g_{t_1} = \mathbf{1},$$

so that property (a) holds. Moreover, in this case, the sequence U_1 is reduced to a single element $u_1 \in \mathcal{U}$ $(u_1 \equiv x_h$ or $u_1 \equiv a_h)$, hence $\lambda(U_1) = 1$. On the other hand, it is obvious that $\alpha(g_{t_1}) = \alpha(\mathbf{1}) = 1$. Thus, for $i = 1$, the assertion (c) holds.

Suppose $i > 1$ and assume that functions g_{t_j} $(j < i)$ are already defined and have one of properties (a), (b) and also property (c).

In this case, the term $t_i \equiv t_i(U_i)$ satisfies one of the conditions (1), (2) above. If it satisfies the condition (1), then we put (as in the case $i = 1$) $g_{t_i} = \mathbf{1}$, and properties (a),(c) hold.

If $t_i \equiv t_i(U_i)$ satisfies the condition (2), we put

$$g_{t_i} = \eta(f_h)(g_{t'_1}, g_{t'_2}, \ldots, g_{t'_k})$$

and property (b) holds. In order to prove that g_{t_i} has the property (c), let us write

$$t'_1 \equiv t'_1(U'_1), \ldots, t'_k \equiv t'_k(U'_k).$$

Taking into account that $t_i(U_i) = f_h(t'_1(U'_1), \ldots, t'_k(U'_k))$ it follows that $U_i \equiv U'_1 \ldots U'_k$ and $\lambda(U_i) = \lambda(U'_1) + \ldots + \lambda(U'_k)$. On the other hand, by the inductive assumption,

$$\alpha(g_{t'_1}) = \lambda(U'_1), \ldots, \alpha(g_{t'_k}) = \lambda(U'_k),$$

hence $\alpha(g_{t_i}) = \alpha(g_{t'_1}) + \ldots + \alpha(g_{t'_k}) = \lambda(U'_1) + \ldots + \lambda(U'_k)$ therefore, $\alpha(g_{t_i}) = \lambda(U_i)$ and property (c) is proved. From the induction principle it follows that all functions g_{t_i} $(1 \le i \le r)$, in particular, the function $g_T = g_{t_r}$, are defined and all have property (c) and one of the properties (a), (b).

The uniqueness of the correspondence $T \longmapsto g_T$ follows easily by routine arguments. Indeed, suppose that

$$T \longmapsto g'_T \text{ and } T \longmapsto g''_T$$

are two correspondences with properties (a), (b). Consider an arbitrary term T and let

$$t_1, t_2, \ldots, t_r \ (r \ge 1),$$

$t_r \equiv T$, be a formal construction of T. The term t_1 is necessarily a variable or a constant and by property (a) $g'_{t_1} = g''_{t_1} = 1$. Taking $i > 1$ and assuming that $g'_{t_j} = g''_{t_j}$ for all $j < i$, we get easily (using the definition of t_i and properties (a), (b)), that $g'_{t_i} = g''_{t_i}$. From the induction principle it follows that $g'_{t_i} = g''_{t_i}$ for all i $(1 \le i \le r)$. In particular $g'_T = g'_{t_r} = g''_{t_r} = g''_T$. Hence $g'_T = g''_T$ for each term T and the two correspondences coincide. ∎

Let us remark that each term T regarded as a formal word in $\mathbf{X}$, is a finite sequence in $\mathbf{X}$ containing only variables, constants, functors, parentheses and commas. If we denote by $\mathcal{Z}$ the set $\mathbf{V} \bigcup \mathbf{F_0} \bigcup \mathbf{F} \bigcup \{(\} \bigcup \{)\} \bigcup \{,\}$, then $T \equiv \sigma_1 \sigma_2 \ldots \sigma_m$ is a finite sequence in $\mathcal{Z}$. The formal word $\tilde{T} \equiv \tilde{\sigma}_1 \tilde{\sigma}_2 \ldots \tilde{\sigma}_m$ obtained from T by putting the symbol $\square$ instead of each variable or constant is called the **skeleton** of T. If T' and T'' are two terms with the same skeleton, then we shall say that T' and T'' are **congruent**.

The congruence of terms is an equivalence relation in the set of terms. (i.e., it is reflexive, symmetric and transitive).

The following result can be proved by induction following the same idea as in Proposition 10.1.

PROPOSITION 10.2

Let T, T' be terms. If T' and T'' are congruent, then $g_{T'} = g_{T''}$.

REMARK 10.1   From the above proposition it follows that the function g_T depends in fact only on the skeleton $\tilde{T}$ of T. ∎

If $T \equiv T(U)$ is a term and $U = u_1 u_2 \ldots u_q$ is the sequence of the arguments of T, then obviously, for each i $(1 \leq i \leq q)$ the element $\hat{u}_i$ is the interpretation of u_i. Therefore, we can say that $\hat{U} \equiv \hat{u}_1 \hat{u}_2 \ldots \hat{u}_q$ is the interpretation of U.

DEFINITION 10.5 *Let* $I_0 = (\mathcal{D}, \xi, \eta_0, \eta, \zeta_0, \zeta)$ *be an interpretation of the literal symbols (of* $\mathbf{X}$*). If* $T \equiv T(U)$ *is a term, then the element* $I(T) = g_T(\hat{U}) \in \mathcal{D}$ *is called the* **interpretation of** T. *The correspondence*

$$I : T \longmapsto I(T)$$

which associates to each term T *the interpretation* $I(T)$ *of* T *is called the* **interpretation of terms generated by** I_0.

REMARK 10.2 The above definition shows that the interpretation I of terms depends only on the domain $\mathcal{D}$ of interpretation, and on the interpretations ξ, η_0, η of variables, constants and functors. ∎

Now let $T \equiv T(U)$ be a term containing the variable x, that is $U \equiv U(x)$; in such a case we shall sometimes write $T(U(x)) \equiv T(x)$. If $T(x)$ is given and $t \equiv t(V)$ is an arbitrary term, where $V \equiv v_1 v_2 \ldots v_p$ is the sequence of the arguments of t, then we can construct the formal word $T(t|x)$ obtained from $T(x)$ by putting the term $t(V)$ instead of each occurrence of x in $T(x)$. It is easy to prove by inductive arguments the following result.

PROPOSITION 10.3
If $T \equiv T(x)$ *and* $t \equiv t(V)$ *are terms, then* $T(t|x)$ *is also a term. If* $U(x)$ *is the sequence of the arguments of* T, *then the sequence of the arguments of* $T(t|x)$ *coincides obviously with the finite sequence* $U(V)$ *in* $\mathcal{U}$.

The following technical result can also be proved without difficulty by inductive considerations.

PROPOSITION 10.4
If $T \equiv T(x)$ *and* $t \equiv t(V)$ *are terms, then*

$$g_{T(t|x)}(\hat{U}(\hat{V})) = g_{T(x)}(\hat{U}(g_t(\hat{V}))).$$

Now we shall pass to the interpretation of formulas. For this purpose we begin with some preliminary considerations.

As in the case of terms, for each formula $F \in \mathcal{L}(\mathbf{X})$ we introduce the sequence U_F of the arguments of F. We recall that every formula F is actually a formal word, i.e., a finite sequence in **X**

$$F \equiv \sigma_1 \sigma_2 \ldots \sigma_m \ (m \geq 1),$$

having special properties (Definitions 9.6, 9.7). This sequence may contain, among other symbols, individual variables, individual constants, squares and clips. If the formula $F \equiv \sigma_1 \sigma_2 \ldots \sigma_m$ is provided with clips, then each clip connects two distinct squares (i.e., having distinct positions (occurrences) in F). The individual variables contained in the sequence $F \equiv \sigma_1 \sigma_2 \ldots \sigma_m$ are called the free variables of F. In abbreviated notations of formulas could exist bound variables which actually are not contained in the formula F. For instance, in the formula $G \equiv \exists x \ (A(x,y))$, the variable x is bound, while y is a free variable of G. If we write the formula G without abbreviation using only the symbols of **X**, then

$$G \equiv \exists \overline{\square \ (A(\square,y))}$$

and is apparent that y is a free variable of G and the variable x does not exist in G; however, in the subformula $A(x,y)$ of G, the individual variable x is a free one.

Let $F \equiv \sigma_1 \sigma_2 \ldots \sigma_m$ be a formula (possibly with clips, which, for simplicity, are not marked in this notation) and let us denote by $U_F \equiv u_1 u_2 \ldots u_q$ the sequence of all individual variables and constants contained (possibly with repetitions) in the sequence $F \equiv \sigma_1 \sigma_2 \ldots \sigma_m$ and written in the order of their occurrence in F. Thus, $U \equiv u_1 u_2 \ldots u_q$ is a finite sequence in $\mathcal{U}$ which will be called the **sequence of the arguments** of F. In the following we shall often use the notation $F \equiv F(U) \equiv F(U_F)$, where $U \equiv U_F$ is the sequence of the arguments of F.

In the first part of this section we have seen that for any term $T \equiv T(U)$ the sequence U of the arguments of T is always an authentic one, i.e., is nonempty $(\lambda(U) \geq 1)$. By contrast with the situation of terms, the sequence of the arguments of a formula F might be empty. For instance, if F is a propositional variable, or if F does not contain individual constants and all individual variables are quantified, then the sequence of the arguments of F is empty (F does not contain elements of $\mathcal{U}$).

Consider, for example, the (elementary) formula $G \equiv P_2(x_3)$, where $P_2 \in$ **F** is a predicator with $\alpha(P_2) = 1$. It is clear that the sequence of the arguments of G is $V \equiv x_3$, $(\lambda(V) = 1)$ and we can write $G \equiv G(V) \equiv G(x_3)$. Applying to G the quantifier $\exists$ with respect to x_3 we get the formula $F \equiv \exists x_3(G(x_3)) \equiv \exists x_3(P_2(x_3))$ which does not contain either variables or constants. In such a case we shall say that the sequence of the arguments is **the null sequence**.

If $F \equiv F(U)$ is a formula and x an individual variable contained in U_F, then we shall write $U_F \equiv U_F(x)$ and sometimes the formula F will be denoted by $F(U_F(x))$ or even by $F(x)$. In such a case, for each term

$t = t(V)$ we shall denote by $F(t|x)$ the formal word obtained from $F(x) \equiv \sigma_1\sigma_2 \ldots \sigma_m$ by putting $t(V)$ instead of each occurrence of x in $F(x)$. We have seen that if $T(x) \equiv T(U(x))$ is a term, then the formal word $T(t|x)$ (having $U(V)$ as sequence of arguments) is also a term.

The corresponding assertion for formulas is also true and can be proved (as in the case of terms) by induction using the recursive definition of formulas and the similar result for terms.

PROPOSITION 10.5

If $F \equiv F(x)$ *is a formula and if* $t \equiv t(V)$ *is a term, then* $F(t|x)$ *is also a formula. Moreover, if* $U_F \equiv U_F(x)$ *is the sequence of the arguments of* F, *then* $U_F(V)$ *is the sequence of the arguments of* $F(t|x)$.

REMARK 10.3   If the formula $F \equiv F(U_F)$ does not contain the variable x, then obviously $F(t|x) \equiv F$ and $U_F(V) \equiv U_F$.   ∎

Now analyze the connection between the sequences U_F and U_G of the arguments of formulas $F(x)$ and $G \equiv \exists x(F(x))$. Suppose, for instance, that $P \in \mathbf{F}$ is a predicator with $\alpha(P) = 10$ and consider the (elementary) formula

$$A \equiv P(x_2, a_4, x_1, a_2, x_2, x_3, x_1, x_4, a_3, a_1).$$

We see that the sequence U_A of the arguments of A is

$$U_A \equiv x_2 a_4 x_1 a_2 x_2 x_3 x_1 x_4 a_3 a_1$$

and if we fix our attention on the variable x_1, then we can write

$$U_A \equiv U_1 x_1 U_2 x_1 U_3,$$

where $U_1 \equiv x_2 a_4$, $U_2 \equiv a_2 x_2 x_3$, $U_3 \equiv x_4 a_3 a_1$. Taking into account that x_1 is contained in U_A we can write $U_A \equiv U_A(x_1)$, $A \equiv A(x_1) \equiv A(U_A(x_1))$; applying to A the quantifier $\exists$ with respect to x_1 we get a new formula $F \equiv \exists x_1(A(x_1))$ having $U_F \equiv x_2 a_4 a_2 x_2 x_3 x_4 a_3 a_1$ as sequence of arguments, which can obviously be written in the form

$$U_F \equiv U_1 U_2 U_3.$$

Thus, we remark that passing from $A(x_1)$ to $\exists x_1(A(x_1))$ the sequence U_F is obtained from U_A by the operation

$$U_A \equiv U_1 x_1 U_2 x_1 U_3$$
$$\downarrow$$
$$U_F \equiv U_1 U_2 U_3$$

called "dropping the variable x_1".

This remark is true in the general case. Let $\mathbf{S}$ be an arbitrary nonempty set and consider a natural number $n \geq 1$. If $\omega = \{i_1, i_2, \ldots, i_k\}$ is a set of natural numbers such that $1 \leq i_1 < i_2 < \ldots < i_k \leq n$, then each finite sequence $S \equiv s_1 s_2 \ldots s_n \in \mathbf{S}^n$ can be written $S = S_1 s_{i_1} S_2 s_{i_2} \ldots s_{i_k} S_{k+1}$, where some of sequences $S_1, S_2, \ldots, S_{k+1}$ might be null sequences. It is clear that $\lambda(S_1) + \ldots + \lambda(S_{k+1}) = \lambda(S) - k = l \geq 0$. Now we denote $\delta_\omega S = S_1 S_2 \ldots S_{k+1} \in \mathbf{S}^l$. The operation $S \longmapsto \delta_\omega S$ defined for sequences in $\mathbf{S}^n$ is called **dropping operation**. This operation will be used in the definition of the next operation with functions.

Let $h : \mathcal{D}^n \to \Gamma$ $(\Gamma = \{0, 1\})$ be a given function and $\omega = \{i_1, i_2, \ldots, i_k\}$, $1 \leq i_1 < i_2 < \ldots < i_k \leq n$ be a fixed subset of $\{1, 2, \ldots, n\}$. If $D \equiv d_1 d_2 \ldots d_n$ is a sequence in $\mathcal{D}^n$, then D can be uniquely represented in the form $D = D_1 d_{i_1} D_2 d_{i_2} \ldots d_{i_k} D_{k+1}$ as above, hence $\delta_\omega D \equiv D_1 D_2 \ldots D_{k+1} \in \mathcal{D}^l$, $l = n - k$. Now starting from h we define two functions, $\max_\omega h$ and $\min_\omega h$, by the equalities

$$(\max_\omega h)(D_1 D_2 \ldots D_{k+1}) = \max_{d \in \mathcal{D}} h(D_1 d D_2 d \ldots d D_{k+1}),$$

$$(\min_\omega h)(D_1 D_2 \ldots D_{k+1}) = \min_{d \in \mathcal{D}} h(D_1 d D_2 d \ldots d D_{k+1}).$$

It is clear that in this way we get the functions

$$\max_\omega h : \mathcal{D}^l \to \Gamma,$$

$$\min_\omega h : \mathcal{D}^l \to \Gamma.$$

In further considerations we shall need two more operations with functions, namely $\overline{h}$ and $h \vee k$.

If $h : \mathcal{D}^l \to \Gamma$ is a given function, then the function $\overline{h} : \mathcal{D}^l \to \Gamma$ is defined by the equality $\overline{h}(D) = \overline{h(D)}$ for all $D \in \mathcal{D}^n$. If $h : \mathcal{D}^p \to \Gamma$ and $k : \mathcal{D}^q \to \Gamma$ are given functions and if we denote $p + q = n$, then the function $h \vee k : \mathcal{D}^n \to \Gamma$ is defined as follows: for each $D \in \mathcal{D}^n$ we write (uniquely) $D = D'D''$, where $D' \in \mathcal{D}^p, D'' \in \mathcal{D}^q$, and put

$$(h \vee k)(D) = (h \vee k)(D'D'') = \max\{h(D'), k(D'')\}.$$

Now we are ready to construct a special correspondence $F \longmapsto h_F$ which associates to each formula $F \equiv F(U_F) \in \mathcal{L}(\mathbf{X})$ a uniquely determined function $h_F : \mathcal{D}^m \to \Gamma$, where $m = \lambda(U_F)$. The function $\zeta(P_j)$, where $P_j \in \mathbf{P}$, will be denoted often by $\hat{P}_j$.

PROPOSITION 10.6

If $I_0 = (\mathcal{D}, \xi, \eta_0, \eta, \zeta_0, \zeta)$ is an interpretation of the literal symbols of $\mathbf{X}$, then there exists a unique correspondence

$$F \longmapsto h_F$$

which associates to each formula $F \equiv F(U_F)$ a function $h_F : \mathcal{D}^m \to \Gamma$, where $m = \lambda(U_F)$, having the following properties:

(a) if $F \equiv p_j \in \mathbf{P}_0$, then h_F is the constant function

$$h_F = \xi_0(p_j) = \hat{p}_j \in \Gamma;$$

if $F \equiv P_j(T_1, T_2, \ldots, T_n)$ where $P_j \in \mathbf{P}$, with $\alpha(P_j) = n$, and $T_1, T_2, \ldots, T_n$ are terms, then

$$h_F = \zeta(P_j)(g_{T_1}, g_{T_2}, \ldots, g_{T_n}) = \hat{P}_j(g_{T_1}, g_{T_2}, \ldots, g_{T_n}).$$

(b) if $F \equiv \neg(G)$, where $F \equiv F(U)$ (and obviously $G \equiv G(U)$), then

$$h_F = \overline{h_G};$$

(c) if $F \equiv L \vee M$, where $L \equiv L(U'), M \equiv M(U'')$ (and obviously $F \equiv F(U)$ with $U \equiv U'U''$), then

$$h_F = h_L \vee h_M;$$

(d) if $F \equiv \exists x(A(x))$, $F \equiv F(U_F)$, with $U_F \equiv \delta_\omega U_A$, where $\omega = \{i_1, i_2, \ldots, i_k\}$ is the set of occurrences of x in $U_A(x)$, then

$$h_F = \max_\omega h_A, \quad h_F : \mathcal{D}^m \to \Gamma, \quad m = \lambda(U_F).$$

More precisely, if $A \equiv A(x) \equiv A(U_A(x))$, then denoting $\omega = \omega(x; U_A(x))$ it follows that

$$U_A(x) \equiv U_1 x U_2 x \ldots x U_{k+1}$$

and

$$U_F \equiv \delta_\omega U_A \equiv U_1 U_2 \ldots U_{k+1}.$$

Therefore,

$$h_F(\widehat{\delta_\omega U_A}) = h_F(\hat{U}_1 \hat{U}_2 \ldots \hat{U}_{k+1}) = \max_{d \in \mathcal{D}} h_A(\hat{U}_1 d \hat{U}_2 d \ldots d \hat{U}_{k+1}).$$

It is clear that

$$m = \lambda(\delta_\omega U_A) = \lambda(U_1) + \ldots + \lambda(U_{k+1}) = \lambda(\delta_\omega \hat{U}_A) = \alpha(h_F).$$

The proof of this proposition is similar to the one given in Proposition 10.1 for terms (using inductive arguments, the recursive definition of formulas and the correspondence $T \longmapsto g_T$).

DEFINITION 10.6   Let $I_0 = (\mathcal{D}, \xi, \eta_0, \eta, \zeta_0, \zeta)$ be an interpretation of the literal symbols of **X**. If $F \equiv F(U_F) \in \mathcal{L}(\mathbf{X})$ is a formula, then the element $I(F) \in \Gamma$ defined by the equality

$$I(F) = h_F(\hat{U}_F)$$

is called **the interpretation of** F. The correspondence $F \longmapsto I(F)$, which associates to each formula $F \in \mathcal{L}(\mathbf{X})$ the element $I(F) \in \Gamma$, is called **the interpretation generated by** I_0 **of the formulas**.

The correspondence $F \longmapsto h_F$ has the important property below; the proof is analogous to that for the similar property for $T \longmapsto g_T$.

PROPOSITION 10.7

Let $I_0 = (\mathcal{D}, \xi, \eta_0, \eta, \zeta_0, \zeta)$ be an interpretation of the literal symbols of **X**. If $F \equiv F(x) \equiv F(U_F(x))$ is a formula containing x and $t \equiv t(V)$ is a term free for x in $F(x)$, then

$$h_{F(t|x)}(\hat{U}_F(\hat{V})) = h_{F(x)}(\hat{U}_F(g_t(\hat{V}))).$$

From this proposition we get the following consequence concerning the interpretation I of formulas.

PROPOSITION 10.8

Let I be the interpretation of $\mathcal{L}(\mathbf{X})$ generated by $I_0 = (\mathcal{D}, \xi, \eta_0, \eta, \zeta_0, \zeta)$. If $F \equiv F(x) \equiv F(U_F(x))$ is a formula containing x and $t \equiv t(V)$ is a term free for x in $F(x)$, then

$$I(F(t|x)) = h_F(\hat{U}_F(I(t))).$$

PROOF   Let $F \equiv F(x) \equiv F(U_F(x))$ be a formula containing x and $t \equiv t(V)$ be a term free for x in $F(x)$. Taking into account the definition of I and the Proposition 10.7 we have $I(F(t|x)) = h_{F(t|x)}(\hat{U}_F(\hat{V})) = h_F(\hat{U}_F(g_t(\hat{V}))) = h_F(\hat{U}_F(I(t)))$.

In the last equality we have used the definition $g_t(\hat{V}) = I(t)$.   ∎

Example 10.4

Let us return to the interpretation I_0 of literal symbols considered in Example 10.1 and as an illustration let us find the interpretation of the terms

$$T_1 \equiv f_1(x_2, a_1) \,;\; T_2 \equiv f_2(x_1, a_2) \,;\; T_3 \equiv f_1(f_2(x_2, a_1), f_1(x_1, a_2))$$

and of the formulas

$$F \equiv P_1(T_2, T_1) \,;\; G \equiv P_2(T_1, T_2, T_3) \,;\; L \equiv \exists x_1(G).$$

The sequences of the arguments of T_1, T_2, T_3 are

$$U_{T_1} \equiv x_2 a_1 \; ; \; U_{T_2} \equiv x_1 a_2 \; ; \; U_{T_3} \equiv x_2 a_1 x_1 a_2.$$

The sequences of arguments of F, G are

$$U_F \equiv x_1 a_2 x_2 a_1 \; ; \; U_G \equiv x_2 a_1 x_1 a_2 x_2 a_1 x_1 a_2.$$

In order to write the sequence U_L of the arguments of L let us remark that

$$U_G \equiv U'_G x_1 U''_G x_1 U'''_G \, ,$$

where $U'_G \equiv x_2 a_1$, $U''_G \equiv a_2 x_2 a_1$, $U'''_G \equiv a_2$. So that $\omega(x_1; G) = \{3, 7\}$ and

$$U_L \equiv \delta_\omega U_G \equiv U'_G U''_G U'''_G \, .$$

Now we have to compute the functions $g_{T_1}, g_{T_2}, g_{T_3}, h_F, h_G, h_L$. Using the definitions of these functions we get:

$$g_{T_1} = \hat{f}_1 \, , \; g_{T_2} = \hat{f}_2 \, , \; g_{T_3} = \hat{f}_1(\hat{f}_2, \hat{f}_1);$$

similarly

$$h_F = \hat{P}_1(\hat{f}_2, \hat{f}_1) \, , \; h_G = \hat{P}_2(g_{T_1}, g_{T_2}, g_{T_3}) \, , \; h_L = \max_\omega h_G,$$

where

$$\hat{f}_1 = \eta(f_1) \, , \; \hat{f}_2 = \eta(f_2) \, , \; \hat{P}_1 = \zeta(P_1) \, , \; \hat{P}_2 = \zeta(P_2).$$

Now we are ready to compute the interpretations of the above terms and formulas. From the corresponding definitions we have

$$I(T_1) = g_{T_1}(\hat{U}_{T_1}) = \hat{f}_1(\hat{x}_2, \hat{a}_1) = \hat{f}_1(2, 4) = 2 + 4 = 6;$$

$$I(T_2) = g_{T_2}(\hat{U}_{T_2}) = \hat{f}_2(\hat{x}_1, \hat{a}_2) = \hat{f}_2(1, 0) = 1 \cdot 0 = 0;$$

$$I(T_3) = g_{T_3}(\hat{U}_{T_3}) = \hat{f}_1(\hat{f}_2, \hat{f}_1)(\hat{U}_{T_3});$$

writing $U_{T_3} \equiv U'_{T_3} U''_{T_3}$ where $U'_{T_3} \equiv x_2 a_1$, $U''_{T_3} \equiv x_1 a_2$, it follows $\hat{U}_{T_3} \equiv \hat{U}'_{T_3} \hat{U}''_{T_3}$, hence

$$I(T_3) = \hat{f}_1(\hat{f}_2, \hat{f}_1)(\hat{U}'_{T_3} \hat{U}''_{T_3}) = \hat{f}_1(\hat{f}_2(\hat{U}'_{T_3}), \hat{f}_1(\hat{U}''_{T_3})) =$$
$$= \hat{f}_1(\hat{f}_2(\hat{x}_2, \hat{a}_1), \hat{f}_1(\hat{x}_1, \hat{a}_2)) = \hat{f}_1(\hat{f}_2(2, 4), \hat{f}_1(1, 0)) =$$
$$= \hat{f}_1(8, 1) = 8 + 1 = 9 \, .$$

Now, let us pass to the interpretations of formulas.

$$I(F) = h_F(\hat{U}_F) = \hat{P}_1(\hat{f}_2, \hat{f}_1)(\hat{U}'_F \hat{U}''_F) = \hat{P}_1(\hat{f}_2(\hat{U}'_F), \hat{f}_1(\hat{U}''_F)) \, ,$$

where $U_F' \equiv x_1 a_2$, $U_F'' \equiv x_2 a_1$.

Therefore

$$I(F) = \hat{P}_1(\hat{f}_2(\hat{x}_1, \hat{a}_2), \hat{f}_1(\hat{x}_2, \hat{a}_1)) = \hat{P}_1(0, 6) = 1.$$

Similarly,

$$I(G) = \hat{P}_2(g_{T_1}, g_{T_2}, g_{T_3})(\hat{x}_2 \hat{a}_1 \hat{x}_1 \hat{a}_2 \hat{x}_2 \hat{a}_1 \hat{x}_1 \hat{a}_2) =$$
$$= \hat{P}_2(g_{T_1}(\hat{x}_2, \hat{a}_1), g_{T_2}(\hat{x}_1, \hat{a}_2), g_{T_3}(\hat{x}_2, \hat{a}_1, \hat{x}_1, \hat{a}_2)) =$$
$$= \hat{P}_2(6, 0, 9) = 0,$$
$$I(L) = h_L(\hat{U}_L) = (\max_\omega h_G)(\hat{U}_L) =$$
$$= \max_\omega h_G(\delta_\omega \hat{U}_G(d)) = \max_{d \in \mathbf{N}} h_G(\hat{U}_G(d)),$$

where $U_G \equiv U_G(x_1)$, $\hat{U}_G(d) \equiv \hat{U}_G' d \hat{U}_G'' d \hat{U}_G'''$. It follows that

$$I(L) = \max_{d \in \mathbf{N}} h_G(\hat{U}_G' d \hat{U}_G'' d \hat{U}_G''')$$
$$= \max_{d \in \mathbf{N}} h_G(\hat{x}_2 \hat{a}_1 d \hat{a}_2 \hat{x}_2 \hat{a}_1 d \hat{a}_2) =$$
$$= \max_{d \in \mathbf{N}} \hat{P}_2(g_{T_1}, g_{T_2}, g_{T_3})(24d024d0) =$$
$$= \max_{d \in \mathbf{N}} \hat{P}_2(g_{T_1}(2, 4), g_{T_2}(d, 0), g_{T_3}(24d0)) =$$
$$= \max_{d \in \mathbf{N}} \hat{P}_2(6, 0, \hat{f}_1(\hat{f}_2, \hat{f}_1)(24d0));$$

taking into account that $\hat{f}_1(\hat{f}_2, \hat{f}_1)(24d0) = \hat{f}_1(\hat{f}_2(2, 4), \hat{f}_1(d, 0)) = \hat{f}_1(8, d) = 8 + d$ we get

$$I(L) = \max_{d \in \mathbf{N}} \hat{P}_2(6, 0, 8 + d).$$

Now it is easy to see that

$$\hat{P}_2(6, 0, 8 + d) = \begin{cases} 1, & \text{if } d = 4 \\ 0, & \text{otherwise} \end{cases}$$

and it follows $\max_{d \in \mathbf{N}} \hat{P}_2(6, 0, 8 + d) = 1$, hence $I(L) = 1$.   ▯

REMARK 10.4   Taking into account that in $I_0 = (\mathcal{D}, \xi, \eta_0, \eta, \zeta_0, \zeta)$ the set $\mathcal{D}$ and the functions $\xi, \eta_0, \eta, \zeta_0, \zeta$ are arbitrary, it follows that the set of all interpretations I of $\mathcal{L}(\mathbf{X})$ is infinite.   ∎

10.2   Semantic Deduction in Predicate Logic

In this section, using the concept of interpretation, we shall introduce the truth structure on $\mathcal{L}$ in the semantic version. Thus we shall define universally true (valid) formulas, universally false (inconsistent) formulas, consistent formulas and contingent formulas; we shall define also the logical equivalence of formulas and the notion of logical (semantical) consequence, etc.

DEFINITION 10.7   *A formula $F \in \mathcal{L}$ is called*

(a) **valid** *or* **universally true**, *if $I(F) = 1$ for each interpretation I;*

(b) **inconsistent**, *if $I(F) = 0$ for each interpretation I;*

(c) **consistent**, *if $I(F) = 1$ for at least one interpretation I;*

(d) **contingent**, *if there are interpretations I_1 and I_2 such that $I_1(F) = 1$ and $I_2(F) = 0$.*

From this definition we get an immediate consequence.

PROPOSITION 10.9
Let F be a formula in $\mathcal{L}$.

(a) F is inconsistent if and only if $\neg F$ is valid;

(b) F is contingent if and only if $\neg F$ is contingent;

(c) F is consistent if and only if F is valid or contingent.

In the following, the assertion "F is a valid formula" will be denoted by $\models F$.

DEFINITION 10.8   *Let F, G be formulas.*

(a) We shall say that F, G are **logically equivalent** *and will write $F \approx G$ if $F \Leftrightarrow G$ is a valid formula, for short*

$$F \approx G \text{ if and only if } \models F \Leftrightarrow G;$$

(b) The formula G is said to be a **logical consequence** *of the formula F and we write $F \models G$, if $F \Rightarrow G$ is a valid formula;*

(c) The formula G is said to be a logical consequence of formulas $F_1, F_2, \ldots, F_k$ and we write $F_1, F_2, \ldots, F_k \models G$ if $F_1 \wedge F_2 \wedge \ldots \wedge F_k \models G$.

PROPOSITION 10.10

(a) $F \approx G$ if and only if $I(F) = I(G)$ for each interpretation I;

(b) $F \models G$ if and only if $I(F) \leq I(G)$ for each interpretation I;

(c) $F_1, \ldots, F_k \models G$ if and only if G is true in all interpretations in which $F_1, F_2, \ldots, F_k$ are all true.

PROOF   The assertion (a) is obvious and the assertion (c) follows from (b). In order to prove (b), let us consider two formulas F, G such that $F \models G$, i.e., $\models F \Rightarrow G$. It follows that for any interpretation I we have $I(F \Rightarrow G) = 1$. Therefore,

$$I = I(F \Rightarrow G) = I(\neg F \vee G) = \overline{I(F)} + I(G).$$

If $I(F) = 1$, then $\overline{I(F)} = 0$ and from the above equality we get $I(G) = 1$, hence $I(F) = I(G)$. If $I(F) = 0$, then obviously $I(F) \leq I(G)$.

Conversely, suppose that $I(F) \leq I(G)$ for an interpretation I and let us compute $I(F \Rightarrow G) = \overline{I(F)} + I(G)$. If $I(F) = 1$, then from our inequality it follows $I(G) = 1$, hence $I(F \Rightarrow G) = 1$. If $I(F) = 0$, then $\overline{I(F)} = 1$ and also $I(F \Rightarrow G) = 1$. So $I(F \Rightarrow G) = 1$ in each interpretation if and only if $I(F) \leq I(G)$ in each interpretation. This means that $F \models G$ if and only if $I(F) \leq I(G)$ for each interpretation I.   ∎

The following result is immediate:

PROPOSITION 10.11
If F, G, H are formulas in $\mathcal{L}(\mathbf{X})$, then all formulas listed below are valid:

(a) $F \vee F \Rightarrow F$;

(b) $F \Rightarrow F \vee G$;

(c) $F \vee G \Rightarrow G \vee F$;

(d) $(F \Rightarrow G) \Rightarrow ((H \vee F) \Rightarrow (H \vee G))$.

The inference rule MP has a correspondent in the semantic of predicate logic.

PROPOSITION 10.12
Let F, G be formulas in $\mathcal{L}(\mathbf{X})$. If F and $F \Rightarrow G$ are valid formulas, then the formula G is also valid, i.e.,

$$\text{If} \quad \models F \quad \text{and} \quad \models F \Rightarrow G,$$

$$\text{then} \quad \models G.$$

PROOF   The assertion is obvious; indeed suppose that F and $F \Rightarrow G$ are valid formulas, i.e., $I(F) = 1$ and $I(F \Rightarrow G) = 1$ for each interpretation I. It follows

$$1 = I(F \Rightarrow G) = \overline{I(F)} + I(G) = I(G)$$

(because $\overline{I(F)} = 0$). Therefore $I(G) = 1$ in each interpretation I, hence G is valid. $\blacksquare$

Next, we shall point out the connection existing between the formulas and tautologies in propositional logic and formulas and valid formulas in predicate logic.

For this purpose let us consider a formula Y belonging to the language $\mathcal{L}(\mathbf{X_0})$ of propositional logic which contains all propositional variables X_1, ..., X_N or only a subset of them; consider a fixed arbitrary sequence of formulas in $\mathcal{L}(\mathbf{X})$

$$F_1, F_2, \ldots, F_N$$

belonging to the language $\mathcal{L}(\mathbf{X})$ of predicate logic. If in Y we replace each propositional variable X_i contained in Y by the formula F_i ($1 \leq i \leq N$), then we obtain a formal word in $\mathbf{X}$ denoted by Y'. So we get a correspondence

$$Y \mapsto Y'$$

which associates each formula $Y \in \mathcal{L}(\mathbf{X_0})$ with a formal word Y' in $\mathbf{X}$. We shall prove that for each $Y \in \mathcal{L}(\mathbf{X_0})$, the formal word Y' is a formula in $\mathcal{L}(\mathbf{X})$. Therefore, the correspondence $Y \mapsto Y'$ is a function defined on $\mathcal{L}(\mathbf{X_0})$ with values in $\mathcal{L}(\mathbf{X})$.

PROPOSITION 10.13
Let Y be a formula in $\mathcal{L}(\mathbf{X_0})$: if $F_1, F_2, \ldots, F_N$ are formulas in $\mathcal{L}(\mathbf{X})$, then the formal word Y' is a formula in $\mathcal{L}(\mathbf{X})$.

PROOF   Taking into account that $Y \in \mathcal{L}(\mathbf{X_0})$, there exists a formal construction in $\mathbf{X_0}$,

$$Y_1, Y_2, \ldots, Y_r \qquad (r \geq 1),$$

with $Y_r \equiv Y$, such that for each index i ($1 \leq i \leq r$), Y_i satisfies one of the conditions:

(a) Y_i is a propositional variable;

(b) $Y_i \equiv \neg(Y_j)$ with $j < i$;

(c) $Y_i \equiv (Y_h) \vee (Y_k)$ with $h, k < i$.

Consider now the sequence

$$Y_1', Y_2', \ldots, Y_r'$$

where the formal words Y_i' $(1 \leq i \leq r)$ are constructed by the above correspondence $Y \mapsto Y'$. We shall show by induction that for each i $(1 \leq i \leq r)$, Y_i' is a formula in $\mathcal{L}(\mathbf{X})$.

For $i = 1$, the formula $Y_1 \in \mathcal{L}(\mathbf{X}_0)$ is necessarily a propositional variable X_l (condition (a)). It follows that $Y_1' \equiv F_l$ is a formula in $\mathcal{L}(\mathbf{X})$.

Suppose $i > 1$ and assume that, for all $j < i$, Y_j' are formulas in $\mathcal{L}(\mathbf{X})$. The formula Y_i satisfies one of conditions (a), (b), (c). If Y_i satisfies condition (a), then Y_i' is a formula in $\mathcal{L}(\mathbf{X})$ as above. If $Y_i \equiv \neg(Y_j)$ with $j < i$, then by the inductive assumption Y_j' is a formula in $\mathcal{L}(\mathbf{X})$, hence $Y_i' \equiv \neg(Y_j')$ is a formula in $\mathcal{L}(\mathbf{X})$. Finally, if $Y_i \equiv (Y_h) \vee (Y_k)$ with $h, k < i$, then Y_h', Y_k' are formulas in $\mathcal{L}(\mathbf{X})$ by the inductive assumption, hence $Y_i' \equiv (Y_h') \vee (Y_k')$ is a formula in $\mathcal{L}(\mathbf{X})$.

Thus, from the principle of mathematical induction we get $Y_i' \in \mathcal{L}(\mathbf{X})$ for all i $(1 \leq i \leq r)$; in particular $Y' \equiv Y_r' \in \mathcal{L}(\mathbf{X})$.   ∎

PROPOSITION 10.14
If $Y \in \mathcal{L}(\mathbf{X}_0)$ is a tautology, then Y' is a valid formula in $\mathcal{L}(\mathbf{X})$.

PROOF   Let $Y \in \mathcal{L}(\mathbf{X}_0)$ be a tautology; from the completeness metatheorem of propositional logic it follows that Y is a theorem. Therefore, there exists a formal proof

$$Y_1, Y_2, \ldots, Y_r \qquad (r \geq 1)$$

in $\mathcal{L}(\mathbf{X}_0)$ such that $Y_r \equiv Y$. We shall show by induction on i $(1 \leq i \leq r)$, that all formulas Y_i' of the sequence

$$Y_1', Y_2', \ldots, Y_r'$$

are valid.

For $i = 1$, Y_1 is necessarily an axiom of the system $\mathbf{H}$ of propositional logic. Suppose, for instance, that

$$Y_1 \equiv (U \Rightarrow V) \Rightarrow [(W \vee U) \Rightarrow (W \vee V)]$$

(axiom IV), where $U, V, W \in \mathcal{L}(\mathbf{X}_0)$. Taking into account that if U', V', W' are formulas in $\mathcal{L}(\mathbf{X})$ it follows that

$$Y_1' \equiv (U' \Rightarrow V') \Rightarrow ((W' \vee U') \Rightarrow (W' \vee V'))$$

is a valid formula in $\mathcal{L}(\mathbf{X})$ (see Proposition 10.11, (d)). For the other axioms the arguments are similar. So, Y_1' is a valid formula.

Suppose now $i > 1$ and assume that for all $j < i$, Y_j' is a valid formula. The formula Y_i in the formal proof $Y_1, Y_2, \ldots, Y_r$ is either an axiom or it is obtained from Y_h, $Y_k \equiv Y_h \Rightarrow Y_i$ $(h, k < i)$ by the MP-rule. If Y_i is an axiom, then, as above, Y_i' is a valid formula. In the second case from the inductive assumption we have $\models Y_h'$ and $\models Y_h' \Rightarrow Y_i'$. From Proposition 10.12 one gets $\models Y_i'$. Thus, the assertion follows from the principle of mathematical induction; in particular $\vdash Y'$ (because $Y' \equiv Y_r'$). ∎

REMARK 10.5   The Proposition 10.14 shows that from a semantic point of view, the predicate logic is an extension of propositional logic in the sense that any tautology Y of propositional logic furnishes valid formulas Y' of predicate logic. ∎

In order to prove the validity of some remarkable formulas, which cannot be obtained from the tautologies of propositional logic, we shall establish certain simple properties of Γ-valued functions defined on an arbitrary set.

Remember that in the Boolean algebra $\Gamma = \{0, 1\}$ the operations $x + y$ and xy are defined by the equalities:

$$x + y = \max\{x, y\} \text{ and } xy = \min\{x, y\},$$

where x, y are arbitrary elements in Γ. In this section, for reasons of convenience, we shall use the notations $x + y = x \vee y$, $xy = x \wedge y$; hence, for any $x, y \in \Gamma$,

$$x \vee y = \max\{x, y\} \text{ and } x \wedge y = \min\{x, y\}.$$

If $\mathbf{S}$ is an arbitrary set and $w : \mathbf{S} \to \Gamma : s \mapsto w(s)$ is a given Γ-valued function defined on $\mathbf{S}$, then, taking into account that Γ is a finite set (Γ has two elements), there are always $s_0', s_0'' \in \mathbf{S}$ such that

$$w(s) \leq w(s_0') \text{ and } w(s) \geq w(s_0''),$$

for all $s \in \mathbf{S}$. Although the elements s_0', s_0'' are not uniquely determined, the values $w(s_0')$ and $w(s_0'')$ in Γ are unique. The values $w(s_0')$ and $w(s_0'')$ will be called respectively, the greatest and lowest values of w and will be denoted by

$$\max_{s \in \mathbf{S}} w(s) \text{ and } \min_{s \in \mathbf{S}} w(s).$$

LEMMA 10.1

(a) *If the function* $h : \mathcal{D} \to \Gamma : d \mapsto h(d)$ *and the element* $\gamma \in \Gamma$ *are given,*
   then the following implications hold:
   (i) $(\gamma \geq h(d)$ *for all* $d \in \mathcal{D})$ *implies* $(\gamma \geq \max_{d} h(d))$;
   (ii) $(\gamma \leq h(d)$ *for all* $d \in \mathcal{D})$ *implies* $(\gamma \leq \min_{d} h(d))$.

(b) *If* $h_1, h_2 : \mathcal{D} \to \Gamma$ *are given functions, then*
   (i) $\max_{d} (h_1(d) \vee h_2(d)) = (\max_{d} h_1(d)) \vee (\max_{d} h_2(d))$;
   (ii) $\min_{d} (h_1(d) \wedge h_2(d)) = (\min_{d} h_1(d)) \wedge (\min_{d} h_2(d))$.

(c) *If* $k : \mathcal{D} \times \mathcal{D} \to \Gamma : (d', d'') \mapsto k(d', d'')$ *is a given function, then*
   (i) $\max_{d'} (\max_{d''} k(d', d'')) = \max_{d''} (\max_{d'} k(d', d''))$;
   (ii) $\min_{d'} (\min_{d''} k(d', d'')) = \min_{d''} (\min_{d'} k(d', d''))$;
   (iii) $\max_{d'} (\min_{d''} k(d', d'')) \leq \min_{d''} (\max_{d'} k(d', d''))$;

PROOF   The assertion (a) follows obviously from the definition of $\max_{d} h(d)$ and $\min_{d} h(d)$.

The proofs of assertions (b)(i) and (b)(ii) are simple exercises. Let us prove, for instance, (b)(ii). If we denote by d_0 an element of $\mathcal{D}$ such that

$$h_1(d_0) \wedge h_2(d_0) = \min_{d} (h_1(d) \wedge h_2(d)),$$

then

$$\min_{d} (h_1(d) \wedge h_2(d)) = h_1(d_0) \wedge h_2(d_0) \geq (\min_{d \in \mathcal{D}} (h_1(d)) \wedge (\min_{d \in \mathcal{D}} (h_2(d)).$$

On the other hand, if d_0', d_0'' are the elements of $\mathcal{D}$ with $h_1(d_0') = \min_{d \in \mathcal{D}} h_1(d)$ and $h_2(d_0'') = \min_{d \in \mathcal{D}} h_2(d)$, then

$$\min_{d \in \mathcal{D}} (h_1(d)) \wedge \min_{d \in \mathcal{D}} (h_2(d)) = h_1(d_0') \wedge h_2(d_0'')$$

$$= \begin{cases} h_1(d_0'), & \text{if } h_1(d_0') \leq h_2(d_0'') \\ h_2(d_0''), & \text{otherwise} \end{cases}$$

$$= \begin{cases} h_1(d_0') \wedge h_2(d_0'), & \text{if } h_1(d_0') \leq h_2(d_0'') \\ h_1(d_0'') \wedge h_2(d_0''), & \text{otherwise} \end{cases} \geq$$

$$\geq \min_{d \in \mathcal{D}} (h_1(d) \wedge h_2(d)).$$

From the above two inequalities the required equality follows.

In order to prove assertion (c)(i) let us remark that, for each $d' \in \mathcal{D}$, there is an element $d_0''(d') \in \mathcal{D}$ depending on d' such that

$$\max_{d'' \in \mathcal{D}} k(d', d'') = k(d', d_0''(d'))$$

and for the function $d' \mapsto \max_{d''} k(d', d'')$ there is some $d_0' \in \mathcal{D}$ such that

$$\max_{d' \in \mathcal{D}} (\max_{d'' \in \mathcal{D}} k(d', d'')) = \max_{d'' \in \mathcal{D}} k(d_0', d'').$$

Thus,

$$\max_{d' \in \mathcal{D}} (\max_{d'' \in \mathcal{D}} k(d', d'')) = \max_{d'' \in \mathcal{D}} k(d_0', d'') =$$
$$= k(d_0', d_0''(d_0')) \leq \max_{d' \in \mathcal{D}} k(d', d_0''(d_0')) \leq \max_{d'' \in \mathcal{D}} (\max_{d' \in \mathcal{D}} k(d', d''));$$

hence

$$\max_{d' \in \mathcal{D}} (\max_{d'' \in \mathcal{D}} k(d', d'')) \leq \max_{d'' \in \mathcal{D}} (\max_{d' \in \mathcal{D}} k(d', d'')).$$

The converse inequalities follow similarly, hence equality (c)(i) is proved. Equality (c)(ii) can be established in the same way. Finally, let us prove equality (c)(iii). If we consider an element $d_0' \in \mathcal{D}$ such that

$$\max_{d' \in \mathcal{D}} (\min_{d'' \in \mathcal{D}} k(d', d'')) = \min_{d'' \in \mathcal{D}} k(d_0', d''),$$

then

$$\max_{d' \in \mathcal{D}} (\min_{d'' \in \mathcal{D}} k(d', d'')) = \min_{d'' \in \mathcal{D}} k(d_0', d'') \leq k(d_0', d'') \leq \max_{d' \in \mathcal{D}} k(d', d'')$$

for all $d'' \in \mathcal{D}$. If we denote $\gamma = \max_{d' \in \mathcal{D}} (\min_{d'' \in \mathcal{D}} k(d', d''))$, the last inequality becomes $\gamma \leq \max_{d' \in \mathcal{D}} k(d', d'')$ for all $d'' \in \mathcal{D}$. It follows that

$$\gamma \leq \min_{d'' \in \mathcal{D}} (\max_{d'} k(d', d''))$$

and so (c)(iii) is proved. ∎

Now we are ready to prove a proposition which will furnish a list of remarkable valid formulas. These valid formulas cannot be obtained from tautologies.

PROPOSITION 10.15

Let F, G be formulas containing the individual variable x and L be a formula containing two distinct individual variables x, y. With these notations the following formulas are all valid:

(1) $\exists x(F \vee G) \Leftrightarrow (\exists xF \vee \exists xG)$;

(2) $\exists x(F \wedge G) \Rightarrow (\exists xF \wedge \exists xG)$;

(3) $\forall x(F \wedge G) \Leftrightarrow (\forall xF \wedge \forall xG)$;

(4) $(\forall xF \vee \forall xG) \Rightarrow \forall x(F \vee G)$;

(5) $\forall x(F \Rightarrow G) \Rightarrow (\forall xF \Rightarrow \forall xG)$;

(6) $\forall x(F \Leftrightarrow G) \Rightarrow (\forall xF \Leftrightarrow \forall xG)$;

(7) $\exists x \exists y L(x,y) \Leftrightarrow \exists y \exists x L(x,y)$;

(8) $\forall x \forall y L(x,y) \Leftrightarrow \forall y \forall x L(x,y)$;

(9) $\exists x \forall y L(x,y) \Rightarrow \forall y \exists x L(x,y)$.

In the above formulas, for simplicity, we have dropped some parentheses.

PROOF   Consider an arbitrary interpretation I of $\mathcal{L}(\mathbf{X})$. The validity of formulas, such as $A \Leftrightarrow B$ and $A \Rightarrow B$ will be proved by showing that $I(A) = I(B)$ and $I(A) \leq I(B)$ for all I, respectively.

(1) Let $F \equiv F(U_F(x))$, $G \equiv G(U_G(x))$; then $U_{F \vee G}(x) \equiv U_F(x)U_G(x)$. Taking into account the definition of I and equality (b), (i) from the previous lemma we get:

$$
\begin{aligned}
I[\exists x(F \vee G)] &= \max_\omega (h_F \vee h_G)(\delta_\omega \hat{U}_{F \vee G}) = \\
&= \max_\omega (h_F \vee h_G)(\delta_\omega \hat{U}_F \delta_\omega \hat{U}_G) = \\
&= \max_d (h_F(\hat{U}_F(d)) \vee h_G(\hat{U}_G(d))) = \\
&= (\max_d h_F(\hat{U}_F(d))) \vee (\max_d h_G(\hat{U}_G(d))) = \\
&= (\max_\omega h_F(\delta_\omega \hat{U}_F)) \vee (\max_\omega h_G(\delta_\omega \hat{U}_G)) = \\
&= I(\exists xF) \vee I(\exists xG) = I(\exists xF \vee \exists xG).
\end{aligned}
$$

Hence, (1) is proved. The proof of the validity of (3) is similar.

$$
\begin{aligned}
(2) \quad I(\exists x(F \wedge G)) &= (\max_\omega h_{F \wedge G})(\delta_\omega \widehat{U_F U_G}) = \\
&= (\max_\omega h_{F \wedge G})(\delta_\omega \widehat{U_F} \delta_\omega \widehat{U_G}) = \\
&= \max_d h_{F \wedge G}(\hat{U}_F(d)\hat{U}_G(d)) = \max_d [h_F(\hat{U}_F(d)) \wedge h_G(\hat{U}_G(d))] \leq \\
&\leq (\max_d h_F(\hat{U}_F(d))) \wedge (\max_d h_G(\hat{U}_G(d))) = \\
&= (\max_\omega h_F(\delta_\omega \hat{U}_F)) \wedge (\max_\omega h_G(\delta_\omega \hat{U}_G)) = \\
&= I(\exists xF) \wedge I(\exists xG) = I(\exists xF \wedge \exists xG).
\end{aligned}
$$

Using similar arguments, one can prove the validity of formulas (4), (5), (6). The validity of (7), (8), (9) follows by applying (c), (i), (ii), (iii) from the previous lemma. For instance, let us prove the validity of (9).

Let $L \equiv L(x,y) \equiv L(U_L(x,y))$, where $U_L(x,y)$ is the sequence of the arguments of L and let us denote by ω_1 and ω_2, respectively, the occurrences of x and y in L. If we denote $N \equiv \forall y L(x,y)$, $M \equiv \exists x \forall y L(x,y)$, $R \equiv \exists x L(x,y)$ and $Q \equiv \forall y \exists x L(x,y)$, we have to prove that $I(M) \le I(Q)$ for all I. Let us remark also that $U_Q \equiv \delta_{\omega_2}\delta_{\omega_1}U_L \equiv \delta_{\omega_1}\delta_{\omega_2}U_L \equiv U_M$, $U_N \equiv \delta_{\omega_2}U_L$. If we consider an arbitrary interpretation I on $\mathcal{L}(\mathbf{X})$, then

$$
\begin{aligned}
I(M) &= I(\exists x \forall y L(x,y)) = h_M(\hat{U}_M) = h_M(\delta_{\omega_1}\delta_{\omega_2}\hat{U}_L) = \\
&= (\max_{\omega_1} h_N)(\delta_{\omega_1}\delta_{\omega_2}\hat{U}_L) = \max_{d' \in \mathcal{D}} h_N(\delta_{\omega_2}\hat{U}_L(d')) = \\
&= \max_{d' \in \mathcal{D}}((\min_{\omega_2} h_L)(\delta_{\omega_2}\hat{U}_L(d'))) = \max_{d' \in \mathcal{D}}(\min_{d'' \in \mathcal{D}} h_L(\hat{U}_L(d')(d''))).
\end{aligned}
$$

Applying (c), (iii) from the previous lemma, it follows that

$$
\begin{aligned}
I(M) &\le \min_{d''}(\max_{d'} h_L(\hat{U}_L(d')(d''))) = \min_{d''}[(\max_{\omega_1} h_L)(\delta_{\omega_1}(\hat{U}_L(d'')))] = \\
&= \min_{d''} h_R(\delta_{\omega_1}(\hat{U}_L(d''))) = (\min_{\omega_2} h_R)(\delta_{\omega_2}\delta_{\omega_1}\hat{U}_L) = \\
&= h_Q(\hat{U}_Q) = I(Q)
\end{aligned}
$$

and (9) is proved. ∎

PROPOSITION 10.16

If $A \equiv A(x) \equiv A(U_A(x))$ is a formula containing the variable x and if $t = t(V)$ is a term free for x in A, then

$$
A(t|x) \Rightarrow \exists x A(x),
$$
$$
\forall x\, A(x) \Rightarrow A(t|x)
$$

are valid formulas.

PROOF   Let I be an interpretation of $\mathcal{L}(\mathbf{X})$. Taking into account Proposition 10.7 we have

$$
I(A(t|x)) = h_A(\widehat{U_A}(I(t))).
$$

If we denote $d_0 = I(t)$, then

$$
I(A(t|x)) = h_A(\hat{U}_A(d_0)) \le \max_d h_A(\hat{U}_A(d)) = I(\exists x A(x)).
$$

Therefore, $I(A(t|x)) \le I(\exists x A(x))$ for all I, hence $A(t|x) \Rightarrow \exists x A(x)$ is a valid formula.

The proof of validity of the second formula is similar.   ▮

The proposition below concerns, from a semantic point of view, the inference rules IR_1 and IR_2 mentioned in Chapter 8.

PROPOSITION 10.17
Let $F \equiv F(x)$ be a formula containing the individual variable x and G a formula which does not contain x.

(a) If $\models F \Rightarrow G$, then $\models \exists x F \Rightarrow G$;

(b) If $\models G \Rightarrow F$, then $\models G \Rightarrow \forall x F$.

PROOF   We have to show that if $I(F) \leq I(G)$ for all interpretations I, then $I(\exists x F) \leq I(G)$ for all I. Suppose that $I(F) \leq I(G)$. Taking into account that G does not contain x it follows that $U_G(x) \equiv U_G$, $\hat{U}_G(\hat{x}) = \hat{U}_G$ and $I(G) = h_G(\hat{U}_G)$, which is a constant.

If, on the other hand, we denote $F(x) \equiv F(U_F(x))$, then

$$I(F(x)) = h_F(\hat{U}_F(\hat{x})).$$

For another interpretation I' in which the function ξ is replaced by ξ' with

$$\xi'(u) = \begin{cases} \xi(u) & , u \neq x \\ d & , u = x \end{cases}$$

we have $I'(F(x)) = h_F(\hat{U}_F(d))$ and obviously d is an arbitrary element of $\mathcal{D}$. By assumption, we have $I(F(x)) \leq I(G)$ for all I. That is $h_F(\hat{U}_F(d)) \leq I'(G)$ for all $d \in \mathcal{D}$. But obviously $I'(G) = I(G)$. It follows that

$$\max_{d \in \mathcal{D}} h_F(\hat{U}_F(d)) \leq I(G).$$

From the definition of the interpretation I it follows that $I(\exists x F) \leq I(G)$ and the first assertion is thus proved.

The second assertion can be proved in the same way.   ▮

Before giving the next proposition, we need to introduce a certain convention. If $F \equiv F(U_F)$ is a formula, then as we know, $U_F \equiv u_1 u_2 \ldots u_q$ is the sequence of all arguments, variables and constants of F. Let us denote by $V_F \equiv v_1 v_2 \ldots v_p$ $(p \leq q)$ the sequence of all free individual variables contained in F (possible with repetitions), written in the order of their occurrence in F and let us write $F \equiv F(V_F)$. We stress that $F(U_F)$ and $F(V_F)$ are notations for the same formula F. Of course, V_F might be a null sequence. In such a case, we shall say that F is a **closed formula**.

If $F \equiv F(V_F)$ is a formula with $V_F \equiv v_1 v_2 \ldots v_p$ $(1 \leq p \leq q)$, then the formula $\forall v_p \forall v_{p-1} \ldots \forall v_1 F(V_F)$ will be called the **(universal) closure of** F and will be denoted by $\forall V_F(F(V_F))$. It is obvious that the (universal) closure of a formula F is a closed formula.

PROPOSITION 10.18

Let $F \equiv F(V_F)$ be a formula in which $V_F \equiv v_1 v_2 \ldots v_p$ $(1 \leq p \leq q)$ is the sequence of all free variables of F. The following assertion holds:

F is a valid formula if and only if $\forall V_F(F(V_F))$ is a valid formula, or briefly:

$$\models F \quad \text{iff} \quad \models \forall V_F(F(V_F)).$$

PROOF Let $V_F \equiv v_1 v_2 \ldots v_p$ $(1 \leq p \leq q)$ be the sequence of all distinct variables in F, then $\forall V_F(F(V_F)) \equiv \forall v_p \forall v_{p-1} \ldots \forall v_1 F(v_1 v_2 \ldots v_p)$. We shall prove the above assertion by induction on p.

Let be $p = 1$, $V_F \equiv v_1$, $F(V_F) \equiv F(v_1)$. Suppose that $\models F(v_1)$. That is $I(F(v_1)) = 1$ for all interpretations I. By definition $I(F(v_1)) = h_F(\hat{v}_1)$ for all I. In particular, $I'(F(v_1)) = 1$ for all I' in which the function ξ is replaced by ξ', where

$$\xi'(v) = \begin{cases} \xi(v) & , v \neq v_1 \\ d & , v = v_1 \end{cases},$$

where d is an arbitrary element of $\mathcal{D}$. It follows that $h_F(d) = 1$ for all $d \in \mathcal{D}$. Therefore, $\min_{d \in \mathcal{D}} h_F(d) = 1$, hence $I(\forall v_1 F(v_1)) = 1$ for all I. Thus, $\models \forall v_1 F(v_1)$.

Conversely, suppose that $I(\forall v_1 F(v_1)) = 1$ for all I. This means that $\min_{d \in \mathcal{D}} h_F(d) = 1$. In particular, $I(F(v_1)) = h_F(\xi(v_1))$ for all ξ and we can write $I(F(v_1)) = h_F(\xi(v_1)) \geq \min_{d \in \mathcal{D}} h_F(d) = 1$. Therefore, $I(F(v_1)) = 1$ for all I, hence $\models F(v_1)$. So, the assertion is proved for $p = 1$.

If $p > 1$ and we assume that the assertion is true for all formulas $F' \equiv F'(V_{F'})$ with $\lambda(V_{F'}) < p$, then it follows easily that the assertion is true for all formulas $F \equiv F(V_F)$ with $\lambda(V_F) = p$. Therefore, we can conclude that the proposition is proved. ∎

11

The Syntactic Truth Structure on the Language $\mathcal{L}$ of Predicate Logic

As in the case of propositional logic the syntactic truth structure on the language $\mathcal{L} = \mathcal{L}(\mathbf{X})$ of predicate logic will be based on concepts such as: axioms, inference rules, theorems, syntactic deduction, etc. Thus, the syntactic truth structure on $\mathcal{L}$ will be an axiomatic theory. This theory will have some similarities with the one of propositional logic but, of course, it will be more complex, having many specific characteristics.

11.1   Axioms, Theorems

The set of axioms of predicate logic contains five schemes of axioms and two inference rules. The list of axioms of predicate logic contains the four schemes representing the axioms of propositional logic and a new specific axiom (scheme) involving the quantifier "∃".

Afterwards we shall introduce additionally an axiom and an inference rule concerning the quantifier "∀"; this axiom and inference rule will be proved because they are not independent from the initial set of axioms and inference rules.

DEFINITION 11.1   *For any formulas $F, G, H, L \in \mathcal{L}(\mathbf{X})$ the following formulas are axioms of predicate logic:*

(A1): $F \vee F \Rightarrow F$;

(A2): $F \Rightarrow F \vee G$;

(A3): $F \vee G \Rightarrow G \vee F$;

(A4): $(F \Rightarrow G) \Rightarrow ((H \vee F) \Rightarrow (H \vee G))$;

(A5): $L(t|x) \Rightarrow \exists x\, L(x)$, *where* x *is an individual variable contained in* L *and* t *is a term free for* x *in* $L(x)$.

REMARK 11.1   As we have already mentioned the formulas (A1)–(A5) are in fact schemes of axioms, because F, G, H, L are arbitrary formulas in $\mathcal{L}(\mathbf{X})$. The inference rules of predicate logic (in our approach) will be implicitly given below when we define the notion of formal proof.   ∎

DEFINITION 11.2   *A finite sequence*

$$F_1, F_2, \ldots, F_n \qquad (n \geq 1)$$

of formulas in $\mathcal{L}(\mathbf{X})$ *is called a* **(formal) proof** *if for each index* i $(1 \leq i \leq n)$ *the formula* F_i *satisfies one of the conditions:*

(a) F_i *is an axiom (for short* $F_i \in Ax$*);*

(b)   *there are indices* $h, k < i$ *such that*

$$F_k \equiv F_h \Rightarrow F_i$$

(in this case we shall say that F_i *is obtained by the MP-rule from* F_h *and* F_k*);*

(c)   *there is an index* $j < i$ *and an individual variable* x *such that*

$$F_j \equiv A(x) \Rightarrow C \text{ and } F_i \equiv \exists x A(x) \Rightarrow C$$

where $A(x)$ *is a formula containing* x *and* C *is a formula which does not contain* x *(in this case we shall say that* F_i *is obtained from* F_j *by the rule* $(\exists \Rightarrow)$*, called the rule of introduction of quantifier* $\exists$*).*

DEFINITION 11.3   *A formula* F *in* $\mathcal{L}(\mathbf{X})$ *is called a* **theorem of predicate logic** *and we shall write* $\vdash F$ *if the formula* F *can be inserted in a (formal) proof. In particular if* $F_1, F_2, \ldots, F_r$ *is a formal proof, then all formulas* F_i $(i \leq i \leq r)$ *are theorems and the sequence above is usually called the proof of its last formula* F_r.

REMARK 11.2   From the above definitions it follows that axioms are theorems too.   ∎

In Section 10.2 we have shown that if $A_1, A_2, \ldots, A_N$ are arbitrary formulas in $\mathcal{L}(\mathbf{X})$, then one can define a correspondence $F \mapsto F'$ which associates to each formula $F \in \mathcal{L}(\mathbf{X}_0)$ a unique formula F' in $\mathcal{L}(\mathbf{X})$; the formula F' in $\mathcal{L}(\mathbf{X})$ is obtained from the formula $F \in \mathcal{L}(\mathbf{X}_0)$ by replacing the propositional variables $X_1, X_2, \ldots, X_N$ by $A_1, A_2, \ldots, A_N$ respectively. In order to make the correspondence $F \mapsto F'$ more expressive, we shall use the notations $F = F(X_1, X_2, \ldots, X_N)$ and $F' \equiv F(A_1, A_2, \ldots, A_N)$. Therefore, the above result can be formulated briefly:

$$\text{if } F(X_1, \ldots, X_N) \in \mathcal{L}(\mathbf{X}_0), \text{ then } F(A_1, \ldots, A_N) \in \mathcal{L}(\mathbf{X}).$$

In Section 10.2 we have proved also the result:

$$\text{if } F(X_1, \ldots, X_N) \text{ is a tautology,}$$
$$\text{then } F(A_1, \ldots, A_N) \text{ is a valid formula.}$$

Next, we shall prove the corresponding result in the syntactic version.

PROPOSITION 11.1

If the formula $F \equiv F(X_1 \ldots, X_N) \in \mathcal{L}(\mathbf{X}_0)$ is a theorem of propositional logic (or equivalently, a tautology), then for each sequence of formulas $A_1, A_2, \ldots, A_N$ in $\mathcal{L}(\mathbf{X})$, the formula $F' \equiv F(A_1, \ldots, A_N)$ is a theorem of predicate logic.

PROOF   Suppose that $F \in \mathcal{L}(\mathbf{X}_0)$ is a theorem of propositional logic and let the sequence

$$(*) \qquad\qquad F_1, F_2, \ldots, F_r \qquad (r \geq 1),$$

$F_r \equiv F$, be a syntactic proof of F in propositional logic. By the correspondence $F_1 \mapsto F_1', F_2 \mapsto F_2', \ldots, F_r \mapsto F_r'$ we get the sequence of formulas

$$(**) \qquad\qquad F_1', F_2', \ldots, F_r' \qquad (r \geq 1)$$

in $\mathcal{L}(\mathbf{X})$ with $F_r' \equiv F'$.

We shall show that the sequence $(**)$ is a proof in predicate logic. First, let us remark that, by the above correspondence, axioms $(\mathbf{H1})$–$(\mathbf{H4})$ of propositional logic go in axioms $(A1)$–$(A4)$, respectively, of predicate logic. Second, if in the initial sequence $(*)$ there are indices h, k, i with $h, k < i$ and $F_k \equiv F_h \Rightarrow F_i$, then in the sequence $(**)$, for the same indices h, k, i, we have $F_k' \equiv F_h' \Rightarrow F_i'$. Thus, if in $(*)$ F_i is an axiom, then in $(**)$ F_i' is an axiom and if in $(*)$ F_i is obtained by MP-rule, then in $(**)$ F_i' is also obtained by the MP-rule. This means that the sequence $(**)$ is a proof in predicate logic. Taking into account that $F_r' \equiv F'$ it follows that F' is a theorem of predicate logic.   ∎

REMARK 11.3   Proposition 11.1 furnishes an infinite set of theorems of predicate logic, namely all formulas which are obtained from tautologies by some correspondence $F \mapsto F'$.   ∎

We mention that the derived rule of syllogism is also true in predicate logic having the same proof as in propositional logic.

PROPOSITION 11.2

If $A(x)$ is a formula in $\mathcal{L}(\mathbf{X})$ containing the individual variable x and t is a term free for x in $A(x)$, then $\vdash \forall x A(x) \Rightarrow A(t|x)$.

PROOF   Consider the following formal proof in predicate logic:

(1) $\neg A(t|x) \Rightarrow \exists x(\neg A(x))$; axiom $(A5)$ for the formula $\neg A(x)$;

(2) $(\neg A(t|x) \Rightarrow \exists x(\neg A(x))) \Rightarrow (\neg \exists x(\neg A(x)) \Rightarrow \neg\neg A(t|x))$; theorem obtained from the tautology of contraposition;

(3) $\neg \exists x(\neg A(x)) \Rightarrow \neg\neg A(t|x)$; MP: (1), (2);

(4) $\neg\neg A(t|x) \Rightarrow A(t|x)$; theorem obtained from the tautology $\neg\neg F \Rightarrow F$;

(5) $\neg \exists x(\neg A(x)) \Rightarrow A(t|x)$; syllogism: (3), (4);

(6) $\forall x A(x) \Rightarrow A(t|x)$; the definition: $\forall x(A(x)) \equiv \neg \exists x(\neg A(x))$.

The sequence (1)–(6) being a proof in predicate logic, it follows that its last formula is a theorem.   ∎

REMARK 11.4   The theorem $\forall x A(x) \Rightarrow A(t|x)$ proved in the above proposition represents the additional derived "axiom" concerning the quantifier "$\forall$".   ∎

PROPOSITION 11.3

Let $A(x)$ be a formula containing the variable x and C be a formula which does not contain x.

$$\text{If } \vdash C \Rightarrow A(x),$$
$$\text{then } \vdash C \Rightarrow \forall x A(x).$$

PROOF   The formula $C \Rightarrow A(x)$ being a theorem there is a proof $F_1, F_2, \ldots, F_p$ with $F_p \equiv C \Rightarrow A(x)$. Consider now the following proof in predicate logic

$$\left.\begin{array}{l} F_1 \\ F_2 \\ \vdots \\ F_p \equiv C \Rightarrow A(x) \end{array}\right\} \; ; \text{ the proof of theorem } C \Rightarrow A(x);$$

$F_{p+1} \equiv (C \Rightarrow A(x)) \Rightarrow (\neg A(x) \Rightarrow \neg C)$; theorem obtained from the

tautology of contraposition

$F_{p+2} \equiv \neg A(x) \Rightarrow \neg C$; MP: F_p, F_{p+1};

$F_{p+3} \equiv \exists x(\neg A(x)) \Rightarrow \neg C$; the rule $(\exists \Rightarrow)$: F_{p+2};

$F_{p+4} \equiv (\exists x \neg (A(x)) \Rightarrow \neg C) \Rightarrow (\neg\neg C \Rightarrow \neg\exists x \neg (A(x)))$; theorem obtained

from the tautology of contraposition

$F_{p+5} \equiv \neg\neg C \Rightarrow \neg\exists x(\neg A(x))$; MP: F_{p+3}, F_{p+4};

$F_{p+6} \equiv C \Rightarrow \neg\neg C$; theorem obtained from the tautology $F \Rightarrow \neg\neg F$;

$F_{p+7} \equiv C \Rightarrow \neg\exists x(\neg A(x))$; syllogism F_{p+5}, F_{p+6};

$F_{p+8} \equiv C \Rightarrow \forall x(A(x))$: the definition: $\forall x(A(x)) \equiv \neg\exists x(\neg A(x))$.

The sequence $F_1 - F_{p+8}$ being a proof in predicate logic it follows that $C \Rightarrow \forall x(A(x))$ is a theorem. ∎

REMARK 11.5 Proposition 11.3 represents the inference rule involving the quantifier "$\forall$". This rule is useful, but it is not essentially new because it is derived from the axioms $(A1)$-$(A5)$ by means of the initial inference rules. ∎

11.2 Some Remarkable Metatheorems

In this section we shall prove some important theorems of predicate logic. As in the case of proportional logic, assertions like, "F is a theorem of predicate logic", have to be proved by means of definitions and usual reasoning. Therefore, they are usual theorems about formal theorems. That is why such assertions will be called **metatheorems**. Among other results we shall establish the metatheorem of deduction called Herbrand's metatheorem for predicate logic.

Before presenting the most relevant metatheorems let us recall that the (derived) rule of syllogism, proved in propositional logic, is also true in predicate logic (with the same proof). Similar remarks are true for many metatheorems below.

METATHEOREM 11.1
Let F, G, H be formulas in $\mathcal{L}(\mathbf{X})$.

$$\text{If} \quad \vdash F \Rightarrow G \text{ and } \vdash G \Rightarrow H,$$
$$\text{then} \quad \vdash F \Rightarrow H.$$

METATHEOREM 11.2
Let F, G be formulas in $\mathcal{L}(\mathbf{X})$.

$$\text{If} \quad \vdash G, \quad \text{then} \quad \vdash F \Rightarrow G.$$

METATHEOREM 11.3
For any formula $F \in \mathcal{L}(\mathbf{X})$, $\quad \vdash F \Rightarrow F$.

The next metatheorem follows from the tautology $(A \Rightarrow (B \Rightarrow C)) \Leftrightarrow (B \Rightarrow (A \Rightarrow C))$.

METATHEOREM 11.4
For all formulas $F, G, H \in \mathcal{L}(\mathbf{X})$, the formula

$$(F \Rightarrow (G \Rightarrow H)) \Leftrightarrow (G \Rightarrow (F \Rightarrow H))$$

is a theorem in predicate logic.

From this metatheorem we get the following direct consequence.

METATHEOREM 11.5

$$If \quad \vdash F \Rightarrow (G \Rightarrow H), \text{ then } \vdash G \Rightarrow (F \Rightarrow H).$$

METATHEOREM 11.6
If F, G are formulas, then $\vdash F \Rightarrow (G \Rightarrow (F \wedge G))$.

PROOF   The formula $A \Rightarrow (B \Rightarrow (A \wedge B))$ is a tautology, hence a theorem in propositional logic. Thus, by Proposition 11.1 for any formulas F, G, the formula $F \Rightarrow (G \Rightarrow (F \wedge G))$ is a theorem of predicate logic.   ∎

From this metatheorem the result below (the Metatheorem of Conjunction) follows directly.

METATHEOREM 11.7

$$If \ \vdash F \ and \ \vdash G, \ then \ \vdash F \wedge G.$$

Taking into account that for any $A, B, C, D \in \mathcal{L}(\mathbf{X_0})$, the formulas

$$((A \Rightarrow B) \wedge (C \Rightarrow D)) \Rightarrow ((A \vee B) \Rightarrow (C \vee D)),$$
$$((A \Rightarrow B) \wedge (C \Rightarrow D)) \Rightarrow ((A \wedge C) \Rightarrow (B \wedge D))$$

are tautologies, hence theorems of propositional logic, we get the following result.

METATHEOREM 11.8
For any formulas F, G, H, K *in* $\mathcal{L}(\mathbf{X})$, *the formulas*

(a) $((F \Rightarrow G) \wedge (H \Rightarrow K)) \Rightarrow ((F \vee H) \Rightarrow (G \vee K))$;
(b) $((F \Rightarrow G) \wedge (H \Rightarrow K)) \Rightarrow ((F \wedge H) \Rightarrow (G \wedge K))$

are theorems of predicate logic.

METATHEOREM 11.9
For any formulas $F, G \in \mathcal{L}(\mathbf{X})$ *containing the variable* x *and for any formula* $L \in \mathcal{L}(\mathbf{X})$ *containing two distinct variables* x, y, *the following formulas are theorems of predicate logic:*

(1) $\exists x(F \vee G) \Leftrightarrow (\exists xF \vee \exists xG)$;

(2) $\exists x(F \wedge G) \Rightarrow (\exists xF \wedge \exists xG)$;

(3) $\forall x(F \wedge G) \Leftrightarrow (\forall xF \wedge \forall xG)$;

(4) $(\forall xF \vee \forall xG) \Rightarrow \forall x(F \vee G)$;

(5) $\forall x(F \Rightarrow G) \Rightarrow (\forall xF \Rightarrow \forall xG)$;

(6) $\forall x(F \Leftrightarrow G) \Rightarrow (\forall xF \Leftrightarrow \forall xG)$;

(7) $\exists x \exists y\, L(x, y) \Leftrightarrow \exists y \exists x\, L(x, y)$;

(8) $\forall x \forall y\, L(x, y) \Leftrightarrow \forall y \forall x\, L(x, y)$;

(9) $\exists x \forall y\, L(x, y) \Rightarrow \forall y \exists x\, L(x, y)$.

PROOF   As an illustration we shall prove only that (2) and (9) are theorems.
(2) Consider the following proof:

1. $F \Rightarrow \exists x(F)$; axiom $(A5)$;

2. $G \Rightarrow \exists x(G)$; axiom $(A5)$;

3. $F \wedge G \Rightarrow (\exists x(F) \wedge \exists x(G))$; MTh. 11.8;

4. $\exists x(F \wedge G) \Rightarrow (\exists x(F) \wedge \exists x(G))$; the inference rule $(\exists \Rightarrow)$: 3 (the formula $\exists x(F) \wedge \exists x(G)$ does not contain the variable x).

(9) Consider the following proof:

1. $L(x,y) \Rightarrow \exists x\, L(x,y)$; axiom $(A5)$;

2. $\forall y\, L(x,y) \Rightarrow L(x,y)$; Proposition 11.2;

3. $\forall y\, L(x,y) \Rightarrow \exists x\, L(x,y)$; syllogism: 1, 2;

4. $\exists x \forall y\, L(x,y) \Rightarrow \exists x\, L(x,y)$; the inference rule $(\exists \Rightarrow)$: 3 (because the formula $\exists x\, L(x,y)$ does not contain x);

5. $\exists x \forall y\, L(x,y) \Rightarrow \forall y \exists x\, L(x,y)$; the derived inference rule proved in Proposition 11.3 applied to 4 (because $\exists x \forall y\, L(x,y)$ does not contain y).

∎

Next, we shall define the notion of **formal deduction** from a set of formulas (hypotheses).

DEFINITION 11.4   *Let $\mathcal{H}$ be a set of formulas in $\mathcal{L}(\mathbf{X})$. A finite sequence*

$$F_1, F_2, \ldots, F_r \quad (r \geq 1)$$

*is called a **deduction** from $\mathcal{H}$ if for any index i $(1 \leq i \leq r)$ the formula F_i satisfies one of the conditions below:*

(a) F_i is an axiom $(F_i \in Ax)$;

(b) F_i is an hypothesis $(F_i \in \mathcal{H})$;

(c) there are indices $h, k < i$ such that $F_k \equiv F_h \Rightarrow F_i$ (F_i is obtained by the MP-rule from F_h and F_k);

(d) there is an index $j < i$, and formulas $A(x), C \in \mathcal{L}(\mathbf{X})$, having the property that $A(x)$ contains the variable x and C does not contain x, such that $F_j \equiv A(x) \Rightarrow C$ and $F_i \equiv \exists x\, A(x) \Rightarrow C$. ($F_i$ is obtained from F_j by the inference rule $(\exists \Rightarrow)$.)

Each formula F which can be inserted in a deduction from $\mathcal{H}$ is said to be **deducible from** $\mathcal{H}$ *and we shall write* $\mathcal{H} \vdash F$. *If the* $\mathcal{H}$ *is empty* $(\mathcal{H} = \emptyset)$, *then obviously F is a theorem of predicate logic and instead of* $\emptyset \vdash F$ *we shall write* $\vdash F$.

From this definition we have as immediate consequence:

METATHEOREM 11.10
Let $\mathcal{H}_1, \mathcal{H}_2$ be two sets of formulas with $\mathcal{H}_1 \subseteq \mathcal{H}_2$.

$$If \quad \mathcal{H}_1 \vdash F, \quad then \quad \mathcal{H}_2 \vdash F.$$

In particular if $\vdash F$, then for any set $\mathcal{H}$ of formulas we have $\mathcal{H} \vdash F$.

In propositional logic we have proved the Herbrand's Metatheorem of Deduction, which says that if $\mathcal{H}, A \vdash B$, then $\mathcal{H} \vdash A \Rightarrow B$, where $\mathcal{H}$ is an arbitrary set of formulas in $\mathcal{L}(\mathbf{X}_0)$ and $A, B \in \mathcal{L}(\mathbf{X}_0)$. Without supplementary precautions such a result cannot be proved in predicate logic. In order to give an appropriate formulation of Herbrand's metatheorem in predicate logic we must first give the definition below.

DEFINITION 11.5 *Let $\mathcal{H}$ be a set of formulas in $\mathcal{L}(\mathbf{X})$ and F, G be formulas in $\mathcal{L}(\mathbf{X})$. We shall say that G is **normally deducible** with respect to F from $\mathcal{H}, F$ if there is a deduction $G_1, G_2, \ldots, G_r$ $(r \geq 1)$, with $G_r \equiv G$, of G from $\mathcal{H}, F$, such that for each formula $G_i \equiv \exists x\, A(x) \Rightarrow C$ which is obtained by the rule $(\exists \Rightarrow)$ from a formula $G_j \equiv A(x) \Rightarrow C$ $(j < i)$, one of the following two conditions is fulfilled:*
(a) $\mathcal{H} \vdash G_j$;
(b) the variable x is not contained in the formula F.
In this case, instead of $\mathcal{H}, F \vdash G$, sometimes we shall write $\mathcal{H}, F \vdash_{(nF)} G$.

METATHEOREM 11.11 (The Metatheorem of Herbrand in predicate logic)
Let $\mathcal{H}$ be a set of formulas in $\mathcal{L}(\mathbf{X})$ and F, G be formulas in $\mathcal{L}(\mathbf{X})$. If the formula G is normally deducible with respect to F from $\mathcal{H}, F$, then $F \Rightarrow G$ is deducible from $\mathcal{H}$; in short,

$$if \quad \mathcal{H}, F \vdash_{(nF)} G, \quad then \quad \mathcal{H} \vdash F \Rightarrow G.$$

PROOF   Suppose that the sequence

$$(*) \qquad\qquad G_1, G_2, \ldots, G_n \qquad (n \geq 1)$$

with $G_n \equiv G$ is a deduction of G from $\mathcal{H}, F$, normal with respect to F. We shall prove by induction that for all i $(1 \leq i \leq r)$, $\mathcal{H} \vdash F \Rightarrow G_i$.

Let $i = 1$; the formula G_1 in $(*)$ satisfies one of the conditions:

$$(a) \ \ G_1 \in Ax; \ \ (b) \ \ G_1 \in \mathcal{H}; \ \ (c) \ \ G_1 \equiv F.$$

In case (a) we have $\vdash G_1$, hence by MTh. 11.2, $\vdash F \Rightarrow G_1$. Therefore, $\mathcal{H} \vdash F \Rightarrow G_1$.

In case (b), we have $\mathcal{H} \vdash G_1$, therefore $\mathcal{H} \vdash F \Rightarrow G_1$.

In case (c), $F \Rightarrow G_1 \equiv F \Rightarrow F$; by MTh. 11.3 we get $\mathcal{H} \vdash F \Rightarrow G_1$.

So, in all cases $\mathcal{H} \vdash F \Rightarrow G_1$.

Let be $i > 1$ and assume that for all indices $j < i$ we have $\mathcal{H} \vdash F \Rightarrow G_j$. For the formula G_i in $(*)$ we have the following possibilities:

(a) $G_i \in Ax$;
(b) $G_i \in \mathcal{H}$;
(c) $G_i \equiv F$;
(d) G_i is obtained from some formulas G_h, G_k $(h, k < i)$ by the MP-rule;
(e) G_i is obtained from a formula G_j, $j < i$, by the inference rule $(\exists \Rightarrow)$.

In cases (a), (b), (c) we obtain as above that $\mathcal{H} \vdash F \Rightarrow G_i$.

In case (d) we have $G_k \equiv G_h \Rightarrow G_i$ (MP-rule). By the inductive assumption we have

$$\mathcal{H} \vdash F \Rightarrow G_h \ \text{ and } \ \mathcal{H} \vdash F \Rightarrow (G_h \Rightarrow G_i).$$

From the tautology $(A \Rightarrow (B \Rightarrow C)) \Rightarrow ((A \Rightarrow B) \Rightarrow (A \Rightarrow C))$ by Proposition 11.1 we get

$$\vdash (F \Rightarrow (G_h \Rightarrow G_i)) \Rightarrow ((F \Rightarrow G_h) \Rightarrow (F \Rightarrow G_i)).$$

Therefore,

$$\mathcal{H} \vdash (F \Rightarrow (G_h \Rightarrow G_i)) \Rightarrow ((F \Rightarrow G_h) \Rightarrow (F \Rightarrow G_i)).$$

Taking into account the inductive assumption applying the the MP-rule twice we obtain

$$\mathcal{H} \vdash F \Rightarrow G_i.$$

In case (e) there is $j < i$ such that

$$G_j \equiv A(x) \Rightarrow C \ \text{ and } \ G_i \equiv \exists x \, A(x) \Rightarrow C,$$

where $A(x)$ is a formula containing x and C does not contain x. The deduction $G_1, G_2, \ldots, G_r$ being normal with respect to F we have to analyze two subcases:

(α) $\mathcal{H} \vdash G_j$;
(β) the variable x is not contained in F.

(α) Taking into account that $\mathcal{H} \vdash G_j$, i.e., $\mathcal{H} \vdash A(x) \Rightarrow C$, it follows (by the ($\exists \Rightarrow$) rule) $\mathcal{H} \vdash \exists x \, A(x) \Rightarrow C$ or

$$\mathcal{H} \vdash G_i$$

and by MTh. 11.2 we get $\mathcal{H} \vdash F \Rightarrow G_i$.
(β) In this subcase from MTh. 11.4 we obtain

$$\mathcal{H} \vdash (F \Rightarrow (A(x) \Rightarrow C)) \Rightarrow (A(x) \Rightarrow (F \Rightarrow C)).$$

Taking into account the inductive assumption ($j < i$), it follows that $\mathcal{H} \vdash F \Rightarrow G_j$, $G_j \equiv A(x) \Rightarrow C$, and by the MP-rule we get $\mathcal{H} \vdash (A(x) \Rightarrow (F \Rightarrow C))$. Now we notice that the variable x is contained neither in F nor in C, therefore x is not contained in $F \Rightarrow C$. Applying the ($\exists \Rightarrow$) rule we obtain

$$\mathcal{H} \vdash \exists x \, A(x) \Rightarrow (F \Rightarrow C).$$

Finally, by MTh. 11.4 we get $\mathcal{H} \vdash F \Rightarrow (\exists x \, A(x) \Rightarrow C)$ or

$$\mathcal{H} \vdash F \Rightarrow G_i.$$

Thus, from the principle of mathematical induction it follows that for all i $(1 \leq i \leq r)$ $\mathcal{H} \vdash F \Rightarrow G_i$. In particular, $\mathcal{H} \vdash F \Rightarrow G$ because $G \equiv G_r$.
∎

We end this section with a useful result, the proof of which we leave as a good exercise for the reader.

METATHEOREM 11.12
If $A(x_1, \ldots, x_n)$ is a formula containing the variables $x_1, \ldots, x_n$, then

$$\vdash A(x_1, \ldots, x_n) \quad \text{iff} \quad \vdash \forall x_1 \ldots \forall x_n (A(x_1, \ldots, x_n)).$$

11.3 Soundness, Noncontradiction (Consistency) and Completeness of Predicate Logic

The first result concerning the connection between syntactic and semantic truth structures of predicate logic is the so-called soundness metatheorem which asserts that all (formal) theorems are valid formulas. The proof uses certain facts established in Chapter 10 and is rather simple. As an easy consequence of the soundness metatheorem we will obtain the noncontradiction of predicate logic. The converse of the soundness metatheorem is the completeness metatheorem. The proof of this last important result is more difficult and requires deeper considerations.

METATHEOREM 11.13
Each theorem of predicate logic is a valid formula, i.e.,

$$\text{if } \vdash F, \text{ then } \models F.$$

PROOF   Suppose that the formula $F \in \mathcal{L}(\mathbf{X})$ is a theorem of predicate logic and let the sequence

$$(*) \quad F_1, F_2, \ldots, F_n \ (n \geq 1)$$

with $F_n \equiv F$, be a (formal) proof of F. We know that all axioms of predicate logic are valid formulas (Proposition 10.11 and Proposition 10.16) and that inference rules MP, $\exists \Rightarrow$ preserve validity (Proposition 10.12 and Proposition 10.17). The first formula F_1 in the above sequence $(*)$ is necessarily an axiom, hence it is a valid formula. If we assume that $i > 1$ and that the formulas $F_1, F_2, \ldots, F_{i-1}$ in $(*)$ are valid, then the formula F_i is an axiom or is obtained from the formulas preceding F_i in $(*)$ by some inference rule; in both cases F_i is a valid formula. Therefore, the assertion follows by the principle of mathematical induction. ∎

METATHEOREM 11.14 (Consistency of predicate logic)
The predicate logic is non-contradictory, i.e., there is no formula $F \in \mathcal{L}(\mathbf{X})$ such that

$$\vdash F \quad and \quad \vdash \neg F.$$

PROOF   Suppose that there is a formula $F \in \mathcal{L}(\mathbf{X})$ such that $\vdash F$ and $\vdash \neg F$. From Metatheorem 11.13 it follows that $\models F$ and $\models \neg F$, which is impossible; indeed, for any interpretation I we will have $I(F) = 1$ and $I(\neg F) = 1$, but $I(\neg F) = \overline{I(F)} = 0$. ∎

Now let us go to the problem of completeness of predicate logic. For this purpose we begin with some preliminary considerations. The completeness metatheorem asserts that any valid formula in $\mathcal{L}(\mathbf{X})$ is a theorem of predicate logic (i.e., it is provable in the syntactic truth structure of $\mathcal{L}(\mathbf{X})$). This remarkable result was proved by Kurt Gödel in 1930. After Gödel, many authors (J. Herbrand (1931), L. Henkin (1949), G. Hasenjäger (1953) and others) have analyzed from different points of view the completeness problem and have obtained new interesting proofs of Gödel's result. Our proof below is based on some ideas of A. Lindenbaum, J. Hasenjäger and E. Mendelson.

We know that each alphabet of predicate logic contains literal symbols, logical symbols and auxiliary symbols. It follows that two alphabets $\mathbf{X}$ and $\mathbf{Y}$ coincide if and only if they have the same literal symbols. Two alphabets $\mathbf{X}$ and $\mathbf{Y}$ may coincide or not; in the first case we will use the notation $\mathbf{X} = \mathbf{Y}$, while in the second, we will use the notation $\mathbf{X} \neq \mathbf{Y}$. If all literal symbols of $\mathbf{X}$ are also literal symbols in $\mathbf{Y}$, then we will write $\mathbf{X} \subseteq \mathbf{Y}$; if $\mathbf{X} \subseteq \mathbf{Y}$ and $\mathbf{X} \neq \mathbf{Y}$, we will use the notation $\mathbf{X} \subset \mathbf{Y}$.

If $\mathbf{X}$ and $\mathbf{Y}$ are two alphabets of predicate logic, the corresponding languages will be denoted by $\mathcal{L}(\mathbf{X})$ and $\mathcal{L}(\mathbf{Y})$. It is clear that if $\mathbf{X} \subseteq \mathbf{Y}$, then all terms and all formulas in $\mathbf{X}$ are, respectively, terms and formulas in $\mathbf{Y}$; in such a case we will write $\mathcal{L}(\mathbf{X}) \subseteq \mathcal{L}(\mathbf{Y})$. If $\mathbf{X} \subset \mathbf{Y}$, then obviously $\mathcal{L}(\mathbf{X}) \subseteq \mathcal{L}(\mathbf{Y})$ and $\mathcal{L}(\mathbf{X}) \neq \mathcal{L}(\mathbf{Y})$; in such a case we will write $\mathcal{L}(\mathbf{X}) \subset \mathcal{L}(\mathbf{Y})$.

If $\mathbf{X}, \mathbf{Y}$ are two alphabets of predicate logic, then the syntactic truth structures on $\mathcal{L}(\mathbf{X})$ and $\mathcal{L}(\mathbf{Y})$ will be denoted by $\Lambda(\mathbf{X})$ and $\Lambda(\mathbf{Y})$, respectively. If $\mathbf{X} \subseteq \mathbf{Y}$, hence $\mathcal{L}(\mathbf{X}) \subseteq \mathcal{L}(\mathbf{Y})$, then each axiom of $\Lambda(\mathbf{X})$ is an axiom of $\Lambda(\mathbf{Y})$ too, but obviously $\Lambda(\mathbf{Y})$ may have a larger set of axioms; it is important to note that the inference rules in $\Lambda(\mathbf{X})$ and $\Lambda(\mathbf{Y})$ are the same. If the formula $F \in \mathcal{L}(\mathbf{X})$ is a theorem in $\Lambda(\mathbf{X})$ and the formula $G \in \mathcal{L}(\mathbf{Y})$ is a theorem in $\Lambda(\mathbf{Y})$, then we will write, respectively,

$$\vdash_{\Lambda(\mathbf{X})} F \quad \text{and} \quad \vdash_{\Lambda(\mathbf{Y})} G.$$

Similarly, if $\mathcal{H} \subseteq \mathcal{L}(\mathbf{X})$ and $\mathcal{K} \subseteq \mathcal{L}(\mathbf{Y})$ are two sets of formulas, then if F and G are deducible in $\Lambda(\mathbf{X})$ and $\Lambda(\mathbf{Y})$, respectively, from $\mathcal{H}$ and $\mathcal{K}$, we will write

$$\mathcal{H} \vdash_{\Lambda(\mathbf{X})} F \quad \text{and} \quad \mathcal{K} \vdash_{\Lambda(\mathbf{Y})} G.$$

Let $\mathbf{Z}$ be an arbitrary alphabet of predicate logic and $\mathcal{H}$ be a set (possibly empty) of formulas in $\mathcal{L}(\mathbf{Z})$. If the formula $F \in \mathcal{L}(\mathbf{Z})$ is not deducible in $\Lambda(\mathbf{Z})$ from $\mathcal{H}$, then we will write

$$\text{not}(\mathcal{H} \vdash_{\Lambda(\mathbf{Z})} F).$$

In particular, if the formula $F \in \mathcal{L}(\mathbf{Z})$ is not a theorem in $\Lambda(\mathbf{Z})$, then, instead of writing $\text{not}(\emptyset \vdash_{\Lambda(\mathbf{Z})} F)$ we will use the notation

$$\text{not}(\vdash_{\Lambda(\mathbf{Z})} F).$$

Finally, some remarks about interpretations. Suppose that $\mathbf{X} \subseteq \mathbf{Y}$, consequently $\mathcal{L}(\mathbf{X}) \subseteq \mathcal{L}(\mathbf{Y})$. If $I_{\mathbf{Y}}$ is any interpretation of $\mathcal{L}(\mathbf{Y})$ with the domain $\mathcal{D}$, then it is clear that $I_{\mathbf{Y}}$ is also an interpretation of $\mathcal{L}(\mathbf{X})$ with the same domain $\mathcal{D}$. It follows that any valid formula $F \in \mathcal{L}(\mathbf{X})$ is true for all interpretations $I_{\mathbf{Y}}$ of $\mathcal{L}(\mathbf{Y})$, hence it is a valid formula in $\mathcal{L}(\mathbf{Y})$.

DEFINITION 11.6   *A term T in $\mathcal{L}(\mathbf{Z})$ is a* **closed term** *if T does not contain individual variables.*

DEFINITION 11.7   *Let $\mathcal{H}$ be a set of closed formulas in $\mathcal{L}(\mathbf{Z})$. We say that $\mathcal{H}$ is* **syntactically consistent** *in $\Lambda(\mathbf{Z})$, or, in short, s-consistent in $\Lambda(\mathbf{Z})$, if there is no closed formula F such that*

$$\mathcal{H} \vdash_{\Lambda(\mathbf{Z})} F \quad and \quad \mathcal{H} \vdash_{\Lambda(\mathbf{Z})} \neg F.$$

If such a closed formula F in $\mathcal{L}(\mathbf{Z})$ exists, then the set $\mathcal{H}$ is said to be **syntactically inconsistent**, *in short, s-inconsistent, in $\Lambda(\mathbf{Z})$.*

DEFINITION 11.8   *Let $\mathcal{H}$ be a set of closed formulas in $\mathcal{L}(\mathbf{Z})$. We say that $\mathcal{H}$ is* **syntactically complete**, *in short, s-complete, in $\Lambda(\mathbf{Z})$ if for each closed formula $F \in \mathcal{L}(\mathbf{Z})$ we have*

$$\mathcal{H} \vdash_{\Lambda(\mathbf{Z})} F \quad or \quad \mathcal{H} \vdash_{\Lambda(\mathbf{Z})} \neg F.$$

If the set $\mathcal{H}$ is empty, then we will say that the predicate logic $\Lambda(\mathbf{Z})$ is s-complete.

PROPOSITION 11.4
  Let F be a closed formula in $\mathcal{L}(\mathbf{Z})$ and $\mathcal{H}$ be a nonempty set of closed formulas in $\mathcal{L}(\mathbf{Z})$.
  (a) If $not(\vdash_{\Lambda(\mathbf{Z})} \neg F)$, then the set $\{F\}$ is s-consistent in $\Lambda(\mathbf{Z})$;
  (b) If $not(\mathcal{H} \vdash_{\Lambda(\mathbf{Z})} \neg F)$, then the set $\mathcal{H}' = \mathcal{H} \cup \{F\}$ is s-consistent in $\Lambda(\mathbf{Z})$.

PROOF   (a) Suppose that the closed formula $\neg F$ is not a theorem in $\Lambda(\mathbf{Z})$, i.e.,

$$(*) \quad not(\vdash_{\Lambda(\mathbf{Z})} \neg F)$$

and assume that the set $\{F\}$ is *s*-inconsistent. This means that there is a closed formula $G \in \mathcal{L}(\mathbf{Z})$ such that

$$F \vdash_{\Lambda(\mathbf{Z})} G \quad and \quad F \vdash_{\Lambda(\mathbf{Z})} \neg G.$$

Taking into account that F is a closed formula we can apply the Metatheorem 11.11 and so we obtain the two theorems below:

$$\vdash_{\Lambda(\mathbf{Z})} F \Rightarrow G \quad and \quad \vdash_{\Lambda(\mathbf{Z})} F \Rightarrow \neg G$$

From the first theorem, by contraposition we get $\vdash_{\Lambda(\mathbf{Z})} \neg G \Rightarrow \neg F$ and from this theorem and from the second theorem above, by the rule of syllogism we deduce the theorem

$$(**) \quad \vdash_{\Lambda(\mathbf{Z})} F \Rightarrow \neg F.$$

On the other hand, from the tautology $(X \Rightarrow \neg X) \Rightarrow \neg X$ we obtain the theorem

$$\vdash_{\Lambda(\mathbf{Z})} (F \Rightarrow \neg F) \Rightarrow \neg F,$$

which together with $(**)$, by the MP-rule, gives the theorem

$$\vdash_{\Lambda(\mathbf{Z})} \neg F.$$

But this last result is in contradiction with the assumption $(*)$. It follows that the set $\{F\}$ is s-consistent in $\Lambda(\mathbf{Z})$.

The assertion (b) follows in the same way.   ∎

PROPOSITION 11.5

Let $\mathcal{H}$ be a nonempty set of closed formulas in $\mathcal{L}(\mathbf{Z})$. If $\mathcal{H}$ is s-consistent in $\Lambda(\mathbf{Z})$, then there is a set $\mathcal{K}$ of closed formulas in $\mathcal{L}(\mathbf{Z})$, $\mathcal{K} \supseteq \mathcal{H}$, which is s-consistent and s-complete in $\Lambda(\mathbf{Z})$.

PROOF   Taking into account that $\mathbf{Z}$ is a countable set of symbols, it is easy to see that the set of all formal words in $\mathbf{Z}$ is countable, hence the set $\mathcal{L}(\mathbf{Z})$ is countable. In particular, the set of all closed formulas in $\mathcal{L}(\mathbf{Z})$, being also countable, can be represented (in many ways) as a sequence. Let us denote by

$$(*) \quad G_1, G_2, \ldots, G_n, \ldots$$

such a sequence of all closed formulas in $\mathcal{L}(\mathbf{Z})$. Now we will inductively construct the increasing sequence of sets

$$\mathcal{H}_0 \subseteq \mathcal{H}_1 \subseteq \ldots \subseteq \mathcal{H}_n \subseteq \ldots$$

of closed formulas $\mathcal{H}_n (n \geq 0)$ by the definition

$$\mathcal{H}_0 = \mathcal{H};$$

and for $n \geq 1$,

$$\mathcal{H}_n = \begin{cases} \mathcal{H}_{n-1}, & \text{if } \mathcal{H}_{n-1} \vdash_{\Lambda(\mathbf{Z})} \neg G_n, \\ \mathcal{H}_{n-1} \cup \{G_n\}, & \text{if } not(\mathcal{H}_{n-1} \vdash_{\Lambda(\mathbf{Z})} \neg G_n). \end{cases}$$

The set $\mathcal{H}_0$ is by hypothesis s-consistent; if the sets $\mathcal{H}_0, \mathcal{H}_1, \ldots, \mathcal{H}_{n-1}$ $(n \geq 1)$ are assumed to be s-consistent, then the set $\mathcal{H}_n$ is also s-consistent. Indeed, if $\mathcal{H}_{n-1} \vdash_{\Lambda(\mathbf{Z})} \neg G_n$, then $\mathcal{H}_n = \mathcal{H}_{n-1}$ and $\mathcal{H}_n$ is s-consistent; if $not(\mathcal{H}_{n-1} \vdash_{\Lambda(\mathbf{Z})} \neg G_n)$, then, by the Proposition 11.4, (b), the set $\mathcal{H}_n = \mathcal{H}_{n-1} \cup \{G_n\}$ is s-consistent in $\Lambda(\mathbf{Z})$. Thus, from the principle of mathematical induction, it follows that all sets $\mathcal{H}_n$ $(n \geq 0)$ are s-consistent.

We consider now the set

$$\mathcal{K} = \mathcal{H}_0 \cup \mathcal{H}_1 \cup \ldots \cup \mathcal{H}_n \cup \ldots = \bigcup_{n \geq 0} \mathcal{H}_n$$

and we intend to prove that $\mathcal{K}$ is s-consistent and s-complete.

Let us assume that $\mathcal{K}$ is s-inconsistent. This means that there is a closed formula F in $\mathcal{L}(\mathbf{Z})$ such that $\mathcal{K} \vdash_{\Lambda(\mathbf{Z})} F$ and $\mathcal{K} \vdash_{\Lambda(\mathbf{Z})} \neg F$; it follows that

$$\mathcal{K} \vdash_{\Lambda(\mathbf{Z})} F \wedge \neg F.$$

If the sequence

$$(**) \quad K_1, K_2, \ldots, K_r \equiv F \wedge \neg F \quad (r \geq 1)$$

is a deduction in $\Lambda(\mathbf{Z})$, from $\mathcal{K}$, of the formula $F \wedge \neg F$, then each formula K_i $(1 \leq i \leq r)$ satisfies one of the conditions:

(a) K_i is an axiom;

(b) K_i belongs to $\mathcal{K}$;

(c) K_i is obtained in $(*)$ by some inference rule.

Taking into account that

$$\mathcal{K} = \bigcup_{n \geq 0} \mathcal{H}_n,$$

it follows that for each formula K_i satisfying condition (b) there is an index $n_i \geq 0$ such that

$$K_i \in \mathcal{H}_{n_i}.$$

If we denote by p the greatest index n_i we deduce that each formula K_i in $(**)$, which satisfies condition (b), belongs to $\mathcal{H}_p$. Thus, each formula K_i in $(**)$ satisfies one of the conditions:

(a) K_i is an axiom;

(b) K_i belongs to $\mathcal{H}_p$;

(c) K_i is obtained in $(**)$ by some inference rule;

hence the sequence $(**)$ is a deduction in $\Lambda(\mathbf{Z})$ from $\mathcal{H}_p$ of the formula $F \wedge \neg F$. But this is impossible because $\mathcal{H}_p$ is s-consistent. Therefore the set $\mathcal{K}$ is s-consistent.

Finally, let us prove that $\mathcal{K}$ is s-complete. For this purpose let us consider an arbitrary closed formula $G \in \mathcal{L}(\mathbf{Z})$; the formula G (being closed) belongs necessarily to the sequence $(*)$, hence there is some index $n \geq 1$ such that $G \equiv G_n$. If $\mathcal{H}_{n-1} \vdash_{\Lambda(\mathbf{Z})} \neg G_n$, then

$$\mathcal{K} \vdash_{\Lambda(\mathbf{Z})} \neg G_n,$$

(because $\mathcal{H}_{n-1} \subset \mathcal{K}$); if not($\mathcal{H}_{n-1} \vdash_{\Lambda(\mathbf{Z})} \neg G_n$), then, by definition, $\mathcal{H}_n = \mathcal{H}_{n-1} \cup \{G_n\}$ and $\mathcal{H}_n \vdash_{\Lambda(\mathbf{Z})} G_n$, hence $\mathcal{K} \vdash_{\Lambda(\mathbf{Z})} G_n$. Taking into account that $G_n \equiv F$, it follows $\mathcal{K} \vdash_{\Lambda(\mathbf{Z})} \neg F$ or $\mathcal{K} \vdash_{\Lambda(\mathbf{Z})} F$; therefore, $\mathcal{K}$ is s-complete. ∎

In the following we will fix an alphabet $\mathbf{X}$ of predicate logic and we will denote as usual by $\mathbf{F}_0 = \{a_1, a_2, a_3, \ldots\}$ the set of all individual constants in $\mathbf{X}$; as we know, the set $\mathbf{F}_0$ is finite (possibly empty) or at most countable. Let us consider also a countable set

$$\beta = \{b_1, b_2, b_3, \ldots, b_n, \ldots\}$$

of symbols, different from all symbols of $\mathbf{X}$, and let us denote by $\mathbf{X}_\beta$ the alphabet obtained from $\mathbf{X}$ changing the set of individual constants by $\mathbf{F}_{0\beta} = \mathbf{F}_0 \cup \beta$. Thus the individual constants of $\mathbf{X}_\beta$ are $a_1, a_2, \ldots, b_1, b_2, \ldots, b_n, \ldots$, while the other literal symbols (variables, functors, propositional variables and predicators) of $\mathbf{X}_\beta$ are the same as those of $\mathbf{X}$. It is clear that $\mathbf{X} \subset \mathbf{X}_\beta$ and $\mathcal{L}(\mathbf{X}) \subset \mathcal{L}(\mathbf{X}_\beta)$. Denoting by $\Lambda(\mathbf{X})$ and $\Lambda(\mathbf{X}_\beta)$, respectively, the syntactic truth structures on $\mathcal{L}(\mathbf{X})$ and $\mathcal{L}(\mathbf{X}_\beta)$, it follows that all axioms of $\Lambda(\mathbf{X})$ are axioms in $\Lambda(\mathbf{X}_\beta)$, but the set of axioms of $\Lambda(\mathbf{X}_\beta)$ is larger; indeed, from each axiom F of $\Lambda(\mathbf{X})$ we can obtain infinitely many axioms of $\Lambda(\mathbf{X}_\beta)$ if we replace (a part of, or all) the variables and the constants $a_1, a_2, \ldots$ in F by constants from the set $\beta = \{b_1, b_2, \ldots, b_n, \ldots\}$. It follows that each proof and each deduction in $\Lambda(\mathbf{X})$ are, respectively, a proof and a deduction in $\Lambda(\mathbf{X}_\beta)$; in particular, if the set $\mathcal{H} \subset \mathcal{L}(\mathbf{X})$ is s-consistent in $\Lambda(\mathbf{X}_\beta)$, then $\mathcal{H}$ is s-consistent in $\Lambda(\mathbf{X})$.

PROPOSITION 11.6

Let $\mathcal{H}$ be a set of closed formulas in $\mathcal{L}(\mathbf{X})$. If $\mathcal{H}$ is s-consistent in $\Lambda(\mathbf{X})$, then it is s-consistent in $\Lambda(\mathbf{X}_\beta)$.

PROOF   Suppose that the set $\mathcal{H}$ of closed formulas is s-consistent in $\Lambda(\mathbf{X})$ and assume that $\mathcal{H}$ is s-inconsistent in $\Lambda(\mathbf{X}_\beta)$. This means that there is a closed formula F in $\mathcal{L}(\mathbf{X}_\beta)$ such that

$$\mathcal{H} \vdash_{\Lambda(\mathbf{X}_\beta)} F \quad \text{and} \quad \mathcal{H} \vdash_{\Lambda(\mathbf{X}_\beta)} \neg F,$$

hence

$$\mathcal{H} \vdash_{\Lambda(\mathbf{X}_\beta)} F \wedge \neg F.$$

Let the sequence

$$(*) \qquad K_1, K_2, \ldots, K_r \equiv F \wedge \neg F \quad (r \geq 1),$$

be a deduction from $\mathcal{H}$ in $\Lambda(\mathbf{X}_\beta)$ of the formula $F \wedge \neg F$. In the sequence $(*)$ there are necessarily formulas which contain individual constants belonging to the set $\beta = \{b_1, b_2, \ldots, b_n, \ldots\}$, otherwise the sequence $(*)$ would be a deduction from $\mathcal{H}$ in $\Lambda(\mathbf{X})$ of the formula $F \wedge \neg F$, which is impossible because the set $\mathcal{H}$ is s-consistent in $\Lambda(\mathbf{X})$. Let $b_{m_1}, b_{m_2}, \ldots, b_{m_p}$ be the sequence of all individual constants from β which are contained in $(*)$ and let us take the sequence $x_{n_1}, x_{n_2}, \ldots, x_{n_p}$ of individual variables which are not contained in $(*)$. If we now replace in $(*)$ the individual constants $b_{m_1}, b_{m_2}, \ldots, b_{m_p}$ by the individual variables $x_{n_1}, x_{n_2}, \ldots, x_{n_p}$, respectively, then we get a new sequence

$$(**) \quad K_1', K_2', \ldots, K_r' \equiv F' \wedge \neg F'$$

of formulas belonging to $\mathcal{L}(\mathbf{X})$ (for some indices i $(1 \le i \le r)$ we may have $K_i \equiv K_i'$). Moreover, it is easy to see that the sequence $(**)$ is a deduction from $\mathcal{H}$ in $\Lambda(\mathbf{X})$ of the formula $K_r' \equiv F' \wedge \neg F'$, which is impossible because the set $\mathcal{H}$ is s-consistent in $\Lambda(\mathbf{X})$. Therefore, the set $\mathcal{H}$ is s-consistent in $\Lambda(\mathbf{X}_\beta)$. ∎

DEFINITION 11.9 *Let $\mathbf{Z}$ be an arbitrary alphabet of predicate logic and let $\mathcal{H}$ be a set of formulas in $\mathcal{L}(\mathbf{Z})$.*
(a) An interpretation I of $\mathcal{L}(\mathbf{Z})$ is called a **model** *of (for) $\mathcal{H}$ if each formula $H \in \mathcal{H}$ is true in the interpretation I. In other words, the interpretation I is a model of (for) $\mathcal{H}$ if and only if*

$$H \in \mathcal{H} \quad implies \quad I(H) = 1.$$

(b) A model I of (for) the set $\mathcal{H}$ is said to be **countable** *if the domain $\mathcal{D}$ of the interpretation I is a countable set.*
(c) If there is a model of (for) $\mathcal{H}$, then we say that $\mathcal{H}$ has a model.

By arguments similar to those used in the proof of Metatheorem 11.13 it is easy to establish the proposition below.

PROPOSITION 11.7
If the set $\mathcal{H} \subset \mathcal{L}(\mathbf{Z})$ has a model, then it is s-consistent.

The following deeper result represents the converse of the preceding proposition.

METATHEOREM 11.15
Let $\mathcal{H}$ be a set of closed formulas in $\mathcal{L}(\mathbf{X})$. If $\mathcal{H}$ is s-consistent in $\Lambda(\mathbf{X})$, then $\mathcal{H}$ has a countable model.

PROOF   Consider in $\mathcal{L}(\mathbf{X})$ the sequence $F_1, F_2, \ldots, F_n, \ldots$ of all formulas which contain just one individual (free) variable; denoting by x_{i_n} the unique individual variable contained in the formula F_n $(n \geq 1)$, we can write this sequence in the form:

$$(*) \qquad F_1(x_{i_1}), F_2(x_{i_2}), ..., F_n(x_{i_n}), ...$$

Now we choose the individual constants

$$(**) \qquad b_{j_1}, b_{j_2}, ..., b_{j_n}, ...$$

such that, for each $n \geq 1$, b_{j_n} is not contained in the formulas $F_1(x_{i_1})$, ..., $F_n(x_{i_n})$ and b_{j_n} is distinct from $b_{j_1}, b_{j_2}, ..., b_{j_{n-1}}$. Using the formulas $(*)$ and the individual constants $(**)$, we construct for each $n \geq 1$ the formula

$$A_n \equiv \exists x_{i_n} \neg F_n(x_{i_n}) \Rightarrow \neg F_n(b_{j_n}).$$

Now we define the sets of closed formulas in $\mathcal{L}(\mathbf{X}_\beta)$:

$$\mathcal{H}_0 \subset \mathcal{H}_1 \subset ... \subset \mathcal{H}_n \subset ...$$

where $\mathcal{H}_0 = \mathcal{H}$, and $\mathcal{H}_n = \mathcal{H}_{n-1} \cup \{A_n\}$ for $n \geq 1$. Let us prove by induction that each set $\mathcal{H}_n$ $(n \geq 0)$ is s-consistent in $\Lambda(\mathbf{X}_\beta)$. By hypothesis the set $\mathcal{H}_0 = \mathcal{H}$ is s-consistent in $\Lambda(\mathbf{X})$; by Proposition 11.6 $\mathcal{H}_0$ is s-consistent in $\Lambda(\mathbf{X}_\beta)$, too. Suppose now that $n > 1$ and that $\mathcal{H}_{n-1}$ is s-consistent in $\Lambda(\mathbf{X}_\beta)$; if we assume, however, that $\mathcal{H}_n$ is s-inconsistent in $\Lambda(\mathbf{X}_\beta)$, then there is a closed formula $G \in \mathcal{L}(\mathbf{X}_\beta)$ such that

$$\mathcal{H}_n \vdash_{\Lambda(\mathbf{X}_\beta)} G \quad \text{and} \quad \mathcal{H}_n \vdash_{\Lambda(\mathbf{X}_\beta)} \neg G.$$

But from the tautology $X \Rightarrow (\neg X \Rightarrow Y)$ it follows $\mathcal{H}_n \vdash_{\Lambda(\mathbf{X}_\beta)} G \Rightarrow (\neg G \Rightarrow F)$ for any formula $F \in \mathcal{L}(\mathbf{X}_\beta)$. Using the two deductions above and applying the MP-rule two times we get $\mathcal{H}_n \vdash_{\Lambda(\mathbf{X}_\beta)} F$. So, from $\mathcal{H}_n$ one can deduce the arbitrary formula F. In particular, we have $\mathcal{H}_n \vdash_{\Lambda(\mathbf{X}_\beta)} \neg A_n$. But $\mathcal{H}_n = \mathcal{H}_{n-1} \cup \{A_n\}$, hence $\mathcal{H}_{n-1}, A_n \vdash_{\Lambda(\mathbf{X}_\beta)} \neg A_n$ and taking into account that A_n is a closed formula we get

$$\mathcal{H}_{n-1} \vdash_{\Lambda(\mathbf{X}_\beta)} A_n \Rightarrow \neg A_n.$$

From the tautology $(X \Rightarrow \neg X) \Rightarrow \neg X$ we also have

$$\mathcal{H}_{n-1} \vdash_{\Lambda(\mathbf{X}_\beta)} (A_n \Rightarrow \neg A_n) \Rightarrow \neg A_n$$

and applying the MP-rule we obtain

$$\mathcal{H}_{n-1} \vdash_{\Lambda(\mathbf{X}_\beta)} \neg A_n.$$

But $\neg A_n \equiv \neg[\exists x_{i_n} \neg F_n(x_{i_n}) \Rightarrow \neg F_n(b_{j_n})]$ and using the tautology $[\neg(X \Rightarrow Y)] \Rightarrow (X \wedge \neg Y)$ we obtain

$$\mathcal{H}_{n-1} \vdash_{\Lambda(\mathbf{X}_\beta)} \neg A_n \Rightarrow [\exists x_{i_n} \neg F_n(x_{i_n}) \wedge \neg\neg F_n(b_{j_n})]$$

from which by MP-rule we get

$$\mathcal{H}_{n-1} \vdash_{\Lambda(\mathbf{X}_\beta)} (\exists x_{i_n} \neg F_n(x_{i_n}) \wedge (\neg\neg F_n(b_{j_n}))).$$

Therefore, taking into account the tautology $\neg\neg X \Rightarrow X$ we obtain

(a) $\mathcal{H}_{n-1} \vdash_{\Lambda(\mathbf{X}_\beta)} \exists x_{i_n} \neg F_n(x_{i_n})$ and (b) $\mathcal{H}_{n-1} \vdash_{\Lambda(\mathbf{X}_\beta)} F_n(b_{j_n})$.

From (b) we deduce that there is a deduction

$$(* * *)\quad E_1, E_2, ..., E_s \equiv F_n(b_{j_n})$$

from $\mathcal{H}_{n-1}$ in $\Lambda(\mathbf{X}_\beta)$ of the formula $E_s \equiv F_n(b_{j_n})$. We note that there is no formula in $\mathcal{H}_{n-1}$ containing the individual constant b_{j_n}. Let x_m be an individual variable which is not contained in the deduction $(* * *)$. If we replace in all formulas of $(* * *)$ the constant b_{j_n} by the variable x_m, then it is easy to see that we get a deduction from $\mathcal{H}_{n-1}$ of the formula $F_n(x_m)$ i.e.,

$$\mathcal{H}_{n-1} \vdash_{\Lambda(\mathbf{X}_\beta)} F_n(x_m).$$

Using the Metatheorem 11.12 we get

$$\mathcal{H}_{n-1} \vdash_{\Lambda(\mathbf{X}_\beta)} \forall x_m F_n(x_m)$$

and taking into account that

$$\forall x_m F_n(x_m) \equiv \neg\exists x_m \neg F_n(x_m) \equiv \neg(\exists x_{i_n} \neg F_n(x_{i_n}))$$

we obtain finally

$$\mathcal{H}_{n-1} \vdash_{\Lambda(\mathbf{X}_\beta)} \neg(\exists x_{i_n} \neg F_n(x_{i_n})).$$

This last deduction together with the deduction (a) above is in contradiction with the inductive assumption (that $\mathcal{H}_{n-1}$ is s-consistent in $\Lambda(\mathbf{X}_\beta)$). Thus, from the induction principle it follows that all sets $\mathcal{H}_n$ $(n \geq 0)$ are s-consistent in $\Lambda(\mathbf{X}_\beta)$.

Now it is easy to see that the set

$$\mathcal{H}_\infty = \bigcup_{n \geq 0} \mathcal{H}_n$$

is s-consistent in $\Lambda(\mathbf{X}_\beta)$. Indeed, each deduction from $\mathcal{H}_\infty$ in $\Lambda(\mathbf{X}_\beta)$ (being a finite sequence in $\mathcal{H}_\infty$) is actually a deduction from some $\mathcal{H}_n$ in $\Lambda(\mathbf{X}_\beta)$. It follows that each deduction from $\mathcal{H}_\infty$ of a contradiction is necessarily a deduction from $\mathcal{H}_n$ (for some $n \geq 0$) of a contradiction; but this is impossible because each set $\mathcal{H}_n$ $(n \geq 0)$ is s-consistent in $\Lambda(\mathbf{X}_\beta)$. Therefore, the set $\mathcal{H}_\infty$ is s-consistent in $\Lambda(\mathbf{X}_\beta)$.

From Proposition 11.5 it follows that there is a set $\mathcal{K}$ of closed formulas, $\mathcal{K} \supset \mathcal{H}_\infty$, which is s-consistent and s-complete in $\Lambda(\mathbf{X}_\beta)$.

Finally we will prove that there is an interpretation I of $\mathcal{L}(\mathbf{X}_\beta)$ which is a model of $\mathcal{H}$.

We know that each interpretation I of $\mathcal{L}(\mathbf{X}_\beta)$ is uniquely determined by the corresponding interpretation $I_0 = (\mathcal{D}, \xi, \eta_0, \eta, \zeta_0, \zeta)$ of the literal symbols in $\mathbf{X}_\beta$ (see Definition 10.1, Definition 10.5, Definition 10.6, Proposition 10.1 and Proposition 10.6).

Let us define the interpretation I_0 of literal symbols for the required interpretation I of $\mathcal{L}(\mathbf{X}_\beta)$. The domain $\mathcal{D}$ of the interpretation will be the set of all closed terms in $\mathcal{L}(\mathbf{X}_\beta)$ (see Definition 11.6); ξ will be an arbitrary function from the set $\mathbf{V}$ of individual variables in $\mathbf{X}_\beta$ (hence in $\mathbf{X}$) into the set $\mathcal{D}$; η_0 will be the identity function from $\mathbf{F}_{0\beta}$ onto itself i.e., η_0 associates to each individual constant from $\{a_1, a_2, ...\} \cup \{b_1, b_2, ..., b_n, ...\}$ the same individual constant (which is a closed term, hence belongs to $\mathcal{D}$); η will be a function which associates to each functor $f \in \mathbf{F}$ with $\alpha(f) = n \geq 1$ the function $\eta(f) = \hat{f}$ from $\mathcal{D}^n$ in $\mathcal{D}$, where the function $\hat{f}$ is defined as follows: if $t_1, t_2, ..., t_n \in \mathcal{D}$, then $\hat{f}(t_1, t_2, ..., t_n) = f(t_1, t_2, ..., t_n) \in \mathcal{D}$; ζ_0 is an arbitrary function which associates to each propositional variable $p \in \mathbf{P}_0$ certain element $\zeta_0(p) = \hat{p}$ belonging to $\Gamma = \{0, 1\}$; finally, ζ is a function which associates to each predicator $P \in \mathbf{F}$, $\alpha(P) = n \geq 1$, the function $\zeta(P) = \hat{P}$ from $\mathcal{D}^n$ into Γ defined as follows: if $t_1, t_2, ..., t_n \in \mathcal{D}$, then

$$\hat{P}(t_1, t_2, ..., t_n) = \begin{cases} 1, & \text{if } \mathcal{K} \vdash_{\Lambda(\mathbf{X}_\beta)} P(t_1, t_2, ..., t_n), \\ 0, & \text{if } not(\mathcal{K} \vdash_{\Lambda(\mathbf{X}_\beta)} P(t_1, t_2, ..., t_n)). \end{cases}$$

The interpretation I_0 of literal symbols of $\mathbf{X}_\beta$ generates a unique interpretation I of $\mathcal{L}(\mathbf{X}_\beta)$. We will prove that, for any closed formula $F \in \mathcal{L}(\mathbf{X}_\beta)$,

$$I(F) = 1 \text{ if and only if } \mathcal{K} \vdash_{\Lambda(\mathbf{X}_\beta)} F.$$

We will proceed by induction on $m = m(F) \geq 0$, where $m(F)$ is the number of operators and quantifiers which occur in the closed formula $F \in \mathcal{L}(\mathbf{X}_\beta)$.

If $m(F) = 0$, then F is an elementary closed formula in $\mathcal{L}(\mathbf{X}_\beta)$; this means that $F \equiv P(t_1, t_2, ..., t_n)$, where P is some predicator with $\alpha(P) = n$ and $t_1, t_2, ..., t_n$ are closed terms. In this case, by the definition of $\zeta(P) = \hat{P}$ we have

$$I(F) = 1 \text{ if and only if } \mathcal{K} \vdash_{\Lambda(\mathbf{X}_\beta)} F.$$

Thus, for $m(F) = 0$ the assertion is true.

Suppose now that $m > 1$ and assume that for all closed formulas F' in $\mathcal{L}(\mathbf{X}_\beta)$ with $m(F') < m$ the assertion is true. Now take a closed formula F with $m(F) = m$. We have to analyze three possibilities:

(a) $F \equiv \neg(G)$, where G is a closed formula with $m(G) < m$;

(b) $F \equiv (H) \vee (K)$, where H, K are closed formulas with $m(H) < m$ and $m(K) < m$;

(c) $F \equiv \exists x[A(x)]$, where $A(x)$ is a formula containing just one individual variable and $m(A(x)) < m$.

In the case (a), $I(F) = 1$ if and only if $I(G) = 0$. By the inductive assumption $I(G) = 0$ if and only if $\mathrm{not}(\mathcal{K} \vdash_{\Lambda(\mathbf{X}_\beta)} G)$. But G is a closed formula and $\mathcal{K}$ is s-complete, hence $\mathcal{K} \vdash_{\Lambda(\mathbf{X}_\beta)} \neg G \equiv F$. Thus, in case (a) the assertion is proved.

In case (b), suppose that $I(F) = 1$. It follows that $I(H) = 1$ or $I(K) = 1$. From the inductive assumption it follows that $\mathcal{K} \vdash_{\Lambda(\mathbf{X}_\beta)} H$ or $\mathcal{K} \vdash_{\Lambda(\mathbf{X}_\beta)} K$, hence $\mathcal{K} \vdash_{\Lambda(\mathbf{X}_\beta)} (H) \vee (K) \equiv F$. Suppose now that $I(F) = 0$, hence $I(H) = 0$ and $I(K) = 0$. From the inductive assumption it follows

$$\mathrm{not}(\mathcal{K} \vdash_{\Lambda(\mathbf{X}_\beta)} H) \quad \text{and} \quad \mathrm{not}(\mathcal{K} \vdash_{\Lambda(\mathbf{X}_\beta)} K).$$

Taking into account that H and K are closed formulas and that $\mathcal{K}$ is s-complete in $\Lambda(\mathbf{X}_\beta)$ we obtain

$$\mathcal{K} \vdash_{\Lambda(\mathbf{X}_\beta)} \neg H \quad \text{and} \quad \mathcal{K} \vdash_{\Lambda(\mathbf{X}_\beta)} \neg K$$

and consequently,

$$\mathcal{K} \vdash_{\Lambda(\mathbf{X}_\beta)} (\neg H) \wedge (\neg K) \equiv \neg (H \vee K) \equiv \neg F.$$

But the set $\mathcal{K}$ is s-consistent, therefore

$$\mathrm{not}(\mathcal{K} \vdash_{\Lambda(\mathbf{X}_\beta)} F).$$

This means that $I(F) = 1$ if and only if $\mathcal{K} \vdash_{\Lambda(\mathbf{X}_\beta)} F$, hence the assertion is proved in case (b), too.

In case (c), $F \equiv \exists x A(x)$ where the formula $A(x)$ contains just one individual variable; therefore there is some $n \geq 1$ such that $A(x_{i_n})$ belongs to the sequence (*). It follows that $A(x_{i_n}) \equiv F_n(x_{i_n})$ and we can write

$$F \equiv \exists x_{i_n} F_n(x_{i_n}).$$

Suppose that $I(F) = 1$; from the definition of the interpretations, $1 = I(F) = \max_{d \in \mathcal{D}} h_{F_n}(d)$. It follows that there is some element (closed term) t in $\mathcal{L}(\mathbf{X}_\beta)$ such that $h_{F_n}(\hat{t}) = 1$, where $\hat{t} = I(t) \in \mathcal{D}$. From the Proposition 10.8 it follows that

$$I(F_n(t|x_{i_n})) = h_{F_n(x_{i_n})}(\hat{t}) = 1.$$

Applying the inductive assumption we obtain

$$\mathcal{K} \vdash_{\Lambda(\mathbf{X}_\beta)} F_n(t|x_{i_n})$$

and using the axiom $F_n(t|x_{i_n}) \Rightarrow \exists x_{i_n} F_n(x_{i_n})$ we get

$$\mathcal{K} \vdash_{\Lambda(\mathbf{X}_\beta)} \exists x_{i_n} F_n(x_{i_n}) \equiv F.$$

Suppose now that $I(F) = 0$, hence $\max_{d \in \mathcal{D}} h_{F_n}(d) = 0$. It follows that $h_{F_n}(d) = 0$ for all $d \in \mathcal{D}$. In particular (we know that $b_i \in \mathcal{D}$), we obtain $h_{F_n(b_i|x_{i_n})}(b_i) = 0$ for all $i \geq 1$; this means that $I(F_n(b_i|x_{i_n})) = 0$. Taking into account that $F_n(b_i|x_{i_n})$ denoted by $F_n(b_i)$ is a closed formula with $m(F_n(b_i)) < m$, we may apply the inductive assumption and so we get

$$(1) \qquad \mathcal{K} \vdash_{\Lambda(\mathbf{X}_\beta)} \neg F_n(b_i), \quad \text{for all} \quad i \geq 1.$$

On the other hand, the formula $\neg F_n(x_{i_n})$ also belongs to the sequence $(*)$ because it contains a unique (free) individual variable. Therefore, there is some $p \geq 1$ such that

$$\neg F_n(x_{i_n}) \equiv F_p(x_{i_p}),$$

where $x_{i_p} \equiv x_{i_n}$. With this remark the formula $A_k \equiv \exists x_{i_p} \neg F_p(x_{i_p}) \Rightarrow \neg F_p(b_{j_p})$ is deducible from $\mathcal{H}_\infty$ in $\Lambda(\mathbf{X}_\beta)$, hence

$$\mathcal{K} \vdash_{\Lambda(\mathbf{X}_\beta)} \exists x_{i_p} \neg F_p(x_{i_p}) \Rightarrow \neg F_p(b_{j_p}).$$

But $F_p(x_{i_p}) \equiv \neg F_n(x_{i_n})$, hence we get

$$\mathcal{K} \vdash_{\Lambda(\mathbf{X}_\beta)} \exists x_{i_n} \neg\neg F_n(x_{i_n}) \Rightarrow \neg\neg F_n(b_{j_p})$$

or

$$\mathcal{K} \vdash_{\Lambda(\mathbf{X}_\beta)} \exists x_{i_n} F_n(x_{i_n}) \Rightarrow F_n(b_{j_p}).$$

From this deduction we obtain by contraposition

$$(2) \qquad \mathcal{K} \vdash_{\Lambda(\mathbf{X}_\beta)} \neg F_n(b_{j_p}) \Rightarrow \neg \exists x_{i_n} F_n(x_{i_n}).$$

Taking into account that $\neg \exists x_{i_n} F_n(x_{i_n}) \equiv \neg F$ and putting in (1) $i = j_p$, we get from (1), (2) by the MP-rule

$$\mathcal{K} \vdash_{\Lambda(\mathbf{X}_\beta)} \neg F.$$

So, if $I(F) = 0$, then $\text{not}(\mathcal{K} \vdash_{\Lambda(\mathbf{X}_\beta)} \neg F)$. We have proved that

$$I(F) = 1 \quad \text{if and only if} \quad \mathcal{K} \vdash_{\Lambda(\mathbf{X}_\beta)} F$$

for each closed formula $F \in \mathcal{L}(\mathbf{X}_\beta)$.

Now, if $F \in \mathcal{H}$, then $\mathcal{H} \vdash_{\Lambda(\mathbf{X}_\beta)} F$, hence $\mathcal{K} \vdash_{\Lambda(\mathbf{X}_\beta)} F$; it follows that $I(F) = 1$. This means that the interpretation I is a model for $\mathcal{H}$. The set $\mathcal{D}$ being countable, the model I is countable by definition. ∎

From the previous metatheorem we obtain the following result.

METATHEOREM 11.16 (Completeness of predicate logic)
Each valid formula in $\mathcal{L}(\mathbf{X})$ is a theorem in $\Lambda(\mathbf{X})$. In other words,

$$\text{if} \quad \models F, \quad \text{then} \quad \vdash F.$$

PROOF   Let $F \in \mathcal{L}(\mathbf{X})$ be a valid formula. Taking into account that a formula F is valid if and only if its closure is valid (Proposition 10.18) and that F is a theorem in $\Lambda(\mathbf{X})$ if and only if its closure is a theorem (Metatheorem 11.12), we can suppose that the valid formula F is closed. If we assume that

$$not(\vdash_{\Lambda(\mathbf{X})} F),$$

then, by Proposition 11.4, the set $\{\neg F\}$ is s-consistent in $\Lambda(\mathbf{X})$. From Metatheorem 11.15, it follows that $\{\neg F\}$ has a (countable) model, i.e., there is an interpretation I of $\mathcal{L}(\mathbf{X})$ such that $I(\neg F) = 1$ or $I(F) = 0$. On the other hand, the formula (being valid) is true in each interpretation, in particular $I(F) = 1$. This contradiction shows that $\vdash_{\Lambda(\mathbf{X})} F$. ∎

REMARK 11.6   Usually, if the alphabet $\mathbf{X}$ of predicate logic is fixed and no confusion is possible, the assertion "F is a theorem of predicate logic" is denoted by $\vdash F$. But in this section we have changed the alphabet, taking sometimes $\mathbf{Z}$ or $\mathbf{X}_\beta$ instead of $\mathbf{X}$ and, consequently, it was necessary to specify the languages $\mathcal{L}(\mathbf{Z})$ or $\mathcal{L}(\mathbf{X}_\beta)$ and the corresponding syntactic truth structures $\Lambda(\mathbf{Z}), \Lambda(\mathbf{X}_\beta)$ on these languages. That is why for assertions like "F is a theorem of predicate logic on $\mathcal{L}(\mathbf{Z})$" and "F is a theorem of predicate logic in $\mathcal{L}(\mathbf{X}_\beta)$" we have used, respectively, more specific notations like $\vdash_{\Lambda(\mathbf{Z})} F$ and $\vdash_{\Lambda(\mathbf{X}_\beta)} F$. Therefore, if the alphabet $\mathbf{X}$ is given and no confusion is possible, then the notations $\vdash F$ and $\vdash_{\Lambda(\mathbf{X})} F$ have the same meaning. ∎

From Metatheorem 11.13 and Metatheorem 11.16 we get, as a corollary, the following important result, called the completeness metatheorem of K. Gödel (1930).

METATHEOREM 11.17
A formula $F \in \mathcal{L}(\mathbf{X})$ is a theorem of predicate logic if and only if F is a valid formula, or, in short

$$\vdash F \quad \text{iff} \quad \models F.$$

Another consequence of Metatheorem 11.15 is the following result (L. Löwenheim (1915) and T. Skolem (1919)).

METATHEOREM 11.18
If the set $\mathcal{H} \subset \mathcal{L}(\mathbf{X})$ has a model, then it has a countable model.

PROOF   If $\mathcal{H}$ has a model, then, by Proposition 11.7, it is s-consistent, hence (by Metatheorem 11.15) $\mathcal{H}$ has a countable model.   ∎

12

Elements of Fuzzy Predicate Logic

The contents of this chapter are a natural continuation of Chapter 7 in Part I. As we already have mentioned, our treatment of this subject follows closely the ideas of the Czech school of logic, mainly those in Novák's paper [36].

Thus, we shall keep the language $\mathcal{L}(\mathbf{X})$ of crisp predicate logic with the difference that we shall replace the primitive logical operator $\vee$ by $\Rightarrow$ and we shall introduce additionally the derived logical operators ∇ and $\&$ (Lukasiewicz's disjunction and conjunction). The notion of interpretation will be adapted; as in the case of propositional logic, the values of each interpretation will belong to a fixed finite MV-algebra $\mathcal{A}$ (which will play the role of the set of truth values). The truth structure in semantic and syntactic versions will briefly be discussed.

12.1 The Language of Fuzzy Predicate Logic

We shall denote by $\mathbf{X}_f$ the alphabet of the language of fuzzy predicate logic. The alphabet $\mathbf{X}_f$ is obtained from the alphabet of crisp predicate logic, but taking $\Rightarrow$ as a primitive operator instead of $\vee$ (the other symbols remaining unchanged). With this change, the general notions of formal words, terms and formulas remain the same; in particular, we keep the notion of elementary formula. For instance, the definition of formulas will be the following:

DEFINITION 12.1 *A finite sequence of (formal) words in $\mathbf{X}_f$*

$$F_1, F_2, \ldots, F_r \quad (r \geq 1)$$

is called **construction of formulas** *(in $\mathbf{X}_f$) if for each index i, $(1 \leq i \leq r)$ the word F_i satisfies one of the conditions below:*

(a) F_i is an elementary formula;

(b) there is an index j $(j < i)$ such that $F_i \equiv \neg(F_j)$;

(c) there are indices h, k $(h, k < i)$ such that $F_i \equiv (F_h) \Rightarrow (F_k)$;

(d) there is an index j $(j < i)$ and an individual (free) variable x contained in F_j such that $F_i \equiv \exists x(F_j)$.

Every word which can be inserted in a construction of formulas is called a formula in $\mathbf{X}_f$. *The set of all formulas in* $\mathbf{X}_f$ *will be called* **the (formal) language of fuzzy predicate logic** *and it will be denoted by* $\mathcal{L}_f = \mathcal{L}(\mathbf{X}_f)$.

REMARK 12.1   In many papers concerning formal fuzzy logic, the set $\mathcal{A}$ of truth values is included "isomorphically" in the alphabet $\mathbf{X}_f$. This approach has the drawback that the alphabet $\mathbf{X}_f$ becomes, at least at first glance, an uncountable set (therefore, the language of fuzzy predicate logic will be also uncountable) but, at the same time, such inclusion presents some technical advantages (in particular, the negation $\neg$ can be introduced as a derived operation). On the other hand, some other connectives c_j are often included in the alphabet. For our limited needs we shall consider only two additional (derived) connectives, called, respectively, Łukasiewicz's conjunction and Łukasievicz's disjunction.   ∎

The following assertion is a direct consequence of Definition 12.1.

PROPOSITION 12.1
If F, G are arbitrary formulas in $\mathcal{L}(\mathbf{X}_f)$, then the formal words below

$$(F) \vee (G) \equiv ((F) \Rightarrow (G)) \Rightarrow (G) \qquad \text{(disjunction)};$$

$$(F) \wedge (G) \equiv \neg(((F) \Rightarrow (G)) \Rightarrow \neg(F)) \qquad \text{(conjunction)};$$

$$(F) \Leftrightarrow (G) \equiv ((F) \Rightarrow (G)) \wedge ((G) \Rightarrow (F)) \qquad \text{(equivalence)};$$

$$(F)\&(G) \equiv \neg((F) \Rightarrow \neg(G)) \qquad \text{(Łukasiewicz's conjunction)};$$

$$(F)\nabla(G) \equiv \neg(\neg(F)\&\neg(G)) \qquad \text{(Łukasiewicz's disjunction)};$$

$$\forall x(F) \equiv \neg(\exists x(\neg(F))) \qquad \text{(universal quantification)};$$

are also formulas in $\mathcal{L}(\mathbf{X}_f)$.

Of course, we shall keep the conventions adopted in Chapter 10 concerning the sequences of arguments of a formula or of a term and the sequence of free variables contained in a formula F. In particular, we keep the following definition.

DEFINITION 12.2   *The term $t \equiv t(V)$ is said to be* **free for x in the formula** $F \equiv F(x)$ *if any subformula of F which contains a free occurrence of x is not quantified with respect to a variable contained in t.*

PROPOSITION 12.2

If $F \equiv F(x) \equiv F(U_F(x))$ is a formula containing the variable x and if $t = T(V)$ is a term, then the word $F(t|x)$ obtained from F by putting $t(V)$ instead of each occurrence of x in F is also a formula.

12.2   The Semantic Truth Structure of Fuzzy Predicate Logic

As in the case of predicate logic (as well as in the case of fuzzy propositional logic) the semantic truth structure on $\mathcal{L}_f = \mathcal{L}(\mathbf{X}_f)$ is defined by means of the notion of **interpretation**. Thus, we have to introduce the appropriate notion of interpretation in our case. Taking into account that for fuzzy logic the set of truth values will be an MV-algebra $\mathcal{A} = \{0, \frac{1}{M}, \ldots, \frac{M-1}{M}, 1\}$, $M \geq 1$, any interpretation I of formulas in $\mathcal{L}(\mathbf{X}_f)$ must take its values in $\mathcal{A}$ and must have special properties similar to the ones of crisp predicate logic. Among such properties, for any interpretation I of $\mathcal{L}(\mathbf{X}_f)$ it is natural to suppose that $I(F \Rightarrow G) = I(F) \to I(G)$, where the symbol "$\to$" designates residuation in $\mathcal{A}$. It is clear that if $M = 1$, then $\mathcal{A} = \Gamma = \{0, 1\}$ and we get the crisp interpretations discussed in Chapter 10; that is why in what follows we shall suppose $M \geq 2$. The existence of $\mathcal{A}$-valued interpretations of $\mathcal{L}(\mathbf{X}_f)$ follows by almost identical inductive arguments with those used in Chapter 3. In other words, the construction of $\mathcal{A}$-valued interpretations of $\mathcal{L}(\mathbf{X}_f)$ can be performed, step by step, starting with an interpretation $I_0 = (\mathcal{D}, \xi, \eta_0, \eta, \zeta_0, \zeta)$ of literal symbols, in which the set Γ is replaced by the set $\mathcal{A}$. The correspondence $T \longmapsto g_T$ will be the same as in Chapter 10 (this correspondence does not depend on the set $\mathcal{A}$). The correspondence $F \longmapsto h_F$ does not coincide with the similar one constructed in Chapter 10 because the functions h_F are valued in $\mathcal{A}$. However, the existence and the uniqueness of this correspondence can be proved directly by inductive reasoning following the same line of ideas. For this reason we give without proof the corresponding assertion.

PROPOSITION 12.3

If $I_0 = (\mathcal{D}, \xi, \eta_0, \eta, \zeta_0, \zeta)$ is an interpretation of literal symbols of $\mathbf{X}_f$, then there is a unique correspondence

$$F \longmapsto h_F,$$

which associates to each formula $F \equiv F(U_F) \in \mathcal{L}(\mathbf{X}_f)$ a function $h_F : \mathcal{D}^m \to \mathcal{A}$ where $m = \lambda(U_F)$, with the following properties:

(a) *if $F \equiv p_j \in \mathbf{F}_0$, then h_F is the constant function $h_F = \zeta_0(p_j) = \hat{p}_j \in \mathcal{A}$; if $F \equiv P_j(T_1, \dots, T_n)$, where $P_j \in \mathbf{P}, \alpha(P_j) = n \geq 1$ and $T_1, \dots, T_n$ are terms, then*

$$h_F = \zeta(P_j)(g_{T_1}, \dots, g_{T_n});$$

(b) *if $F \equiv \neg(G)$, then $h_F = \overline{h_G}$;*

(c) *if $F \equiv (L) \Rightarrow (M)$, then*

$$h_F = h_L \to h_M;$$

(d) *if $F \equiv \exists x A(x)$, then*

$$h_F = \max_\omega h_A,$$

where ω is the set of occurrences of x in $A(x)$.

REMARK 12.2 From the above proposition it obviously follows that for each $F \in \mathcal{L}(\mathbf{X}_f)$ the function h_F takes values in $\mathcal{A}$. We have to mention additionally that in property (c) we have used the notation $h_L \to h_M$. The function $h_L \to h_M$ is defined in a similar way as the function $h_L \vee h_M$ in Chapter 10, Proposition 10.5. More precisely, if $h' : \mathcal{D}^p \to \mathcal{A}$ and $h'' : \mathcal{D}^q \to \mathcal{A}$ are given functions, then the function $h' \to h''$ has the arity $\alpha(h' \to h'') = \alpha(h') + \alpha(h'') = l$, $h' \to h'' : \mathcal{D}^l \to \mathcal{A}$ and is defined by the equality $(h' \to h'')(D) = (h' \to h'')(D'D'') = h'(D') \to h''(D'')$, where $D' \in \mathcal{D}^p, D'' \in \mathcal{D}^q$. In what concerns the function $\max_\omega h_A$ it is defined in the same way as in the case of crisp predicate logic with the difference that the element $\max_{d \in \mathcal{D}} h_A(\hat{U}_A(d))$ is computed in MV-algebra $\mathcal{A}$. ∎

DEFINITION 12.3 *For each term $T \equiv T(U)$ and each formula $F \equiv F(U_F)$, the elements $I(T) = g_T(\hat{U}) \in \mathcal{D}$ and $I(F) = h_F(\hat{U}_F) \in \mathcal{A}$ are called, respectively, the **interpretation** of T and the **interpretation** of F. The correspondence $F \longmapsto I(F)$ is called the $\mathcal{A}$-**interpretation** (generated by $I_0 = (\mathcal{D}, \xi, \eta_0, \eta, \zeta_0, \zeta)$) of the language $\mathcal{L}(\mathbf{X}_f)$ of fuzzy predicate logic.*

The following proposition also holds.

PROPOSITION 12.4

If $T \equiv T(U_T)$, which contains the variable x, $t \equiv t(V)$ are terms, then

$$I(T(t|x)) = g_T(\hat{U}_T(I(t)));$$

if additionally the term t is free for x in the formula $F \equiv F(x) \equiv F(U_F(x)))$, then

$$I(F(t|x)) = h_F(\hat{U}_F(I(t))).$$

REMARK 12.3   Taking into account that the set $\Gamma = \{0,1\}$ is a MV-subalgebra of $\mathcal{A}$ it follows that each interpretation $I_0 = (\mathcal{D}, \xi, \eta_0, \eta, \zeta_0, \zeta)$ of literal symbols of $\mathbf{X}_f$ in which the (constant) functions $\zeta_0(p_j)$ and the functions $\zeta(P_j)$ are Γ-valued, generates a Γ-interpretation of $\mathcal{L}(\mathbf{X}_f)$. Identifying in a natural way the Γ-interpretations of $\mathcal{L}(\mathbf{X}_f)$ with the interpretations of the language $\mathcal{L}(\mathbf{X})$ of crisp predicate logic we see that any interpretation of $\mathcal{L}(\mathbf{X})$ is also an $\mathcal{A}$-interpretation of $\mathcal{L}(\mathbf{X}_f)$.   ∎

In what follows we shall denote by $\mathcal{V}$ the set of all interpretations of crisp predicate logic and by $\mathcal{W}$ the set of all interpretations of fuzzy predicate logic. Taking into account the above remark, we can write $\mathcal{V} \subset \mathcal{W}$.

Next, we shall define the semantic deduction in $\mathcal{L}(\mathbf{X}_f)$, i.e., the notion of deduction of a formula F from a fuzzy set $\mathcal{H}$ of hypotheses in fuzzy predicate logic. For this purpose we recall that, as in Part I, Chapter 6, we shall deal exclusively with $\mathcal{A}$-fuzzy sets in the "universe" $\mathcal{L}(\mathbf{X}_f)$. In other words, all our fuzzy sets will be usual $\mathcal{A}$-valued functions defined on $\mathcal{L}(\mathbf{X}_f)$. In particular, it follows that each $\mathcal{A}$-interpretation I of $\mathcal{L}(\mathbf{X}_f)$ may be considered an $\mathcal{A}$-fuzzy set. In the following we shall use the term fuzzy set instead of "$\mathcal{A}$-fuzzy set". On the other hand, speaking about a semantic consequence from a fuzzy set $\mathcal{H}$ of hypotheses, we shall use the notion of consequence in some degree $\mathbf{a}$ from $\mathcal{H}$, where $\mathbf{a}$ is an element of $\mathcal{A}$. Subsequently, instead of the notation $\mathcal{H} \models F$, we shall indicate this by $\mathcal{H} \models_a F$.

DEFINITION 12.4   *The fuzzy set of all semantic consequences from a fuzzy set $\mathcal{H}$ (of hypotheses) will de denoted by $C^{sem}(\mathcal{H})$ and will be defined by the equalities*

$$C^{sem}(\mathcal{H}) = \bigcap \{I \mid I \in \mathcal{W}, I \sqsupseteq \mathcal{H}\} = \bigwedge \{I \mid I \in \mathcal{W}, I \geq \mathcal{H}\}$$

or

$$C^{sem}(\mathcal{H}) = \bigcap_{I \in \mathcal{W}, I \sqsupseteq \mathcal{H}} I = \bigwedge_{I \in \mathcal{W}, I \geq \mathcal{H}} I$$

or equivalently, for each $F \in \mathcal{L}(\mathbf{X}_f)$,

$$C^{sem}(\mathcal{H})(F) = \bigwedge \{I(F) | I \in W, I \sqsupseteq \mathcal{H}\} =$$
$$= \bigwedge \{I(F) | I \in W, I \geq \mathcal{H}\} =$$
$$= \bigwedge_{I \in W, I \geq \mathcal{H}} I(F).$$

REMARK 12.4   In the previous definition we have implicitly used the fact that the MV-algebra $\mathcal{A}$ is complete with respect to infinite operations $\bigwedge_{\alpha} a_\alpha$. This follows from the fact that $\mathcal{A}$ is a finite MV-algebra. Moreover, for the same reason, $\mathcal{A}$ is complete with respect to infinite operations $\bigvee_{\alpha} a_\alpha$. Actually, even in the general case when $\mathcal{A}$ is an infinite MV-algebra, $\mathcal{A}$ is supposed to be complete under the two mentioned operations. ∎

DEFINITION 12.5

(a) *If $C^{sem}(\mathcal{H})(F) = a$, then we shall say that F is a **semantic consequence** from $\mathcal{H}$ **in degree a** and we shall write $\mathcal{H} \models_a F$.*

(b) *If $\mathcal{H} \simeq \emptyset$, then instead of $\emptyset \models_a F$ we shall write simply $\models_a F$ and we shall say that F is an **a-tautology**.*

(c) *If $\models_1 F$, then we shall write usually $\models F$ and we shall say that F is a **fuzzy tautology**.*

Let us denote by $\mathcal{W}_\mathcal{H}$ the set of all interpretations $I \in W$ such that $I \sqsupseteq \mathcal{H}$ ($I(F) \geq \mathcal{H}(F)$ for all $F \in \mathcal{L}(\mathbf{X}_f)$), i.e.,

$$\mathcal{W}_\mathcal{H} = \{\mathcal{I} \in W \mid \mathcal{I} \geq \mathcal{H}\}.$$

Using this notation, we can write

$$C^{sem}(\mathcal{H}) = \bigcap_{I \in \mathcal{W}_\mathcal{H}} I = \bigwedge_{I \in \mathcal{W}_\mathcal{H}} I$$

and

$$C^{sem}(\mathcal{H})(F) = \bigwedge_{I \in \mathcal{W}_\mathcal{H}} I(F).$$

From these remarks we get the following obvious result.

PROPOSITION 12.5

If $\mathcal{H}$ is an arbitrary fuzzy set of hypotheses and $F \in \mathcal{L}(\mathbf{X}_f)$, then $\mathcal{H} \models_a F$ if and only if $I(F) \geq a$ for all $I \in \mathcal{W}_{\mathcal{H}}$ and there is at least one $I' \in \mathcal{W}_{\mathcal{H}}$ such that

$$I'(F) = a.$$

In particular, F is a fuzzy tautology if and only if $I(F) = 1$ for all $I \in \mathcal{W}$.

Taking into account that a formula F from crisp predicate logic ($F \in \mathcal{L}(\mathbf{X})$) is valid if and only if $I(F) = 1$ for all $I \in \mathcal{V}$ and that $\mathcal{V} \subset \mathcal{W}$ we immediately get the result below.

PROPOSITION 12.6

Any fuzzy tautology is a valid formula.

REMARK 12.5   The converse of the previous proposition is not true. Indeed, the formula $T \equiv (p \Rightarrow \neg p) \Rightarrow \neg p$, where $p \in \mathbf{P}_0$ is a valid formula (being a tautology), but it is not a fuzzy tautology because there is an interpretation $I' \in \mathcal{W}$ such that $I'(T) \neq 1$. The interpretation I' can be obtained in a similar way as the interpretation φ, constructed in fuzzy propositional logic for the same purpose (Section 6.3). More precisely, let $\mathcal{A} = \{0, \frac{1}{2}, 1\}$ and consider an interpretation of literal symbols $I'_0 = (\mathcal{D}', \xi', \eta'_0, \eta', \zeta'_0, \zeta')$, where $\mathbf{P}_0$ is not empty and contains the propositional variable p. The only condition imposed on I'_0 is that $\zeta_0(p) = \frac{1}{2}$. Denoting by I' the interpretation of $\mathcal{L}(\mathbf{X}_f)$ generated by I'_0, we have obviously $I'(T) = (\frac{1}{2} \to \overline{\frac{1}{2}}) \to \overline{\frac{1}{2}} = \frac{1}{2}$, hence $I'(T) \neq 1$. Therefore, T is not a fuzzy tautology.

12.3 The Syntactic Truth Structure of Fuzzy Predicate Logic

After the construction of the syntactic version of fuzzy **propositional** logic (mainly through the contributions of J. Pavelka [37] and J.A. Goguen [13]), in the last decade many authors have successfully tackled the syntactic and semantic versions of fuzzy **predicate** logic (fuzzy first-order logic). Thus, today fuzzy predicate logic represents a mature and deep theory with non-trivial applications (among others, in the foundations of approximate reasoning). These achievements are essentially due to the crucial results of V. Novák [34], [35], [36], P. Hájek [17], [18] and also to many other contributions (see the list of references at the end of the book).

In this section we shall sketch some basic aspects of the subject. Thus, we shall consider a fuzzy set of axioms and we shall define the notion of syntactic consequence; we shall also discuss some connections between the syntactic and semantic truth structures of fuzzy predicate logic.

Let us consider the following schemes of formulas.

(C1): $F \Rightarrow (G \Rightarrow F)$;

(C2): $(F \Rightarrow G) \Rightarrow ((G \Rightarrow H) \Rightarrow (F \Rightarrow H))$;

(C3): $(\neg G \Rightarrow \neg F) \Rightarrow (F \Rightarrow G)$;

(C4): $((F \Rightarrow G) \Rightarrow G) \Rightarrow ((G \Rightarrow F) \Rightarrow F)$;

(C5): $A(t|x) \Rightarrow \exists x(A(x))$.

where F, G, H are arbitrary formulas in $\mathcal{L}(\mathbf{X}_f)$, $A(x)$ is a formula containing the variable x and t is a term free for x in $A(x)$.

It is visible that schemes (C1)-(C4) coincide with the axioms of the system **C** of Lukasiewicz and Tarski discussed in Part I, Chapter 5, Section 5.3. The scheme (C5) coincides with the axiom ($A5$) of crisp predicate logic.

In what follows we shall denote by **Ax** the set of all formulas in $\mathcal{L}(\mathbf{X}_f)$ which can be obtained by means of schemes (C1)-(C5). Each formula belonging to **Ax** will be called an axiom. We notice that the (crisp) set **Ax** can be identified with its own characteristic function, i.e., can be regarded as a function

$$\mathbf{Ax} : \mathcal{L}(\mathbf{X}_f) \to \mathcal{A}$$

with

$$\mathbf{Ax}(F) = \begin{cases} 1 \,, & \text{if } F \text{ is an axiom} \\ 0 \,, & \text{otherwise.} \end{cases}$$

Thus **Ax** is a special fuzzy set having only two membership degrees. It is easy to see, by similar arguments to those used in Part I, Section 6.4, that each axiom is a fuzzy tautology; this means that for each interpretation $I \in \mathcal{W}$ we have $\mathbf{Ax} \leq I$. Therefore, $\mathbf{Ax} \leq \bigwedge_{I \in \mathcal{W}} I$. In the following instead of **Ax** we shall also consider a slightly generalized fuzzy set **ax**, which will be called the fuzzy set of axioms. The only condition imposed on **ax** will be the following

$$\mathbf{Ax} \leq \mathbf{ax} \leq \bigwedge_{I \in \mathcal{W}} I.$$

In fuzzy predicate logic, just as in crisp predicate logic, we shall make use of the rule (MP) and of the rule ($\exists \Rightarrow$) and we shall denote by $r = \{ (\text{MP}) , (\exists \Rightarrow) \}$ the set of these two inference rules.

DEFINITION 12.6   *A fuzzy set U is called **r-closed** if it has the following two properties:*

(a) $U(F) \odot U(F \Rightarrow G) \leq U(G)$, for all formulas $F, G \in \mathcal{L}(\mathbf{X}_f)$;

(b) $U(A(x) \Rightarrow C) \leq U(\exists x(A(x) \Rightarrow C))$, for all formulas $A(x), C \in \mathcal{L}(\mathbf{X}_f)$ such that $A(x)$ contains the variable x and C does not contain x.

REMARK 12.6   As in Part I, Section 6.4, we shall say that a fuzzy set U satisfying condition (a) is MP-closed. ∎

PROPOSITION 12.7
Any interpretation $I \in \mathcal{W}$, as a fuzzy set, is r-closed.

PROOF   Let $I \in \mathcal{W}$ be an arbitrary interpretation of $\mathcal{L}(\mathbf{X}_f)$. The fact that I, as a fuzzy set, is MP-closed follows easily from the definition of interpretations, like the corresponding result for the interpretations of fuzzy propositional logic. In order to prove that I has property (b), let us denote by ω the set of occurrences of x in $A(x)$ (i.e., in $U_A(x)$). From the definition of interpretations we get

$$I(\exists x A(x) \Rightarrow C) = I(\exists x A(x)) \rightarrow I(c) = [\max_{d \in \mathcal{D}} h_A(\hat{U}_A(d))] \rightarrow I(C).$$

Taking into account that the residuation "$\rightarrow$" is monotonic in the first argument we obtain

$$[\max_{d \in \mathcal{D}} h_A(\hat{U}_A(d))] \rightarrow I(C) \geq h_A(\hat{U}_A(\hat{x})) \rightarrow I(C) =$$
$$= I(A(x)) \rightarrow I(C) \, , \text{ where } \hat{x} = \xi(x).$$

Therefore, we have $I(\exists x A(x) \Rightarrow C) \geq I(A(x) \Rightarrow C)$, which means that I has property (b). ∎

Next, we shall introduce the fuzzy set of syntactic consequences from a given fuzzy set $\mathcal{H}$ of hypotheses. Consider a fuzzy set $\mathcal{H}$, called the fuzzy set of hypotheses, and denote by $\mathcal{U}_\mathcal{H}$ the (crisp) set of all fuzzy sets Z having the following three properties:

(a) Z is r-closed;

(b) $Z \geq \mathbf{ax}$;

(c) $Z \geq \mathcal{H}$.

If $\mathcal{H} \simeq \emptyset$, then instead of $\mathcal{U}_\emptyset$ we shall write simply $\mathcal{U}$.

DEFINITION 12.7   *Let $\mathcal{H}$ be a given fuzzy set of hypotheses. The fuzzy set*

$$C^{syn}(\mathcal{H}) = \bigcap_{Z \in \mathcal{U}_\mathcal{H}} Z = \bigwedge_{Z \in \mathcal{U}_\mathcal{H}} Z$$

will be called **the fuzzy set of syntactic consequences from $\mathcal{H}$.** *For any formula F the element*

$$a = C^{syn}(\mathcal{H})(F) = \bigwedge_{Z \in \mathcal{U}_\mathcal{H}} Z(F)$$

will be called **the degree of consequence of F from $\mathcal{H}$.** *In this case we shall write $\mathcal{H} \vdash_a F$ and we shall say that F is a* **consequence in degree a** *from $\mathcal{H}$. If $\mathcal{H} \simeq \emptyset$, then instead of $\emptyset \vdash_a F$ we shall write simply $\vdash_a F$ and we shall say that F is a* **fuzzy theorem in degree a.** *In particular, if $\vdash_1 F$ we shall say that F is a* **fuzzy theorem** *and we shall write $(f) \vdash F$.*

PROPOSITION 12.8
For any fuzzy set $\mathcal{H}$
$$C^{syn}(\mathcal{H}) \leq C^{sem}(\mathcal{H}).$$

PROOF   By definition

$$C^{syn}(\mathcal{H}) = \bigwedge_{Z \in \mathcal{U}_\mathcal{H}} Z \quad \text{and} \quad C^{sem}(\mathcal{H}) = \bigwedge_{I \in W, I \geq \mathcal{H}} I.$$

It is clear that each interpretation $I \in W$ with $I \geq \mathcal{H}$ belongs to $\mathcal{U}_\mathcal{H}$, hence the set $\mathcal{U}_\mathcal{H}$ is larger than the set of all interpretations $I \in W$ with $I \geq \mathcal{H}$. Therefore, we have immediately

$$\bigwedge_{Z \in \mathcal{U}_\mathcal{H}} Z \leq \bigwedge_{I \in W, I \geq \mathcal{H}} I,$$

which proves our assertion.   ∎

From the proposition above we get the following result.

PROPOSITION 12.9
$C^{syn}(\emptyset) \leq C^{sem}(\emptyset)$; *in other words, each fuzzy theorem is a fuzzy tautology.*

An important corollary of the above proposition is the following soundness result.

PROPOSITION 12.10

There are not formulas $F \in \mathcal{L}(\mathbf{X}_f)$, such that

$$(f) \vdash F \quad and \quad (f) \vdash \neg F.$$

PROOF   Suppose that there is a formula F such that F and $\neg F$ are fuzzy theorems. From Proposition 12.9 it follows that F and $\neg F$ are fuzzy tautologies. This means that for any interpretation $I \in \mathcal{W}$ we have $I(F) = 1$ and $I(\neg F) = 1$, which is impossible. ∎

Now we shall go to the definition of syntactic proofs (and deductions) in fuzzy predicate logic. As we have seen in Part I, in fuzzy logic we have to deal with so-called evaluated proofs. Therefore, we begin with the definition of evaluated deductions and proofs.

DEFINITION 12.8   Let $\mathcal{H}$ be a fuzzy set of hypotheses. A finite sequence of pairs

$$\pi : \qquad [F_1; a_1], [F_2; a_2], \ldots, [F_r; a_r]; \quad (r \geq 1)$$

where F_i are formulas in $\mathcal{L}(\mathbf{X}_f)$ and a_i are elements in $\mathcal{A}$ is called an **evaluated deduction from** $\mathcal{H}$ if for any index i $(1 \leq i \leq r)$ the pair $[F_i; a_i]$ satisfies one of the conditions below:

(1) if there are indices h, k $(h, k < i)$ such that $F_k \equiv F_h \Rightarrow F_i$, then

$$a_i = a_h \odot a_k \quad or \quad a_i = \mathbf{ax}(F_i) \quad or \quad a_i = \mathcal{H}(F_i);$$

(2) if there is an index j $(j < i)$ such that $F_j \equiv A(x) \Rightarrow C$ and $F_i \equiv \exists x(A(x)) \Rightarrow C$, where $A(x)$ contains the individual variable x and C does not contain x, then

$$a_i = a_j \quad or \quad a_i = \mathbf{ax}(F_i) \quad or \quad a_i = \mathcal{H}(F_i);$$

(3) in all other cases, $a_i = \mathbf{ax}(F_i)$ or $a_i = \mathcal{H}(F_i)$.

The sequence π is called an **evaluated deduction from** $\mathcal{H}$ of F_r; the element $a_r \in \mathcal{A}$ is called the **value of the deduction** π and we shall write $a_r = val_{\mathcal{H}}(\pi)$. If $\mathcal{H} \simeq \emptyset$, then π is called an **evaluated proof** of F_r.

As in Part 1, Section 7.4 we shall compare $val_{\mathcal{H}}(\pi)$ with $\mathcal{C}^{syn}(\mathcal{H})$. For this purpose, for any fuzzy set $\mathcal{H}$ and any formula F we shall denote by $\Pi_{\mathcal{H}}(F)$ the (crisp) set of all evaluated deductions π from $\mathcal{H}$ of the formula F and we shall consider the fuzzy set $\mathcal{Z}_{\mathcal{H}}$ defined by the equality

$$\mathcal{Z}_{\mathcal{H}}(F) = \bigvee_{\pi \in \Pi_{\mathcal{H}}(F)} val_{\mathcal{H}}(\pi),$$

where $\bigvee_{\pi \in \Pi_{\mathcal{H}}(F)} val_{\mathcal{H}}(\pi)$ is the l.u.b. (lowest upper bound) of all elements $val_{\mathcal{H}}(\pi) \in \mathcal{A}$, when π runs over all $\pi \in \Pi_{\mathcal{H}}(F)$. Taking into account that our MV-algebra is linearly ordered and finite, it follows that this l.u.b. exists and is an element of $\mathcal{A}$ (this property also holds in the case of infinite complete MV-algebras).

PROPOSITION 12.11

Let $\mathcal{H}$ be a fuzzy set of hypotheses and $F \in \mathcal{L}(\mathbf{X}_f)$ be a formula. If we denote by $\Pi_{\mathcal{H}}$ the (crisp) set of all evaluated deductions π of F from $\mathcal{H}$, then

$$\bigvee_{\pi \in \Pi_{\mathcal{H}}} val_{\mathcal{H}}(\pi) = \mathcal{C}^{syn}(\mathcal{H})(F).$$

PROOF   Let $\pi \in \Pi_{\mathcal{H}}(F)$ be an evaluated proof

$$\pi : \quad [F_1; a_1], \ldots, [F_r; a_r]$$

from $\mathcal{H}$ of the formula $F_r \equiv F$. It is easy to establish by induction along π (in a similar way as in Part I, Proposition 7.10) that

$(*)$ $$\mathcal{Z}_{\mathcal{H}}(F) \leq \mathcal{C}^{syn}(\mathcal{H})(F).$$

For the converse inequality we shall show that the fuzzy set $\mathcal{Z}_{\mathcal{H}}$ is an element of $\mathcal{U}_{\mathcal{H}}$. For this we have to show that $\mathcal{Z}_{\mathcal{H}}$ has the following three properties:

(a) $\mathcal{Z}_{\mathcal{H}} \geq \mathbf{ax}$;

(b) $\mathcal{Z}_{\mathcal{H}} \geq \mathcal{H}$;

(c) $\mathcal{Z}_{\mathcal{H}}$ is r-closed.

Let F be an arbitrary formula. It is obvious that the sequences

$$\pi' : \quad [F; \mathbf{ax}(F)] \quad \text{and} \quad \pi'' : \quad [F; \mathcal{H}(F)]$$

reduced to a single pair are evaluated deductions from $\mathcal{H}$ of F with $val_{\mathcal{H}}(\pi') = \mathbf{ax}(F)$ and $val_{\mathcal{H}}(\pi'') = \mathcal{H}(F)$. It follows that $\mathcal{Z}_{\mathcal{H}}(F) \geq \mathbf{ax}(F)$ and $\mathcal{Z}_{\mathcal{H}}(F) \geq \mathcal{H}(F)$; thus, $\mathcal{Z}_{\mathcal{H}}$ has the properties (a) and (b). Let us prove now that the fuzzy set $\mathcal{Z}_{\mathcal{H}}$ is closed with respect to the rules MP and $(\exists \Rightarrow)$.

Let F, G be arbitrary formulas and consider, respectively, two evaluated deductions π' and π'' from $\mathcal{H}$ of the formulas F and $F \Rightarrow G$:

$$\pi' : \quad [F_1'; a_1'], \ldots, [F_r'; a_r'];$$

$$\pi'' : \quad [F_1''; a_1''], \ldots, [F_s''; a_s''],$$

where $F'_r \equiv F$, $F''_s \equiv F \Rightarrow G$. From π' and π'' we get the following evaluated deduction

$$\pi: \quad [F'_1; a'_1], \dots, [F'_r; a'_r], [F''_1; a''_1], \dots, [F''_s; a''_s], [G; a'_r \odot a''_s]$$

from $\mathcal{H}$ of the formula G; let us denote $\pi = (\pi', \pi'')$. Thus, each pair $\pi' \in \Pi_{\mathcal{H}}(F)$ and $\pi'' \in \Pi_{\mathcal{H}}(F \Rightarrow G)$ provides an evaluated deduction $\pi \in \Pi_{\mathcal{H}}(G)$; in other words, we can identify the set of all deductions $\pi = (\pi', \pi'')$ with a subset $\mathcal{S}$ of $\Pi_{\mathcal{H}}(G)$. It follows that

$$\bigvee_{\pi \in \mathcal{S}} val_{\mathcal{H}}(\pi) \leq \bigvee_{\pi \in \Pi_{\mathcal{H}}(G)} val_{\mathcal{H}}(\pi) = \mathcal{Z}_{\mathcal{H}}(G).$$

With these remarks let us prove that $\mathcal{Z}_{\mathcal{H}}$ is MP-closed. If F, G are arbitrary formulas, then

$$\mathcal{Z}_{\mathcal{H}}(F) \odot \mathcal{Z}_{\mathcal{H}}(F \Rightarrow G) = \left(\bigvee_{\pi' \in \Pi_{\mathcal{H}}(F)} val_{\mathcal{H}}(\pi') \right) \odot \left(\bigvee_{\pi'' \in \Pi_{\mathcal{H}}(F \Rightarrow G)} val_{\mathcal{H}}(\pi'') \right)$$

$$= \bigvee_{(\pi', \pi'') \in \mathcal{S}} (val_{\mathcal{H}}(\pi') \odot val_{\mathcal{H}}(\pi'')) \leq \bigvee_{\pi \in \mathcal{S}} val_{\mathcal{H}}(\pi) \leq \mathcal{Z}_{\mathcal{H}}(G);$$

hence, $\mathcal{Z}_{\mathcal{H}}$ is MP-closed.

In order to prove that $\mathcal{Z}_{\mathcal{H}}$ is closed with respect to the rule $(\exists \Rightarrow)$, consider an arbitrary formula $A(x)$ containing the variable x and a formula C which does not contain x. If π' is an evaluated deduction from $\mathcal{H}$ of the formula $A(x) \Rightarrow C$,

$$\pi': \quad [F_1; a_1], \dots, [F_r; a_r],$$

where $F_r \equiv A(x) \Rightarrow C$, then obviously the sequence

$$\pi: \quad [F_1; a_1], \dots, [F_r; a_r], [\exists x A(x) \Rightarrow C; a_r]$$

is an evaluated deduction from $\mathcal{H}$ of the formula $\exists A(x) \Rightarrow C$ and $val_{\mathcal{H}}(\pi') = val_{\mathcal{H}}(\pi) \leq \bigvee_{\pi \in \Pi_{\mathcal{H}}(\exists x A(x) \Rightarrow C)} val_{\mathcal{H}}(\pi) = \mathcal{Z}_{\mathcal{H}}(\exists x A(x) \Rightarrow C)$; hence

$$\mathcal{Z}_{\mathcal{H}}(A(x) \Rightarrow C) = \bigvee_{\pi' \in \Pi_{\mathcal{H}}(A(x) \Rightarrow C)} val_{\mathcal{H}}(\pi') \leq \mathcal{Z}_{\mathcal{H}}(\exists x A(x) \Rightarrow C).$$

The last inequality proofs that $\mathcal{Z}_{\mathcal{H}}$ is closed with respect to $(\exists \Rightarrow)$. Being also MP-closed, the fuzzy set $\mathcal{Z}_{\mathcal{H}}$ is r-closed and therefore, has properties (a), (b), (c) above. This means that $\mathcal{Z}_{\mathcal{H}}$ belongs to $\mathcal{U}_{\mathcal{H}}$, hence

$$\mathcal{Z}_{\mathcal{H}} \geq \mathcal{C}^{syn}(\mathcal{H}).$$

From this inequality and from $(*)$ we get the required equality. ∎

The proposition above shows that the definition of deducibility by means of the abstract (and more artificial) fuzzy set $C^{syn}(\mathcal{H})$ is conceptually equivalent with the definition of provability by means of evaluated deductions (which is more intuitive).

FINAL REMARKS.

We could bring more results in fuzzy predicate logic, but this is not our goal. We could, for instance, discuss problems such as completeness, special topics about soundness and so on. As we mentioned in Part I, Chapter 6, the reader can find details in the papers [33], [34], [35], [36], [37], [17], [18], [13], [16].

13

Further Applications of Logic in Computer Science

Logic provides important formal systems that are used in various branches of computer science. Logic is the basis of a general purpose, problem-solving language, Prolog, in **logic programming**. Logic is a foundation for the study of deductive **knowledge bases**. Logic is essential in **software engineering** for the construction, verification and transformation of computer programs. Logic is also of primary importance in **artificial intelligence**. It is used as a knowledge representation language in expert systems, and it is crucial in automated theorem proving. Logical inference represents a fundamental tool of computation which, together with expert systems, is used for the construction of **automated reasoning systems**.

The focus of this chapter is to present some elements of two important applications of logic in computer science: the logic programming language Prolog and approximate reasoning for expert system design. In order to prepare for understanding how Prolog works, elements of a general theory of resolution are discussed.

13.1   Elements of the Theory of Resolution

Until now we have discussed several methods of proof in logic, namely truth tables and syntactic deductions. In this section we present some elements of the so-called resolution method of proof in logic and the particular form in which it has been applied in Prolog: SLD-resolution for Horn clauses. Our approach is along the lines discussed in [48] and [1]. We shall prove that the resolution method is sound (it should not prove false formulas) and complete (all logical consequences are provable). Therefore, this method is useful in practical applications where some essential questions are to be addressed: Is there a computer algorithm for finding all consequences of a set of axioms? Are all the solutions true? Will all the solutions be found?

Resolution without variables (in the context of the propositional calculus)

Resolution proofs are a form of proof by contradiction (refutation). In order to prove that a formula is a logical consequence of a set of formulas, we have to show the inconsistency of a new set of formulas which contains the initial formulas and the negation of the expected consequent (conclusion or answer). The essential idea of the resolution method is to check whether this new set of formulas contains, after some transformations, an inconsistent formula. Therefore, resolution can be viewed as a special inference rule that can generate new formulas from a given set of formulas. In order to apply the resolution method, the formulas have first to be transformed into special forms called **clausal forms**.

We have seen that F is a semantic consequence of $\mathcal{H} = \{H_1, H_2, ..., H_n\}$ (we denote this by $\mathcal{H} \models F$) if $\varphi(H_1) = \varphi(H_2) = ... = \varphi(H_n) = 1$ implies $\varphi(F) = 1$.

We have shown (Section 2.2) that if $H \equiv H_1 \wedge H_2 \wedge ... \wedge H_n$, then we have: $\mathcal{H} \models F$ iff $H \models F$.

Let us consider now a finite set of formulas $U = \{H_1, H_2, ..., H_n\}$. We say that U is consistent if there exists an interpretation φ such that $\varphi(H_i) = 1$ for every i. The set U is said to be inconsistent if for every interpretation at least one of the formulas H_i is false. It is obvious that the system (or set) U is consistent (inconsistent) iff the formula $H_1 \wedge H_2 \wedge ... \wedge H_n$ is consistent (inconsistent). Taking into account that all inconsistent formulas are logically equivalent, an inconsistent formula will be designated by the symbol θ and called the empty clause. Therefore, the formula H is inconsistent iff $H \approx \theta$ where "$\approx$" stands for logical equivalence. The well-known principle of contradiction is expressed by the following result.

PROPOSITION 13.1

$H_1, H_2, ..., H_n \models F$ *iff the system* $K = \{H_1, H_2, ..., H_n, \neg F\}$ *is inconsistent.*

PROOF   Let us assume that $H_1, H_2, ..., H_n \models F$ and consider $\varphi \in \mathcal{V}$. We have to analyze two situations:
a) $\varphi(H_1 \wedge H_2 \wedge ... \wedge H_n) = 1$;
b) $\varphi(H_1 \wedge H_2 \wedge ... \wedge H_n) = 0$.
In the first case we get: $\varphi(H_1) = \varphi(H_2) = ... = \varphi(H_n) = 1$ and by our assumption we obtain: $\varphi(F) = 1$, hence $\varphi(\neg F) = 0$. Therefore, $\varphi(H_1 \wedge$

$H_2 \wedge ... \wedge H_n \wedge \neg F) = 0$. In the second case we obtain:

$$\varphi(H_1 \wedge H_2 \wedge ... \wedge H_n \wedge \neg F) = \min\{(\varphi(H_1 \wedge ... \wedge H_n), \varphi(\neg F)\} = 0.$$

Therefore, $\varphi(H_1 \wedge H_2 \wedge ... \wedge H_n \wedge \neg F) = 0$ for all $\varphi \in V$; hence the system K is inconsistent.

Conversely, let us suppose that K is inconsistent and let us consider $\varphi \in V$ such that $\varphi(H_1) = \varphi(H_2) = ... = \varphi(H_n) = 1$. Because K is inconsistent we get:

$$0 = \varphi(H_1 \wedge H_2 \wedge ... \wedge H_n \wedge \neg F) = \min\{\varphi(H_1 \wedge H_2 \wedge ... \wedge H_n), \varphi(\neg F)\} =$$

$$= \min\{1, \varphi(\neg F)\} = \varphi(\neg F).$$

Therefore, $\varphi(\neg F) = 0$, hence $\varphi(F) = 1$. ∎

Based on the above result the problem of deduction of F from the set $H_1, H_2, ..., H_n$ is reduced to the problem of inconsistency of the set $K = \{H_1, H_2, ..., H_n, \neg F\}$ which is equivalent to the problem of determining the inconsistency status of the formula $H_1 \wedge H_2 \wedge ... \wedge H_n \wedge \neg F \equiv G$.

Consequently, we have to solve the inconsistency of a given formula G. This problem is symbolized by "$G \approx \theta$". We know that $G \approx \theta$ means $G \models \theta$ and $\theta \models G$. The deduction $\theta \models G$ is always true (because $\theta \Rightarrow G$ is a tautology). Thus our problem is to show:

$(\mathcal{P})$ $G \models \theta$.

This problem will be tackled by the so-called **resolution method**. In order to achieve this, the formula G will first be represented in an equivalent normal conjunctive form

$$G \approx T , \; T \equiv \bigwedge_{j=1}^{l} C_j$$

where the formulas C_j, called **disjunctive clauses** or simply **clauses**, are disjunctions of literals; this means that, for every j, $(1 \leq j \leq l)$

$$C_j \equiv L_1^j \vee L_2^j \vee ... L_k^j$$

where each L_i^j is a propositional variable or the negation of a propositional variable. Therefore, the inconsistency of G is reduced to the same problem for T, i.e., to the inconsistency of the set $T = \{C_1, ..., C_l\}$ of all clauses in T.

After reducing the problem to the inconsistency of T, the resolution method provides a procedure to successively transform the system T in other larger systems of clauses

$$T \equiv T_0 \rightarrow T_1 \rightarrow ... \rightarrow T_q$$

such that if the system T is inconsistent, then the last system contains the empty clause, i.e., $\theta \in T_q$.

Hence, the problem $(\mathcal{P})$ $G \models \theta$ can be reduced to the following two subproblems:

$(\mathcal{P}_1)$ Reduce the given formula G to the logically equivalent clausal form T;

$(\mathcal{P}_2)$ Test the inconsistency of the system T using the resolution method, which is based on Proposition 13.1.

Let us consider briefly these two problems.

$(\mathcal{P}_1)$ Reduction to clausal form (conjunctive normal form)

We recall that every formula $F \in \mathcal{L}_0$ is either a propositional variable, the negation of a formula, or the disjunction of two formulas (since the other connectives can be expressed with the help of $\neg$ and $\vee$, i.e., $H \wedge K \approx \neg(\neg H \vee \neg K), H \Rightarrow K \approx \neg H \vee K, H \Leftrightarrow K \approx (H \Rightarrow K) \wedge (K \Rightarrow H)$). We are now going to define the notion of **subformula** of a given formula F, i.e., a consecutive sequence of symbols from F which is itself a formula.

DEFINITION 13.1   *Let $F \in \mathcal{L}_0$ be a given formula. Then:*

(j) if F is $\neg G$, then G is a subformula of F;

(jj) if F is $H \vee K$ or $H \wedge K$ or $H \Rightarrow K$ or $H \Leftrightarrow K$, then H, K are subformulas of F;

(jjj) every subformula of a subformula of F is a subformula of F (F is itself considered a subformula of F);

(jv) propositional variables do not have subformulas other than the variable itself.

It is easy to prove the following two results which are the basis for "the principle of substitution" in formulas.

PROPOSITION 13.2
Let L be a subformula of F. If $L \approx L'$, then $F \approx F'$ where F' is obtained by replacing L by L' in F.

PROPOSITION 13.3
If F is a tautology and F contains as a subformula the propositional variable X, then by replacing X with an arbitrary formula L in F we obtain a new formula F' which is also a tautology.

Let us now consider a formula $F \in \mathcal{L}_0$, $\sigma \in \{0, 1\}$, and recall the definition of the formula F^σ:

$$F^\sigma = \begin{cases} F & \text{if } \sigma = 1 \\ \neg F & \text{if } \sigma = 0 \end{cases}$$

If F is a propositional variable X, then the formula X^σ is called a literal. Consequently, there exists 2N literals in $\mathcal{L}_0$:

$X_1, ..., X_N$, (positive literals) and $\neg X_1, ..., \neg X_N$, (negative literals).

DEFINITION 13.2

(i) *A* **disjunctive clause** *C is a disjunction of literals, i.e.,*

$$C \equiv L_1 \vee L_2 \vee ... \vee L_k.$$

(ii) *A formula T is in* **clausal form** *(conjunctive normal form or simply CNF) if it is a conjunction of clauses, i.e., it has the form:*

$$T \equiv C_1 \wedge C_2 \wedge ... \wedge C_l$$

where C_i $(1 \leq i \leq l)$ are clauses. Usually, we write T as $T = \{C_1, C_2, ..., C_l\}$.

REMARK 13.1 It is obvious that the formula T (in clausal form) is true for an interpretation φ if and only if all the clauses C_i are true for this interpretation. ∎

The following result allows the reduction of a formula F to a clausal form.

PROPOSITION 13.4
For every formula F, there exists a clausal form F' (not unique) such that $F \approx F'$, i.e., $\models F \Leftrightarrow F'$.

PROOF This result has been proved in Part I, Chapter 3. Therefore, we shall only sketch the algorithm to obtain F' and we shall show that it works. The steps of the algorithm are as follows:

(I) Eliminate $\Leftrightarrow$. If $M \equiv H \Leftrightarrow K$ is a subformula of F, then replace M with $M' \equiv (H \Rightarrow K) \wedge (K \Rightarrow H)$. Doing that for all occurrences of $\Leftrightarrow$ we obtain a new formula F_1.

(II) Eliminate $\Rightarrow$. If $M \equiv H \Rightarrow K$ is a subformula of F_1, then replace M with $M' \equiv \neg H \vee K$. Doing that for all occurrences of $\Rightarrow$ we get a new formula F_2.

(III) Move negations inside parentheses. If $M_1 \equiv \neg(H \wedge K)$ or $M_2 \equiv \neg(H \vee K)$ is a subformula of F_2, then replace M_1 and M_2 with $M'_1 \equiv \neg H \vee \neg K$ and $M'_2 \equiv \neg H \wedge \neg K$. Repeating this operation as many times as possible on F_2 we obtain a new formula F_3.

(IV) Eliminate multiple negations. If $M \equiv \neg\neg H$ is a subformula of F_3, then replace M with $M' \equiv H$. Doing that everywhere in F_3 we get a new formula F_4.

(V) Move $\vee$ inward and $\wedge$ outward simultaneously. If $M_1 \equiv H \vee (K \wedge L)$ or $M_2 \equiv (H \wedge K) \vee L$ are subformulas of F_4, then replace M_1 and M_2 with $M'_1 \equiv (H \vee K) \wedge (H \vee L)$ and $M'_2 \equiv (H \vee L) \wedge (K \vee L)$. Repeating this operation as many times as possible on F_4 we obtain a new formula F_5 which is logically equivalent to F and is in clausal form. ∎

REMARK 13.2   The above result states that for every formula $F \in \mathcal{L}_0$ there exists (at least) a system of clauses $S = \{C_1, C_2, ..., C_h\}$ such that $F \approx S$. ∎

Example 13.1

Let us consider the formula $(X_1 \wedge (X_2 \Rightarrow X_3)) \Rightarrow X_4$. To obtain its clausal form we get successively:

$$(X_1 \wedge (X_2 \Rightarrow X_3)) \Rightarrow X_4 \approx (X_1 \wedge (\neg X_2 \vee X_3)) \Rightarrow X_4$$

$$\approx \neg(X_1 \wedge (\neg X_2 \vee X_3)) \vee X_4 \approx (\neg X_1 \vee \neg(\neg X_2 \vee X_3)) \vee X_4$$

$$\approx (\neg X_1 \vee (X_2 \wedge \neg X_3)) \vee X_4 \approx ((\neg X_1 \vee X_2) \wedge (\neg X_1 \vee \neg X_3)) \vee X_4$$

$$\approx (X_4 \vee (\neg X_1 \vee X_2)) \wedge (X_4 \vee (\neg X_1 \vee \neg X_3))$$

$$\approx (\neg X_1 \vee X_2 \vee X_4) \wedge (\neg X_1 \vee \neg X_3 \vee X_4) \equiv C_1 \wedge C_2.$$

▯

($\mathcal{P}_2$) **Resolution method**

We have shown that the formula $T \equiv \bigwedge_{j=1}^{l} C_j$ entails the formula F if the set of formulas $\{C_1, C_2, \ldots, \neg F\}$ is inconsistent. In order to deal only with clauses, first we reduce the formula $\neg F$ to its normal conjunctive form $\neg F \equiv \bigwedge_{j=1}^{m} C'_j$, so that T entails F if and only if the system of clauses $\{C_1, \ldots, C_l, C_1', \ldots, C_m'\}$ is inconsistent. So, to simplify the discussion we shall consider the problem of inconsistency of an arbitrary set $U = \{C_1, C_2 \ldots, C_n\}$ of clauses.

In the following we shall adopt two useful conventions. Namely, if $T \equiv \bigwedge_{j=1}^{l} C_j$ is a formula in normal conjunctive form, then we shall write also $T = \{C_1, C_2, \ldots, C_l\}$ and we shall say that T is a system (finite set) of clauses. However, we have to keep in mind that T is actually a conjunction of all clauses contained in the system $\{C_1, C_2, \ldots, C_l\}$. The same convention will be adopted for clauses; for instance, if $C \equiv L_1 \lor L_2 \lor \ldots \lor L_k$, we shall often write $C \equiv \{L_1, L_2 \ldots, L_k\}$, i.e., we shall represent C as a system of literals, also keeping in mind that C actually is the disjunction of all literals contained in $\{L_1, L_2, \ldots, L_k\}$. Therefore, we shall use set-theoretical notations such as $C_2 \in T$, $L_3 \in C$, etc. Both systems of notations will be used equivalently depending of our needs.

Let us now return to a general system U of clauses. To simplify the discussion we shall suppose that our system U has the following properties:
(a) all clauses of U are distinct;
(b) in each clause of U, all literals are distinct.
A formula which is in clausal form and satisfies (a) and (b) is called a **pure formula**. The fact that any formula in clausal form is logically equivalent to its pure form is an immediate consequence of the following logical equivalences:

$$H \land H \land K \approx H \land K$$

$$H \lor H \lor K \approx H \lor K.$$

It is not difficult to prove the following result.

PROPOSITION 13.5

For every $H, F \in \mathcal{L}_0$, $H \models F$ iff $H \land F \approx H$; in a particular case if R is a clause and $T = \{C_1, C_2, \ldots, C_l\}$ is a system of clauses, then $T \models R$ iff $T \approx T \cup \{R\}$.

The resolution method of proof is based on the result below.

PROPOSITION 13.6

For any formulas $F, H, K \in \mathcal{L}_0$, from $H \lor F$ and $K \lor \neg F$, we can infer $H \lor K$, i.e.,

$$\{H \lor F, K \lor \neg F\} \models H \lor K.$$

PROOF   Let $\varphi \in \mathcal{V}$ be an interpretation and assume that $\varphi(H \lor F) = 1$ and $\varphi(K \lor \neg F) = 1$. Then, we have to verify that $\varphi(H \lor K) = 1$. We should consider two situations: (a) $\varphi(F) = 1$ and (b) $\varphi(F) = 0$. In the first case we get:

$$1 = \varphi(K \vee \neg F) = \max\{\varphi(K), \varphi(\neg F)\} = \max\{\varphi(K), 0\} = \varphi(K).$$

Thus, $\varphi(H \vee K) = \max\{\varphi(H), \varphi(K)\} = 1$.
In the second case we obtain:

$$1 = \varphi(H \vee F) = \max\{\varphi(H), \varphi(F)\} = \max\{\varphi(H), 0\} = \varphi(H).$$

Hence, $\varphi(H \vee K) = \max\{\varphi(H), \varphi(K)\} = 1$. This means that

$$\{H \vee F, K \vee \neg F\} \models H \vee K . \quad \blacksquare$$

The above result furnishes the following corollary often used in automatic deduction and logic programming.

PROPOSITION 13.7

(Resolution rule) For every two clauses C_1, C_2 and every propositional variable X, one has $\{C_1 \vee X, C_2 \vee \neg X\} \models C_1 \vee C_2$, i.e.,

$$\models (C_1 \vee X) \wedge (C_2 \vee \neg X) \Rightarrow C_1 \vee C_2.$$

For example: $\{X_1, \neg X_1 \vee X_2\} \models X_2$, $\{X_1 \vee X_2, \neg X_1 \vee X_3\} \models X_2 \vee X_3$, $\{\neg X_1 \vee X_2 \vee X_3, \neg X_2 \vee X_4\} \models \neg X_1 \vee X_3 \vee X_4$.
Applying the resolution rule one obtains:

PROPOSITION 13.8

Let U be a system of clauses and $C_1, C_2 \in U$. If C_1 contains the literal (propositional variable) X and C_2 contains the literal $\neg X$ (complementary of X), then from U one can infer the clause R which is the disjunction of the remaining literals in C_1 and C_2, i.e., $R \equiv (C_1 \setminus \{X\}) \cup (C_2 \setminus \{\neg X\})$.

Therefore, the clause $R = C_1' \cup C_2'$ is inferred from U, where $C_1' = C_1 \setminus \{X\}$ and $C_2' = C_2 \setminus \{\neg X\}$, i.e., R is obtained by deleting X and $\neg X$ from C_1 and C_2.

DEFINITION 13.3   *The clause $R = C_1' \cup C_2'$ is called the **resolvent** of C_1 and C_2.*

From Proposition 13.5 and Proposition 13.6 we can deduce the following fundamental result:

PROPOSITION 13.9

Let U be a system of clauses and $C_1, C_2 \in U$. If the clause R is the resolvent of C_1 and C_2, then $U \approx U \cup \{R\}$.

The following question naturally arises: Does an arbitrary system of clauses contain all possible resolvents? If not, how are we to deal with this problem? The answers are given below.

DEFINITION 13.4   A system of clauses $U = \{C_1, C_2, ..., C_h\}$ is **complete** if U contains the resolvent of all pairs of clauses. If there exists a pair of clauses $C_i, C_j \in U$ such that the resolvent R of C_i and C_j does not belong to U, then the system $U' \approx U \cup \{R\}$ is called **the direct extension of** U.

REMARK 13.3   It is obvious that $U \subseteq U'$ and $U \approx U'$.   ∎

If the system U is not complete, then starting with U we construct successively the direct extensions of U. Consequently, we obtain a **finite** sequence of systems:

(∗) $U \equiv U_0, U_1, U_2, ..., U_m$ with $m \geq 1$ such that:

(a) U_i is a direct extension of U_{i-1}, for $1 \leq i \leq m$;
(b) U_m is a complete system of clauses.
It is not difficult to establish that sequence (∗) must be a finite sequence of systems. First, we can assume that each term of the sequence is a pure system of clauses. Otherwise, we can replace it with another term which is pure and logically equivalent. We note easily that the set $\sum$ of all pure systems of clauses constructed with the help of distinct literals from U is finite. Taking into account that the terms of the sequence (∗) are distinct members of $\sum$, it follows that this sequence is finite. Hence, we can state the following result.

PROPOSITION 13.10

For each system of clauses U, there exists a complete system U_c such that $U \approx U_c$.

The fundamental property of complete systems (resolution theorem) is expressed by the next theorem presented without proof in the book of A. Thaysse [48]. For the proof, A. Thaysse refers to the book of C. L. Chang and R. C. T. Lee (Symbolic Logic and Mechanical Theorem Proving, Academic Press, New York, 1973).

In order to facilitate the considerations below we introduce some conventions and notations. We know that the set of propositional variables is supposed to be finite (actually this restriction is not essential) and it is denoted by $\{X_1, X_2, ..., X_N\}$ ($N \geq 1$); the language of propositional logic generated by $X_1, X_2, ..., X_N$ is denoted by $\mathcal{L}_0$. If $Z_1, Z_2, ..., Z_n$ ($1 \leq n \leq N$) are distinct variables belonging to $\{X_1, X_2, ..., X_N\}$, maybe written in some different order, then the language $\mathcal{L}'_0$ generated by $Z_1, Z_2, ..., Z_n$ is said to be a sublanguage of $\mathcal{L}_0$. Taking into account that each Boolean interpretation φ of $\mathcal{L}_0$ ($\varphi \in \mathcal{V}$) is uniquely determined by the values $\varphi(X_1), \varphi(X_2), ..., \varphi(X_N)$, it follows that any Boolean interpretation ψ of $\mathcal{L}'_0$ can be extended to a Boolean interpretation φ of $\mathcal{L}_0$ in many ways (if $n < N$); on the other hand, the restriction to $\mathcal{L}'_0$ of any Boolean interpretation of $\mathcal{L}_0$ is obviously a Boolean interpretation of $\mathcal{L}'_0$.

Now, if φ is an interpretation of $\mathcal{L}_0$ and $V = \{v_1, v_2, ..., v_h\}$ is a system of clauses, then we will use the notation

$$\varphi(V) = \varphi(v_1 \wedge v_2 \wedge ... \wedge v_h).$$

If W is also a system of clauses, then, by definition, we will write

$$\varphi(V \cup W) = \varphi(V) \wedge \varphi(W).$$

With these specifications and notations we can pass to the fundamental property of complete systems.

THEOREM 13.1
A complete system U of clauses is inconsistent if and only if a propositional variable Z exists such that

$$U \ni Z \quad and \quad U \ni \neg Z,$$

i.e., if and only if U contains the empty clause.

PROOF   It is obvious that if U contains Z and $\neg Z$, then U is inconsistent. The converse assertion is more difficult and will be proved below. For each system U of clauses, we denote by $\nu(U)$ the number of distinct propositional variables in the clauses of U. The proof will be performed by induction on $\nu(U) \geq 1$.

Suppose $\nu(U) = 1$; in this case a propositional variable Z exists such that, for each $d_i \in U$, we have either $d_i \equiv Z$ or $d_i \equiv \neg Z$. If $d_i \equiv Z$ for all $d_i \in U$ (respectively, $d_i \equiv \neg Z$ for all $d_i \in U$), then each interpretation $\varphi \in \mathcal{V}$ with $\varphi(Z) = 1$ (respectively, with $\varphi(Z) = 0$) is a model of U, which is impossible because U has no model. Therefore, there are two clauses $d_i, d_j \in U$ such that $d_i \equiv Z$ and $d_j \equiv \neg Z$; hence, in the case $\nu(U) = 1$, the assertion is true.

Suppose now $\nu(U) = n > 1$ and assume that for every inconsistent and complete system U', with $\nu(U') < n$, the assertion is true. Consider an inconsistent and complete system of clauses

$$U = \{d_1, d_2, ..., d_s\},$$

with $\nu(U) = n$ and let us denote by $Z_1, Z_2, ..., Z_n$ $(1 \leq n \leq N)$ the propositional variables occuring in the clauses of U. We decompose the set U in three disjoint subsets, $U = A \cup B \cup C$, where:
A is the set of all $d_i \in U$, such that $d_i \ni Z_1$;
B is the set of all $d_i \in U$, such that $d_i \ni \neg Z_1$;
C is the set of all $d_i \in U$, such that $d_i \not\ni Z_1$ and $d_i \not\ni \neg Z_1$.
The sets A, B, C are obviously disjoint because we can suppose that clauses containing Z_h and $\neg Z_h$, for some h $(1 \leq h \leq n)$, are eliminated (we suppose also that the clauses $d_1, d_2, ..., d_s$ in U are all distinct). We will denote

$$A = \{a_1, a_2, ..., a_p\}, \ B = \{b_1, b_2, ..., b_q\}, \ C = \{c_1, c_2, ..., c_r\},$$

where $p \geq 0$, $q \geq 0$, $r \geq 0$, $p + q > 0$ and $p + q + r = n$.
We have to consider the following four possibilities:

(1) $A \ni Z_1$ and $B \ni \neg Z_1$; (2) $A \not\ni Z_1$ and $B \ni \neg Z_1$;

(3) $A \ni Z$ and $B \not\ni \neg Z_1$; (4) $A \not\ni Z_1$ and $B \not\ni \neg Z_1$.

In case (1), the system U contains the clauses Z_1 and $\neg Z_1$, hence the assertion is true.
In case (2), we can write

$$A = \{Z_1 \vee a_1', ..., Z_1 \vee a_p'\}$$

$$B = \{\neg Z_1 \vee b_1', ..., \neg Z_1 \vee b_{q-1}', \neg Z_1\}$$

where the clauses $a_1', ..., a_p', b_1', ..., b_{q-1}'$ do not contain the literals Z_1 and $\neg Z_1$. Denoting

$$A' = \{a_1', ..., a_p'\}, \quad B' = \{b_1', ..., b_{q-1}'\}$$

we will show that the system $A' \cup C$ is inconsistent and complete. Suppose that the system is consistent, hence it has a model. Taking into account that the clauses of A' and C may contain only literals involving the variables $Z_2, Z_3, ..., Z_n$, it follows that there is an interpretation ψ of the language $\mathcal{L}_0'$, generated by $Z_2, Z_3, ..., Z_n$, such that $\psi(A' \cup C) = 1$, hence

$$\psi(A') = 1 \ \text{and} \ \psi(C) = 1.$$

Now, let us define an interpretation φ of the language $\mathcal{L}_0$, taking $\varphi(Z_1) = 0$, $\varphi(Z_h) = \psi(Z_h)$ for all h $(2 \leq h \leq n)$ and putting $\varphi(X) = \gamma$ for each variable $X \in \{X_1, ..., X_N\} \setminus \{Z_1, ..., Z_n\}$, where γ is an arbitrary element of $\Gamma = \{0, 1\}$. For this interpretation φ we have:

$$\varphi(U) = \varphi(A \cup B \cup C) = \varphi(A \cup B) \wedge \varphi(C) = \varphi(A \cup B) \wedge \psi(C) =$$

$$= \varphi(A \cup B) \wedge 1 = \varphi(A \cup B) = \varphi(A) \wedge \varphi(B).$$

But $A = \{Z_1 \vee a'_1, ..., Z_1 \vee a'_p\}$ and $B = \{\neg Z_1 \vee b'_1, ..., \neg Z_1 \vee b'_{q-1}, \neg Z_1\}$, therefore, taking into account that $\varphi(Z_1) = 0$ and $\varphi(a'_i) = \psi(a'_i)$, we get

$$\varphi(A) = \varphi(Z_1 \vee a'_1) \wedge ... \wedge \varphi(Z_1 \vee a'_p) = \psi(A') = 1.$$

Similarly, since $\varphi(\neg Z_1) = 1$, it follows that

$$\varphi(B) = \varphi(\neg Z_1 \vee b'_1) \wedge ... \wedge \varphi(\neg Z_1 \vee b'_{q-1}) \wedge \varphi(\neg Z_1) = 1.$$

Thus, $\varphi(U) = \varphi(A) \wedge \varphi(B) = 1$, which is impossible because U is an inconsistent system. It follows that the system $A' \cup C$ is inconsistent.

The system $A' \cup C$ is also complete. Indeed, denoting by $R(u, v)$ the resolvent of clauses u and v, we have to prove that

$$(u \in A' \cup C \quad \text{and} \quad v \in A' \cup C) \quad \text{implies} \quad (R(u, v) \in A' \cup C).$$

If $u, v \in C$, then the resolvent $R(u, v)$ belongs to U because U is complete; on the other hand, the clause $R(u, v)$ does not contain literals Z_1 and $\neg Z_1$, because the system C itself has this property. So, $R(u, v) \in C$, hence $R(u, v) \in A' \cup C$.

If $u \in A'$ and $v \in C$, then we can write $u \equiv a'_i$ and $v \equiv c_j$; if the resolvent $R(u, v) \equiv R(a'_i, c_j)$ exists, then $a'_i \equiv l'_h \vee a''_i$, $c_j \equiv l''_h \vee c'_j$, where l'_h, l''_h are opposite literals involving certain variables Z_h $(2 \le h \le n)$, hence $R(a'_i, c_j) \equiv a''_i \vee c'_j$. By the definition of A we have $a_1 \equiv Z_1 \vee a'_i$ or $a_i \equiv Z_1 \vee l'_h \vee a''_i \in A$. Taking into account that U is complete, it follows $R(a_i, c_j) \in U$ or $Z_1 \vee a''_i \vee c'_j \in U$. By the definition of A' we get $a''_i \vee c'_j \in A'$ or $R(u, v) \equiv R(a'_i, c_j) \in A'$, hence $R(u, v) \in A' \cup C$. Finally, if $u, v \in A'$, then $u \equiv Z_1 \vee a'_i$, $v \equiv Z_1 \vee a'_j$ with $a'_i, a'_j \in A'$. If the resolvent $R(u, v)$ exists, then $a'_i \equiv l'_h \vee a''_i$ and $a'_j \equiv l''_h \vee a''_j$, where l'_h, l''_h are opposite literals involving certain variable Z_h $(2 \le h \le n)$ and $R(u, v) \equiv a''_i \vee a''_j$. On the other hand, the clauses $a_i \equiv Z_1 \vee a'_i$, $a_j \equiv Z_1 \vee a'_j$ belong to A (by the defintion of A'), hence

$$a_i \equiv Z_1 \vee l'_h \vee a''_i \in A \quad \text{and} \quad a_j \equiv Z_1 \vee l''_h \vee a''_j \in A$$

and (since U is complete) $R(a_i, a_j) \in U$, or $(Z_1 \vee a''_i) \vee (Z_1 \vee a''_j) \equiv Z_1 \vee (a''_i \vee a''_j) \in U$. By the definition of A we get $Z_1 \vee (a''_i \vee a''_j) \in A$, hence $R(u, v) \equiv a''_i \vee a''_j \in A'$, which implies $R(u, v) \in A' \cup C$.

In conclusion, we have proved that the system $A' \cup C$ is inconsistent and complete. Taking into account that $\nu(A \cup C) < n$ and applying the inductive assumption, it follows that the system $A' \cup C$ contains two clauses of the form $Z_k, \neg Z_k$ for some k $(2 \leq k \leq n)$. If $Z_k, \neg Z_k \in C$, then obviously $Z_k, \neg Z_k \in U$. If we suppose that $Z_k \in A'$ and $\neg Z_k \in C$, then $Z_1 \vee Z_k \in A$ and we get $R(Z_1 \vee Z_k, \neg Z_k) \equiv Z_1 \in U$, hence, by definition, $Z_1 \in A$ which (in our case (2)) is impossible. Finally, if we suppose that $Z_k, \neg Z_k \in A'$, then $Z_1 \vee Z_k$ and $Z_1 \vee \neg Z_k$ belong to A and

$$R(Z_1 \vee Z_k, Z_1 \vee \neg Z_k) \equiv Z_1 \in A,$$

which is impossible. In conclusion, we have proved that $U \ni Z_k, \neg Z_k$, which means that in case (2) the system U contains the empty clause.

The cases (3), (4) are similar; so, in all cases the system U contains the empty clause. Now the assertion of the theorem follows from the induction principle. ∎

REMARK 13.4   For establishing the inconsistency of a system of clauses $U = \{C_1, C_2, ..., C_n\}$ we can proceed in the following manner: starting with $\{C_1, ..., C_n\}$, we can find resolvents successively until we obtain the empty clause θ. In this situation the system U' which contains U and all the obtained resolvents is logically equivalent to U, and $\theta \in U'$. Hence, U' and consequently U is inconsistent. The crucial idea behind the resolution method, which is a special rule of inference (derivation), for establishing the inconsistency of a system of clauses is to check that the system contains the empty clause. ∎

The next definition expresses in rigorous terms the notion of deducibility by the resolution method.

DEFINITION 13.5   *Let $T = \{C_1', C_2', \ldots, C_l'\}$ be a finite system of clauses (or equivalently a formula in conjunctive normal form) and let F be an arbitrary formula, such that $\{C_1'', C_2'', \ldots, C_n''\}$ is the set of clauses of $\neg F$. Let us consider also the system of clauses $U = \{C_1', \ldots, C_l'\} \cup \{C_1'', \ldots, C_n''\}$. We shall say that F is **deducible from T** by the **resolution method** and we shall write $T \vdash_{\mathcal{R}} F$ if there is a sequence of systems of clauses*

$$U \equiv U_0, U_1, \ldots, U_q$$

such that for every index i $(1 < i \leq q)$ the system U_i is a direct extension of U_{i-1} and $\theta \in U_q$.

In particular, the notation $\vdash_{\mathcal{R}} F$ means that (denoting by $U = \{C_1, C_2, \ldots, C_n\}$ the system of all clauses of $\neg F$ written in normal conjunctive form) there is a sequence of direct extensions of U as above, such that the last term of the sequence contains the empty clause.

Our goal now is to study the relation between the above syntactic notion (the proof method of resolution) and the semantic one (semantic inference).

THEOREM 13.2

The resolution method of proof is sound, i.e., the resolution method preserves truth; formally, if $T \vdash_\mathcal{R} F$, then $T \models F$.

PROOF   Notice first that $T \models F$ if and only if $\models T \Rightarrow F$, that is $\models \neg T \vee F$. Thus, it is sufficient to prove that if $\vdash_\mathcal{R} H$, then $\models H$ for all formulas H. Let us prove this by contradiction, i.e., suppose that $\vdash_\mathcal{R} H$ and not($\models H$).

By the above definition, $\vdash_\mathcal{R} H$ means that denoting by $U = \{C_1, C_2, \ldots, C_n\}$ the system of clauses of $\neg H$ ($\neg H$ being converted in conjunctive normal form), there is a finite sequence

$$U \equiv U_0, U_1, \ldots, U_p$$

of direct extensions of U such that $\theta \in U_p$.

On the other hand, supposing that not($\models H$) it follows that there is an interpretation $\varphi \in V$ such that $\varphi(H) = 0$, therefore, $\varphi(\neg H) = 1$. Now we shall show that any resolvent R of two clauses $C_i, C_j \in U$ is φ-true, i.e., $\varphi(R) = 1$. Indeed, let $C_i = C_i' \cup \{X\}$, $C_j = C_j' \cup \{\neg X\}$ and consequently $R = C_i' \cup C_j'$. Taking into account that $\varphi(C_i) = \varphi(C_j) = 1$ and assuming that $\varphi(R) = 0$, it follows that $\varphi(C_i') = 0$ and $\varphi(C_j') = 0$; thus

$$1 = \varphi(C_i) = \varphi(C_i' \vee X) = \varphi(C_i') \vee \varphi(X) = \varphi(X)$$

or $\varphi(X) = 1$. Similarly, we get $\varphi(\neg X) = 1$, which is impossible. Therefore, we have the following conclusion:

$$1 = \varphi \left(\bigwedge_{C \in U_0} C \right) = \varphi \left(\bigwedge_{C \in U_1} C \right) = \ldots = \varphi \left(\bigwedge_{C \in U_p} C \right)$$

which is impossible because $\theta \in U_p$, hence $\varphi \left(\bigwedge_{C \in U_p} C \right) = 0$. The resulting contradiction proves the assertion.   ∎

Next, we shall give without proof the following result (see, for example, [1]):

THEOREM 13.3

The resolution method of proof is complete, i.e., each formula that is true in every interpretation is provable by resolution; in short, if $S \models F$, then $S \vdash_\mathcal{R} F$.

Algorithm of the resolution method in propositional logic

Let T be a set of formulas and F a formula, where all formulas are in clausal form. For establishing that $\models T \Rightarrow F$ one has to check whether $T' = T \cup \{\neg F\}$ is inconsistent as follows:

A. If T' contains the empty clause, then it is already inconsistent. In a particular case when the system is complete and inconsistent, then necessarily it contains the empty clause (Theorem 13.1).

B. If T' does not contain the empty clause, then one can generate, by resolution, sequences of new systems (each being a direct extension of the preceding one) until we get a sequence such that the last term contains the empty clause. The generation subalgorithm of such systems is as follows:

$$i = 0$$
$$T'_0 = T'; T'_1 = T'$$

While $\theta \notin T'_i$ and $T'_i \neq T'_{i-1}$

- Choose a propositional letter X_i and two clauses C_1, C_2 such that $C_1, C_2 \in T'_i$, $X_i \in C_1$ and $\neg X_i \in C_2$;

- Compute the resolvent R and simplify by eliminating duplicate elements;

- Discard the clause if it contains a propositional letter and its negation;

- If the resolvent is not in T', then replace T' with $T' \cup \{R\}$, i.e.,
  $$T'_{i+1} = T'_i \cup \{R\}$$
  $$i = i + 1$$

This algorithm is nondeterministic due to the use of "choose". We shall illustrate this algorithm below by means of some examples.

Example 13.2

1) Let $U = \{\neg X_1 \vee X_2, \neg X_2, X_1\}$. Then, $U \models \{\neg X_1, X_1\} \models \theta$.

2) Let $U = \{X_1 \vee X_2, \neg X_1 \vee X_2, X_1 \vee \neg X_2, \neg X_1, \neg X_2\}$. Then, $U \models \{X_2, \neg X_2\} \models \theta$.

3) Let $U \equiv \neg((X_1 \Rightarrow X_2) \Rightarrow ((X_3 \vee X_1) \Rightarrow (X_3 \vee X_2)))$.

First, let us convert U to clausal form. We get successively:

1. $\neg(\neg(X_1 \Rightarrow X_2) \vee \neg(X_3 \vee X_1) \vee (X_3 \vee X_2))$;

2. $\neg(\neg(\neg X_1 \vee X_2) \vee \neg(X_3 \vee X_1) \vee (X_3 \vee X_2))$;

3. $\neg\neg(\neg X_1 \vee X_2) \wedge \neg\neg(X_3 \vee X_1) \wedge (\neg(X_3 \vee X_2))$;

4. $(\neg X_1 \vee X_2) \wedge (X_3 \vee X_1) \wedge \neg X_3 \wedge \neg X_2$.

Then, the procedure leads to $U = \{C_1, C_2, C_3, C_4\}$ where $C_1 \equiv \neg X_1 \vee X_2$, $C_2 \equiv X_3 \vee X_1$, $C_3 \equiv \neg X_3$ and $C_4 \equiv \neg X_2$ which is the sought-for clausal form. Moreover, we compute the resolvents:

$C_5 \equiv X_1$; the resolvent of (C_2, C_3);
$C_6 \equiv X_2$; the resolvent of (C_1, C_5);
$C_7 \equiv \theta$; the resolvent of (C_4, C_6).

Because the resolution method is sound and complete, this algorithm has received important practical implementations. The algorithm will be always able to determine the consistency status of an input set of clauses. Using the algorithm we shall find all true solutions of a system of clauses.

SLD-Resolution of Horn Clauses

The SLD-resolution is a special case of general resolution since it dictates the sequence for finding the resolvents (the choice of literal on which to resolve). The logic programming language Prolog is based on this type of resolution for a particular type of clauses called Horn clauses. A **Horn clause** is a clause containing at most one positive literal. In Prolog, as we shall discuss later, a Horn clause with no negative literals is called a fact, a Horn clause with one positive literal and some negative literals is called a rule and a Horn clause with no positive literals is called a **goal clause**. First, assume that we have some algorithm (procedure) for ordering the literals in any goal clause (in Prolog the ordering is from left to right, i.e., one starts from the leftmost literal in the ordered clause). Second, suppose that all goal clauses are ordered using this algorithm. Then, SLD-resolution specifies that resolution proceeds by repeatedly resolving one goal clause to obtain another. The literal to resolve is indicated by the selected ordering algorithm.

Algorithm of SLD-resolution for Horn clauses

Let T be a set of Horn clauses and let G_0 be exactly one ordered goal. To establish that $T \models (\neg G_0)$ one checks whether $T' = T \cup G_0$ is inconsistent as follows:

- For i = 0,1..., select a clause $C_i \in T$ to resolve with the first literal of G_i, thereby yielding a new goal G_{i+1} (do not remove duplicate literals from G_i);

- If there is some choice of C_i that leads to θ, then $\neg G_0$ has been proved by SLD-resolution.

THEOREM 13.4

The SLD-resolution is sound and complete for the propositional calculus of Horn clauses regardless of what ordering algorithm is used for G_0.

Resolution with variables (in the context of predicate logic)

In this section we shall sketch how to extend the method of resolution from propositional logic to predicate logic. In predicate logic, formulas (clauses) may contain variables and quantifiers. Hence, the theoretical issues will be more complicated. The key concepts which make the difference between resolution proofs for propositional logic and resolution proofs for predicate logic are **Skolemization** and **unification** which will be briefly discussed. The emphasis in this section will be on examples.

To carry out a resolution proof in predicate logic we must, first, rewrite a given formula so that the finite set of quantifiers will precede the formula. Second, all quantifiers have to be universal, i.e., we have to eliminate existential quantifiers by Skolemization. Third, we have to convert the formula (without quantifiers) into clausal form. We recall that a formula is in clausal form if it is a conjunction of clauses. A clause is a disjunction of literals. A **literal** in predicate logic is an elementary formula or the negation of an elementary formula.

PROPOSITION 13.11

*In predicate logic, for every formula there is a logically equivalent formula in which the finite set of all quantifiers appears at the beginning (such a formula is said to be in **prenex form**).*

PROOF   We shall prove this result in a constructive manner. The stages to follow are:

- Eliminate the equivalence and the implication connectors (as shown in propositional logic, Section 1).

- Rename in the given formula and in all subformulas the bound variables to avoid any variable having simultaneously free and bound occurrences.

- Remove the quantifiers whose scopes do not contain occurrences of quantified variables.

- Transform all the occurrences of negation by using the following equivalences:

$$\neg(\forall x_1 A) \approx \exists x_1 \neg A;$$
$$\neg(\exists x_1 A) \approx \forall x_1 \neg A;$$
$$\neg(A \wedge B) \approx (\neg A \vee \neg B);$$
$$\neg(A \vee B) \approx (\neg A \wedge \neg B);$$
$$\neg\neg A \approx A.$$

- Move the quantifiers to the beginning of the formula by using the following equivalences (and their duals):

$$\forall x_1 A \wedge \forall x_1 B \approx \forall x_1 (A \wedge B);$$
$$\forall x_1 A \wedge B \approx \forall x_1 (A \wedge B) \quad \text{if B does not contain } x_1;$$
$$A \wedge \forall x_1 B \approx \forall x_1 (A \wedge B) \quad \text{if A does not contain } x_1;$$
$$\exists x_1 A \wedge B \approx \exists x_1 (A \wedge B) \quad \text{if B does not contain } x_1;$$
$$A \wedge \exists x_1 B \approx \exists x_1 (A \wedge B) \quad \text{if A does not contain } x_1.$$

This completes the schema of the proof.

For example, the formula:

$$\forall x_1 (\exists x_2 (P_1(x_1, x_2) \wedge P_2(x_1, x_2)) \Rightarrow P_3(x_1))$$

becomes:

$$\forall x_1 \forall x_2 (\neg P_1(x_1, x_2) \vee \neg P_2(x_1, x_2) \vee P_3(x_1)).$$

DEFINITION 13.6   *A **Skolem form** of a given formula, which is in clausal form preceded by a finite set of quantifiers, is a formula where all existential quantifiers have been eliminated.*

For example, to obtain the Skolem form for the following formula:

$$\exists x_1 \forall x_2 \forall x_3 \exists x_4 P(x_1, x_2, x_3, x_4)$$

we first replace the existential variable x_1 by a constant a and similarly x_4 by a term where f is a new functor $f(x_2, x_3)$ of each universally quantified variable in whose scope (sequence of symbols from the alphabet which form the quantified formula) x_4 appears. Thus, we obtain the following Skolem form associated with the given formula:

$$\forall x_2 \forall x_3 P(a, x_2, x_3, f(x_2, x_3)).$$

We note that by Skolemization we add new constants and functors to the initial alphabet. The constants and the functors used to replace the existential variables in formulas are called Skolem functions. Naturally the questions arise: Is the Skolem form logically equivalent to the given formula? Generally the answer is no; Skolemization of a formula does not, in general, preserve its validity.

PROPOSITION 13.12

For every formula F with no free variables in a given language, there is a universal formula F' (with only universal quantifiers occurring at the beginning) with no free variables, in an expanded language, that has been produced by the addition of new Skolem functions such that $F' \Rightarrow F$ is valid but $F \Rightarrow F'$ need not always hold.

Once all existential quantifiers have been eliminated and the universal quantifiers occur all at the beginning, these universal quantifiers can be dropped and the formula will be in clausal form.

We present now, without proof, the Resolution Theorem for predicate logic:

THEOREM 13.5

Let T be a set of clauses which represents the Skolem form of a given formula F; then F is inconsistent if and only if T is inconsistent.

REMARK 13.5 There is no finite method to establish whether a system of clauses in Skolem form is inconsistent. However, if the system is inconsistent the resolution method can be used to establish a solution in a finite number of steps: a system of clauses is inconsistent iff the empty clause (which denotes inconsistency) is a logical consequence of the given system. On the other hand, if a system is consistent, resolution may never terminate. ∎

The process to obtain a contradiction by means of resolution is similar to that for propositional calculus, except that we must use unification due to the presence of variables. In the unification process we need to use the operation of substitution.

DEFINITION 13.7

*(i) A **substitution** is a function $\sigma: \mathbf{V} \to \mathcal{T}$, i.e., from the set of variables to the set of terms.*

(ii) Two elementary formulas $P_1(x_1)$ and $P_2(x_1)$ are **unifiable** if there is
a substitution (called a unifier) σ such that $P_1(\sigma(x_1))$ and $P_2(\sigma(x_1))$
are identical.

For example, $P_1(x_1)$ and $P_1(a)$ are unifiable with the substitution $\sigma = \{x_1 \swarrow a\}$, i.e., x_1 is replaced by a. Also $P_1(x_1, f(x_2, x_1))$ and $P_1(g(b, x_3),$
$f(x_2, g(b, x_3)))$ are unifiable with the substitution $\sigma = \{x_1 \swarrow g(b, x_3)\}$.
There could be more than one substitution to unify two elementary formulas. Therefore, we need to apply a developed theory of substitutions (which
are functions) for the unification process to succeed. In the context of this
theory the notion of **most general unifier** is important. The most general
unifier is a substitution from which all others can be obtained by further
substitutions (composition of functions), i.e., σ is the most general unifier
for two elementary formulas if for every unifier ϱ there is a substitution τ
such that $\varrho = \sigma\tau$ (or $\varrho = \sigma \circ \tau$ where $\circ$ is the function composition operator). There are various ways to describe an algorithm to find the most
general unifier, but we won't go into this topic here.

DEFINITION 13.8   (Resolution in predicate logic) Let $C_1 = \{L_1, ...\}$,
$C_2 = \{\neg L_2, ...\}$ (supposed without common variables, otherwise rename
them) and R be clauses in predicate logic. Then R is called a **resolvent** of
C_1 and C_2 if there is a most general unifier σ such that $L_1\sigma = L_2\sigma = L$ (are
unifiable) and $R = C_1' \cup C_2'$ where $C_1' = C_1\sigma \setminus \{L\sigma\}$ and $C_2' = C_2\sigma \setminus \{\neg L\sigma\}$.

For example, let $C_1 = P_1(x_1) \vee P_2(x_1)$ and $C_2 = \neg P_1(a) \vee P_3(x_1)$. First,
we rename the variable in C_2. Thus, we get $C_2 = \neg P_1(a) \vee P_3(x_2)$. Then,
$L_1 = P_1(x_1)$, $L_2 = P_1(a)$ and $\sigma = \{x_1 \swarrow a\}$. Thus, we obtain $C_1' = P_2(a)$,
$C_2' = P_3(x_2)$ and $R = P_2(a) \vee P_3(x_2)$.

Using unification we are able to extend the resolution method without
variables to the resolution method with variables. The basic operation in
the unification mechanism is substitution. Hence we replace in clauses all
the occurrences of a variable by a term. Unification is the fundamental
operation in Prolog. Unification permits us to determine if two terms are
identical. It allows the manipulation and the transfer of data. It also
permits the expression of constraints on data through the use of several occurrences of the same variable in a clause. Finally, unification allows us to
access different subparts of a term by using variables which can be unified
with a part of the term. Since in the context of predicate logic we are in
the presence of variables and quantifiers, the formulas have to be converted
first to Skolem form.

Algorithm for resolution proof in predicate logic:

Let Σ be a finite set of formulas. In order to prove the inconsistency of

Σ, we have to do the following:

A. Replace the formulas by their universal Skolem forms.

B. Find a set of clauses T such that Σ is inconsistent iff T is inconsistent.

C. Apply the resolution rule until the empty clause is derived. The non-deterministic generation subalgorithm of such systems is as follows:

While $\theta \notin T$

- Choose two elementary formulas P_1, P_2 and two clauses C_1, C_2 such that $P_1 \in C_1$, $P_2 \in C_2$ and P_1, P_2 are unifiable.
- Compute the resolvent R
- Replace T by $T \cup \{R\}$.

We conclude this section with an important result concerning soundness and completeness:

THEOREM 13.6
The resolution method for predicate logic is sound and complete.

Logic programming is based on the resolution method with variables limited to the particular case of Horn clauses. Some details concerning the SLD-resolution for Horn clauses will be given in the following section regarding Prolog.

Example 13.3
A. Consider the following formulas:

$$F_1 : \forall x_1 (P_1(x_1) \Rightarrow (P_2(x_1) \wedge P_3(x_1)));$$
$$F_2 : \exists x_1 (P_1(x_1) \wedge P_4(x_1));$$
$$F_3 : \exists x_1 (P_4(x_1) \wedge P_3(x_1)).$$

To show that F_3 is a logical consequence of F_1 and F_2, i.e., the system $\{F_1, F_2, \neg F_3\}$ is inconsistent, we first have to convert the given formulas into their Skolem form. Thus we obtain the following clauses:

$$C_1 : \neg P_1(x_1) \vee P_2(x_1);$$
$$C_2 : \neg P_1(x_1) \vee P_3(x_1);$$
$$C_3 : P_1(a);$$
$$C_4 : P_4(a);$$
$$C_5 : \neg P_4(x_1) \vee \neg P_3(x_1).$$

Then we compute:

$$C_6 : P_3(a) \text{ resolvent of } C_2 \text{ and } C_3$$
$$C_7 : \neg P_3(a) \text{ resolvent of } C_4 \text{ and } C_5$$
$$C_8 : \theta \text{ resolvent of } C_6 \text{ and } C_7.$$

B. Consider the following sentences:

Richard likes all courses he follows at Glendon College.
"Logic" is one of these courses.
"Calculus" is another one of these courses.
A course is a series of educational events in which the student participates and that deal with a specific area of knowledge.
All we see happening in a classroom that arouse our interest: lectures, discussions, quizzes, tests, exams, etc. are some components of a course.
Alain likes the "Data Structures" course and he is interested.
Nathalie likes following the same courses as Alain.

  (a) Translate these sentences into logical formulas in predicate logic.

  (b) Convert the obtained formulas into Skolem form.

  (c) Use the resolution method to answer the question: "What courses does Nathalie follow?

  (d) Prove using resolution that Richard likes the course "Data Structures".

(a) We obtain successively:

  (1) $\forall x_1(\text{course}(x_1) \Rightarrow \text{likes}(\text{richard}, x_1))$.

  (2) course(logic).

  (3) course(calculus).

  (4) $\forall x_1(\exists x_2(\text{likes}(x_2, x_1) \wedge \text{interested}(x_2)) \Rightarrow \text{course}(x_1))$.

  (5) likes(alain,data structures).

  (6) interested(alain).

  (7) $\forall x_1(\text{likes}(\text{alain}, x_1) \Rightarrow \text{likes}(\text{nathalie}, x_1))$.

(b) Applying the theory we get:

  (1) $\neg\text{course}(x_1) \vee \text{likes}(\text{richard}, x_1)$.

  (2) course(logic).

(3) course(calculus).

(4) ¬likes(x_4, x_3) ∨ ¬interested(x_4) ∨ course(x_3).

(5) likes(alain,data structures).

(6) interested(alain).

(7) ¬likes(alain, x_7) ∨ likes(nathalie, x_7).

(c) We compute by resolution adding the clause

(8) ¬likes(nathalie, x_8) to our system of clauses.
Then, we infer successively:

(9) ¬likes(alain, x_7), resolvent of (7) and (8) with $\sigma = \{x_8 \swarrow x_7\}$

(10) θ, resolvent of (9) and (5) with $\tau = \{x_7 \swarrow$ data structures$\}$.

Therefore, Nathalie likes the course "Data Structures". This is shown by the following resolution tree:

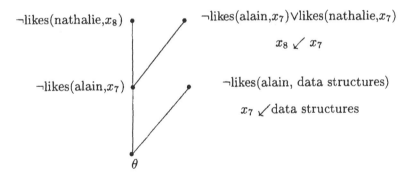

(d) By adding the clause

(11) ¬likes(richard, data structures)
a deduction of empty clause is given by:

(12) ¬course(data structures), resolvent of (11) and (1) with $\varrho = \{x_1 \swarrow$ data structures$\}$

(13) $\neg\text{likes}(x_4, \text{data structures}) \vee \neg\text{interested}(x_4)$, resolvent of (12) and (4) with $\delta = \{x_3 \swarrow \text{data structures}\}$

(14) $\neg\text{interested}(\text{alain})$, resolvent of (13) and (5) with $\gamma = \{x_4 \swarrow \text{alain}\}$

(15) θ, resolvent of (14) and (6).

The following resolution tree illustrates the algorithm:

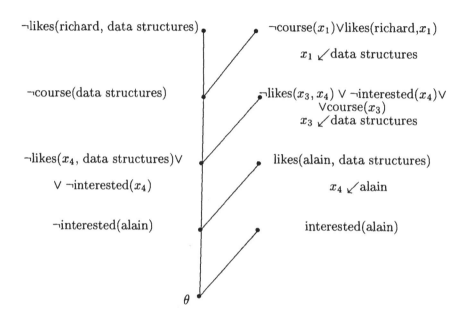

$\square$

13.2   Elements of Logical Foundations of Prolog

The programming language Prolog is based on mathematical logic. Prolog transforms declarative and procedural sentences into pieces of software. The process of "reasoning" implemented in Prolog is based on a subset of the predicate calculus called Horn clauses treated using a particular type of resolution called SLD-resolution. This section is not intended to be a

Prolog manual. We limit ourselves to discussing some predicate logic aspects of Prolog. We study, from the logical point of view, how programs are made up in Prolog. Finally, we present a particular case of the foundations of logic programming along the lines of [23]. To illustrate these issues some simple examples will be given.

Prolog Programs

We recall that a formula P is a logical consequence of the finite set of formulas $Q_1, Q_2, ..., Q_n$ if P is true under all interpretations for which $Q_1 \wedge Q_2 \wedge ... \wedge Q_n$ is true, i.e., P is a logical consequence of $Q_1, Q_2, ..., Q_n$ if and only if the formula $Q_1 \wedge Q_2 \wedge ... \wedge Q_n \Rightarrow P$ is valid. Also, P is a logical consequence of $Q_1, Q_2, ..., Q_n$ if and only if the formula $Q_1 \wedge Q_2 \wedge ... \wedge Q_n \wedge \neg P$ is inconsistent.

We recall that in predicate logic a **literal** is an elementary formula or the negation of an elementary formula. A positive literal is an elementary formula. A negative literal is the negation of an elementary formula. A clause is a disjunction of literals. For example let us consider the clause: $P_1 \vee \neg P_2 \vee \neg P_3$. To obtain the Prolog notation we transform it into $P_1 \vee \neg(P_2 \wedge P_3)$ or $P_2 \wedge P_3 \Rightarrow P_1$. This clause is denoted by $P_1 \leftarrow P_2, P_3$ or $P_1 : -P_2, P_3$ in Prolog notation.

The first step is to transform formulas into clausal forms and more specifically into Horn clausal forms.

DEFINITION 13.9   *If $P_1, P_2, ..., P_k$ are elementary formulas, then Horn clauses are clauses which contain at most one positive literal, i.e., they can have the following forms:*

    *(1) $P_1 :- P_2, P_3, ..., P_k$. (This is called a **rule** as the head and the body are not empty and the head is composed of only one elementary formula)*

    *(2) P_1. (This is called a **fact**, as the body is empty, i.e., no negative literals)*

    *(3) $:- P_2, P_3, ..., P_k$. (This is called a **goal**, as the head is empty, i.e., no positive literal)*

    (4) θ (it is the empty clause, which denotes a contradiction).

DEFINITION 13.10   **A Prolog program** *Pr is composed of a finite set of Horn clauses of type (1) and of type (2) (called definite Horn clauses or positive clauses). A Horn clause of type (3) (called negative clause) represents the goal needed to trigger the program, while a Horn clause of type (4) stops the program.*

A Prolog program Pr composed of a set of Horn clauses $Pr = \{C_1, C_2,$ ..., $C_m\}$ can be considered as a logic formula $Pr = C_1 \wedge C_2 \wedge ... \wedge C_m$ (thus, it can be viewed as a set of axioms).

A Prolog rule of the form $P_1 : -P_2, P_3, ..., P_k$ corresponds to the logic formula $\forall x_1 ... \forall x_m ((P_2 \wedge P_3 \wedge ... \wedge P_k) \Rightarrow P_1)$ where $x_1, ..., x_m$ are all the variables which appear in the formula. Thus, the formula is closed (no free variables).

A Prolog fact P_1 corresponds to the logic formula $\forall x_1 ... \forall x_n P_1$ where $x_1, ..., x_n$ are all free variables in P_1.

We would like to check that a Prolog goal is a logical consequence of the program, i.e.,

$$Pr \Rightarrow \exists x_1 \exists x_2 ... \exists x_m \ (P_2 \wedge P_3 \wedge ... \wedge P_k) \approx$$

$$\neg Pr \vee (\exists x_1 \exists x_2 ... \exists x_m \ (P_2 \wedge P_3 \wedge ... \wedge P_k)) \approx$$

$$\neg (Pr \wedge \neg (\exists x_1 \exists x_2 ... \exists x_m \ (P_2 \wedge P_3 \wedge ... \wedge P_k))).$$

Thus, a goal corresponds to the following formula:

$$\neg (\exists x_1 \exists x_2 ... \exists x_m \ (P_2 \wedge P_3 \wedge ... \wedge P_k)).$$

The proof procedure has to prove that this formula together with axioms of types (1) and (2) result in a contradiction, i.e., represents an inconsistent set. From this it follows that the formula:

$$\exists x_1 \exists x_2 ... \exists x_m \ (P_2 \wedge P_3 \wedge ... \wedge P_k)$$

is entailed by the axioms of types (1) and (2) in the program. This method is called proof by refutation: to prove that a set of axioms entails a formula, we show that the set of axioms together with the negation of the formula is inconsistent.

Example 13.4
In the following program:

likes(alain,tennis).
likes(claude, football).
likes(nathalie,x_1):- likes(alain,x_1).

we have x_1 as a variable (uppercase string in Prolog) which can take the values **tennis, football**; the constants (lowercase strings in Prolog) in our example are functors with 0 arguments and can take the values **alain, tennis, claude, football, nathalie** and the only predicators (functional lowercase notation in Prolog) of arity two is **likes** which is true for the following pairs **(alain, tennis), (claude, football), (nathalie, tennis)**. A goal can be

:- *likes*(**nathalie, x$_1$**). which has the meaning $\neg(\exists x_1 \text{ likes(nathalie, } x_1))$.
□

Let us consider now a particular case in predicate calculus called **datalog** when the alphabet is composed of variables, constants, predicate symbols and connectives (hence, the function symbols are missing, i.e., a variable can be replaced only by a constant).

Example 13.5

Consider the following program:

$P_1(a)$.
$P_1(b)$.
$P_2(x_1)$:- $P_1(x_1)$.

In this situation the alphabet is composed of variables (infinite number), of constants a, b and of the predicators $P_1{}^1$ and $P_2{}^1$ (the superscript 1 shows the arity of the predicators, i.e., the number of arguments they require).
□

There is no algorithm to decide whether a formula from predicate logic is valid or unsatisfiable because the number of its interpretations is infinite. However, J. Herbrand showed that for deciding the unsatisfiability of a formula we can use a particular domain (still infinite but not too large). If the formula takes the false value for all interpretations on this particular domain, then it is unsatisfiable. This particular domain is called **a Herbrand universe**.

We recall that the Herbrand universe of a program Pr without functors, denoted by U_{Pr}, is composed of all constants of the alphabet which appear in Pr.

The **Herbrand base** of a program Pr without functors, denoted by B_{Pr} is the set of all elementary formulas composed of predicators and constants of the alphabet.

Let a goal $G(x_1, x_2, ..., x_n)$ be given, i.e., a list of predicates depending on the variables $x_1, x_2, ..., x_n$. We say that Prolog with a given program Pr solves the goal $G(x_1, x_2, ..., x_n)$ if the goal evaluation finds values for the variables, hence $x_1 = a_1, ..., x_n = a_n$.

The soundness property for Prolog with program Pr would have two parts:

Level 1 soundness: if Prolog solves the goal $G(x_1, x_2, ..., x_n)$ with solution $(a_1, a_2, ..., a_n)$, then this instantiation of the variables makes $G(x_1, x_2, ..., x_n)$ true in the Herbrand universe U_{Pr}.

Level 2 soundness: if Prolog, given $G(x_1, x_2, ..., x_n)$ as a goal, eventually stops, then there is no possible instantiation of the variables which makes $G(x_1, x_2, ..., x_n)$ true in the Herbrand universe.

The completeness property for Prolog with program Pr has two parts, too:

Level 1 completeness: Prolog, given the goal $G(x_1, x_2, ..., x_n)$, should eventually terminate (stop the search for solutions).

Level 2 completeness: Prolog, given the goal $G(x_1, x_2, ..., x_n)$, should eventually list all possible instantiations of the variables which make the goal true in the Herbrand universe.

However, in practice these properties do not hold in all cases. The level 1 soundness property is generally true, but all other properties frequently fail. The Prolog programmer has to write programs carefully in order to benefit from the soundness and completeness properties, if possible.

A Herbrand interpretation of a program Pr is a function which maps its Herbrand base into $\{0, 1\}$:
$f : B_{Pr} \rightarrow \{0, 1\}$, $f^{-1}(\{1\}) = I_{Pr} \subseteq B_{Pr}$.

Example 13.6

In the case of the program of Example 13.5 we have:
$U_{Pr} = \{a, b\}$, $B_{Pr} = \{P_1(a), P_1(b), P_2(a), P_2(b)\}$. Let $I_{Pr} = \{P_1(a), P_1(b)\}$.
The interpretation I_{Pr} of the universal quantifier $(\forall x_1) \, f(x_1)$ is given by $\bigwedge_a f(a), a \in U_{Pr}$.
In the case of Example 13.5 we can have: $\forall x_1 \, (P_1(x_1) \Rightarrow P_2(x_1))$ which in the interpretation I_{Pr} becomes $(P_1(a) \Rightarrow P_2(a)) \wedge (P_1(b) \Rightarrow P_2(b))$.
If we write the program Pr as a formula (conjunction of clauses), then we obtain:
Pr: $P_1(a) \wedge P_1(b) \wedge ((P_1(a) \Rightarrow P_2(a)) \wedge (P_1(b) \Rightarrow P_2(b)))$.
If we consider the expression (goal) $E \equiv \neg P_2(a) \wedge \neg P_2(b) \approx \neg(P_2(a) \vee P_2(b))$, then we get the following table:

$P_1(a)$	$P_1(b)$	$P_2(a)$	$P_2(b)$	Pr	$Pr \wedge E$
1	1	1	1	1	0
1	1	1	0	0	0
1	1	0	1	0	0
1	1	0	0	0	0
1	0	1	1	0	0
1	0	1	0	0	0
1	0	0	1	0	0
1	0	0	0	0	0
0	1	1	1	0	0
0	1	1	0	0	0
0	1	0	1	0	0
0	1	0	0	0	0
0	0	1	1	0	0
0	0	1	0	0	0
0	0	0	1	0	0
0	0	0	0	0	0

DEFINITION 13.11   *The Herbrand interpretation I_{Pr} is a **model** for program Pr, if $Pr \wedge E$ is true under the considered interpretation I_{Pr}. If we are using a truth table, the last entry of the line should contain the Boolean value 1.*

Let us consider as a goal G for the Program Pr the following clause:
:- $P_2(x_1)$., i.e., $\forall x_1 \, \neg P_2(x_1) \approx \neg(\exists x_1 \, P_2(x_1))$.
Then we get the following formula for $Pr \cup \{G\}$:

$$P_1(a) \wedge P_1(b) \wedge (P_1(a) \Rightarrow P_2(a)) \wedge (P_1(b) \to P_2(b)) \wedge (\neg P_2(a) \wedge \neg P_2(b)).$$

From J. Herbrand's results it follows that to run the program Pr with goal G means to find the Herbrand interpretations under which this formula is unsatisfiable. Thus, a set of axioms is satisfiable if and only if it has a Herbrand model.

Example 13.7
For Example 13.6 we get:

$$I_{Pr} = \{P_1(a), P_1(b), P_2(a), P_2(b)\},$$

which is a model for Pr. By inspecting the truth table one can verify that $Pr \cup \{G\}$ is false under I_{Pr} because the last column $Pr \wedge (\neg P_2(a) \wedge \neg P_2(b))$ is 0. Thus, $P_2(a)$ or $P_2(b)$ are true and we obtain as a solution $x_1 = a$ or $x_1 = b$.   ☐

We have denoted by $\mathbf{V}$ the set of all variables, and by $\mathbf{F_0}$ the set of all constants and remind that a finite **substitution** $\sigma : \mathbf{V} \to \mathbf{F_0} \cup \mathbf{V}$ is a function which replaces some variables (a finite number) by terms (constants or other variables).

We denote $\sigma(x_i) = T_i$, $i = 1, 2, ..., n$ or $\sigma = \{x_1 \swarrow T_1, ..., x_n \swarrow T_n\}$. The most general substitution is a substitution from which all others can be obtained by further substitutions.

Two elementary formulas $P_1(x_1)$ and $P_2(x_1)$ are **unifiable** if there exists a substitution σ such that $P_1(\sigma(x_1))$ is identical to $P_2(\sigma(x_1))$.

If two elementary formulas differ just by the names of variables, then they can be unified by changing the names of these variables.

Let G_0 be a goal (query) of the form : $-P_l, P_{l+1}, ..., P_s$, let C_0 be an entry clause (fact or rule) of the form $P_1 : -P_2, P_3, ..., P_k$ and let σ_0 be a substitution.

DEFINITION 13.12   *A* **resolvent** *associated to the triple (G_0, C_0, σ_0) is a clause G_1 which satisfies the following conditions:*

1) *there exists an elementary formula P_i in the goal G_0 such that it is unifiable with the head of the clause P_1 by using the substitution σ_0;*

2) *G_1 is the new goal obtained by replacing P_i by $P_2, P_3, ..., P_k$, hence G_1 is of the form : $-(P_l, P_{l+1}, ..., P_2, P_3, ..., P_k, ...P_s)\sigma_0$.*

Thus, by using C_0 and σ_0 one can construct an edge from the vertex G_0 to the vertex G_1 in a search tree where the vertices represent the goals. The goals need to be "proved" to establish a query. The strategy adopted is a depth-first search. If Prolog obtains an empty goal (θ), then a positive answer to the query has been found.

Let Pr be a Prolog program and let G be a goal.

DEFINITION 13.13   *A* **SLD-resolution** *(Selection-rule-driven Linear resolution for Definite clauses) is composed by:*

1) *a sequence of clauses $G_0, G_1, ...$ starting with $G_0 = G$;*

2) *a sequence of entry clauses $C_0, C_1, ...$ associated to the goals such that each clause C_i is a clause of Pr or is a modified clause of Pr under some substitution σ_i;*

3) *a sequence of substitutions $\sigma_0, \sigma_1, ...$ such that:*

   (i) *G_{i+1} is a resolvent of (G_i, C_i, σ_i), hence $G_i \to G_{i+1}$;*

   (ii) *the clause C_i has no variables in common with $G_0, C_0, ..., C_{i-1}$.*

A SLD-resolution could be finite with success ($\exists G_i = \theta$) or not ($\forall G_i \neq \theta$), or it could be infinite when recursivity exists.

The main differences between resolution proofs in propositional and predicate calculus consist of the need for Skolemization (elimination of existential quantifiers) and unification in predicate calculus. However, for Prolog Skolemization is not required because in the resolution process, the goal, which contains the existential quantifiers, is negated. Therefore, we do not deal in the Prolog context with existential quantifiers.

PROPOSITION 13.13

SLD-resolution for Horn clauses is sound (it should not prove false formulas) and complete (all logical consequences are provable), or formally

$$Pr \vdash_{\mathcal{HR}} G \quad \text{if and only if} \quad Pr \models G.$$

To illustrate resolution one can construct a SLD-tree where the vertices are the goals and the edges are obtained using the substitutions for unifications and the entry clauses.

SLD-resolution for Horn clauses represents a **nondeterministic** algorithm (driven by the goals but with the possibility of backtracking); the programming language Prolog is based on it and allows it to be implemented on the computer: the algorithm is programmed and automatically triggered when an inference process is called for. SLD-resolution is a particular case of resolution because the steps of resolution are specified by the sequence of goals.

Example 13.8

Let Pr be the following program:

$P_1(a)$.	(C_1)
$P_1(b)$.	(C_2)
$P_2(x_1)\text{:- } P_1(x_1)$.	(C_3)

and let the goal G_0 be :- $P_2(x_1)$. In order to construct the associated SLD-tree we consider the first elementary formula of the goal $\exists x_2\ P_2(x_2)$ and the clause C_3 of Pr. They are unifiable because the substitution $\{x_1 \swarrow x_2\}$ exists. Now, the goal becomes $P_1(x_2)$. The new goal can be unified with the clause C_1 using the substitution $\{x_2 \swarrow a\}$ and the SLD_1 resolution step is a success. One can proceed in a similar way for the SLD_2 resolution step. So we obtain:

SLD_1
$C_3:\ P_2(x_2)\text{:- } P_1(x_2)$, by $\sigma_1 = \{x_1 \swarrow x_2\}$
$C_1:\ P_1(x_2) = P_1(a)$, by $\sigma_2 = \{x_2 \swarrow a\}$
hence $G_0 \to G_1 \to \theta$.

SLD_2

$C_3 : P_2(x_2):- P_1(x_2),$ by $\sigma_1 = \{x_1 \swarrow x_2\}$

$C_2 : P_1(x_2) = P_1(b),$ by $\sigma_2 = \{x_2 \swarrow b\}$

hence $G_0 \rightarrow G_1 \rightarrow \theta.$

The solutions are $x_2 = a$ and $x_2 = b$ and the SLD-tree is:

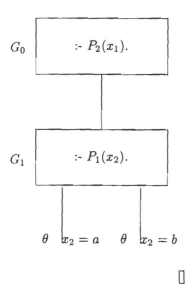

In the following example we present a SLD-resolution with failure and backtracking. By backtracking one means a process whereby Prolog "backs up" in the knowledge base it is searching and finds alternative values for the variables in the goal.

Example 13.9

Let Pr be the following program:

$P_1(a).$	(C_1)
$P_1(b).$	(C_2)
$P_2(x_1):- P_1(x_1), P_3(x_1).$	(C_3)
$P_3(b).$	(C_4)

and let the goal G_0 be $:-P_2(x_1).$

Here is the associated SLD-tree without comments:

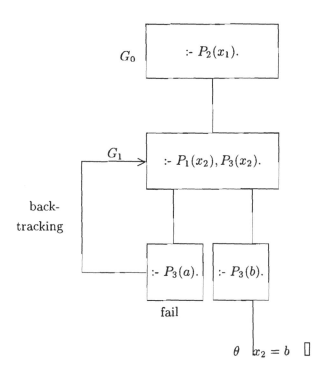

Final Remarks

In order to use a set of logic formulas as a Prolog program, a programmer tries to describe the problem under consideration as a model using declarative sentences. These sentences, which are facts and rules, will be expressed as Horn definite clauses (particular logic formulas). Prolog runs the instructions which are sentences of a particular and elementary logical language, the Horn clause language. The program may have several models, among which is the proposed model. The logical consequences of the program form an infinite set. Hence, the programmer has to propose **goals** for the program. The essential issue is to establish if the program unified with a goal can be satisfied, i.e., if there exists an interpretation for the program and the goal which is a model. The program will be used to obtain conclusions (logical consequences) concerning the considered model. From a procedural point of view, a clause defines a sequence of procedure calls. The programming language Prolog might be considered as a procedural and nondeterministic interpreter (because of the possibility of backtracking for Horn clauses). The program allows variables in the facts and in the rules. In the process of unification the variables will be initialized during the stage of pattern matching. This stage of pattern matching includes an algorithm for unifying elementary formulas. Prolog also contains an inference engine (built-in SLD-resolution) having a depth-first search strategy, built-in predicates (automatically available, i.e., made available in the system rather than defined by the user) and control tools for the execution process which are beyond first-order logic to provide a useful programming language (e.g., negation as failure and the cut mechanism which is a built-in procedure to control generation of solutions, needed not for logical reasons, but for processing efficiency, i.e., that "cuts" paths in which to search for yet additional matches). Some logical limitations of Prolog include the following: Prolog does not allow disjunctive facts or conclusions; does not allow negative facts or conclusions; does not allow existential quantifications in facts, conclusions or rules, and does not allow second-order logic, i.e., does not allow predicate names as variables. However, Prolog is still one of the major artificial intelligence languages because it is: logic-based, rule-based, structured (has a modular top-down approach), declarative, interactive, compact (has a fairly small code), elegant and operational. The latest versions of Prolog, such as Prolog IV, allow meta-programming (allow a program to examine itself, i.e., a program can treat program text, including itself, as data), constraint-based programming (allow the use of constraints in logic programming to solve a set of complex problems), object-oriented programming (programs are constructed by describing classes of objects, which integrate data structure and procedures to work on them), and allow explanations to be given. They yield easy testing and debugging, highly readable, highly modifiable code with a longer life cycle and procure a reusable, stand-alone, application generator.

13.3   Elements of Approximate Reasoning for Expert Systems Design

Approximate reasoning (fuzzy logic in a narrow sense) is a logical system for reasoning with imprecise (vague or fuzzy) knowledge. One of the aims of approximate reasoning is to provide a formal, computationally oriented system of concepts and techniques for dealing with models of reasoning which are approximate. Hence, approximate reasoning helps us draw imprecise (approximate) conclusions starting with imprecise premises, which are stated using linguistic knowledge (e.g., descriptive adjectives from everyday language). The imprecision of knowledge is retained by using translation rules from a part of natural language to the formal language of fuzzy logic. Approximate reasoning constitutes an excellent framework for common sense reasoning and it is a necessary but not a sufficient condition for finding solutions to problems in artificial intelligence and decision-making. As an originator of this theory, L. A. Zadeh pointed out that the key concepts which underline most of the applications of fuzzy logic are **linguistic variables** (variables whose values are not numbers but words or phrases in a natural or artificial language), **canonical forms** (sentence expressing fuzzy constraints on a variable, e.g., x is R) and **interpolative reasoning** (interpolation of fuzzy relations which are partially defined by a collection of fuzzy if-then rules). Linguistic variables play a key role in human communication and may be viewed as a form of data compression (granulation). They can be non-quantifiable (e.g., concerning personal appearance: beautiful, pretty, handsome) or quantifiable but not precisely known (e.g., age: young, old, middle-aged). The principal capabilities of **fuzzy logic** include: fuzzy predicates (tall, small, kind, competent, intelligent, attractive, etc.); fuzzy truth values (true, quite true, not very true, etc.), fuzzy quantifiers (most, few, many, several, frequently, often, some, basically, usually, essentially, typically, etc.); linguistic hedges or predicate modifiers (very, highly, extremely, quite, more or less, roughly, somewhat, rather, almost, etc.); fuzzy probabilities (likely, very likely, highly unlikely, etc.); and fuzzy possibilities (quite possible, more or less possible, almost impossible, etc.). In a wide sense fuzzy logic leads to fuzzy set theory. Fuzzy set theory, which can capture the vagueness of human knowledge as it is expressed in natural language, includes fuzzy sets, fuzzy arithmetic, fuzzy mathematical programming, fuzzy topology, fuzzy data analysis, fuzzy graph theory, and fuzzy logic. Fuzzy set theory is a branch of **soft computing** which also includes **fuzzy neural networks** (learning) and **genetic algorithms** (search algorithms based on the mechanics of natural selection and natural genetics).

In this section we discuss some elements of a formal theory of approximate reasoning (fuzzy logic) along the lines presented in Novák [36] which can be a basis for the design of fuzzy expert systems. Expert systems are computer programs which emulate the reasoning processes of human experts and incorporate their specialized knowledge. They are made up of at least three parts: a knowledge base which contains the expert domain knowledge, an inference engine which makes inferences and draws conclusions and a user interface which allows the communication between the user and the machine. The most frequently used knowledge representation methods are **production rules** (if-then rules), **semantic networks** (graphs where nodes represent objects, concepts or situations in the domain and arcs represent the relations between these nodes) and **frames** (complex data structures with slots that require fillers, used to represent stereotyped objects or situations).

The vagueness in expert systems may arise from translating human expertise expressed in natural language into fuzzy **if-then** rules to produce a knowledge base during the process of knowledge acquisition (which yields rules that may contain fuzzy quantifiers and linguistic hedges), by the partial match (compatibility) between the antecedent patterns and facts which can trigger the firing (activation) of a rule by the inference engine, and by some models of inference engines based on approximate reasoning. The logical part in the inference engine that does the (fuzzy) reasoning may be viewed as producing a proof within the framework of a formal theory of approximate reasoning.

Let us now present a formal theory of approximate reasoning which is a special first-order fuzzy theory given by a fuzzy set of special axioms. The main goal of this logical system will be to achieve a maximal truth value (or provability degree). Formally, we consider a set S of **syntagms** from natural language (well-formed linguistic structures or expressions) and a formal language $\mathcal{L}$ which may contain additional connectives and a sufficiently large set of constants. In the following we shall denote by $\mathcal{T}$ the set of all terms without variables from the language $\mathcal{L}$ under consideration.

The set S contains the following set of natural language syntagms:

1. Simple syntagms have the form
   "$< noun >$ is $[< linguistic\ hedge >] < adjective >$",

   where $< linguistic\ hedge >$ is an intensifying adverb with a narrowing or extending effect (e.g., very, highly, extremely, more or less, roughly, etc.) and $< adjective >$ has a fuzzy meaning (e.g., competent, intelligent, attractive, etc.). Simple syntagms are syntagms.

2. Let $S_i, i = 1, 2, ..., n$ be syntagms. Then:
   S_1 AND S_2 AND...AND S_n
   S_1 OR S_2 OR...OR S_n
   are syntagms.

3. Let S_1 and S_2 be syntagms. Then
   IF S_1 THEN S_2
   is a syntagm.

An example of a simple syntagm is: "the person is very young". Usually the $< noun >$ is replaced by a variable x_j.

Let us now define the **translation rules** from a part of natural language to the formal language of fuzzy logic $\mathcal{L}$.

a) To $< adjective >$ is assigned a predicator $P_i \in \mathcal{L}$

b) To $< linguistic\ hedge >$ is assigned a unary connective $C \in \mathcal{L}$

c) To $< noun >$ is assigned a variable $x_j \in \mathcal{L}$

d) To the syntagm "$< noun > is [< linguistic\ hedge >] < adjective >$"
is assigned a set of evaluated formulas, i.e., a pair $< A(x_1), \underline{A} >$ where

$$A(x_1) \equiv C(P_i(x_1)) \quad \text{and} \quad \underline{A} = \{[A_{x_1}[T]; a_T] \mid T \in \mathcal{T}\}.$$

Here, $A_{x_1}[T]$ is obtained from A by replacing all free occurrences of x_1 by the term T. The terms T are names of objects which have a property $A_{x_1}[T]$ characterized by the linguistic expression $[< linguistic\ hedge >] < adjective >$. Hence, this is how to pass from the vagueness of logical facts to a many-valued logic and thus how to manipulate evaluated formulas. The value $a_T \in [0, 1]$ is a (syntactic) truth value being an evaluation of the formula $A_{x_1}[T]$ in the syntax of fuzzy logic. For example, if the fuzzy set "very young" is given by the discrete function $\{(20, 1), (25, 0.8), (30, 0.6), (40, 0.1), (50, 0)\}$ and Alex is 15 years old, Raoul is 25 and Tim is 50, then we obtain the evaluated formulas [Alex is very young; 1], [Raoul is very young; 0.8], [Tim is very young; 0].

e) The connective AND represents a logical conjunction and is interpreted as a logically fitting **t-norm**. We say that a t-norm t is logically fitting if there are non-zero natural numbers k_1, k_2 such that:

$$(a_1 \leftrightarrow b_1)^{k_1} \odot (a_2 \leftrightarrow b_2)^{k_2} \leq t(a_1, a_2) \leftrightarrow t(b_1, b_2)$$

for every a_1, a_2, b_1, b_2 from [0,1]. Again here:

$$a \to b = \min\{1, 1 - a + b\} \quad \text{and} \quad a \odot b = \max\{0, a + b - 1\}.$$

As a particular case of t-norm we can consider $AND \equiv \wedge$.

f) The connective OR stands for a logical disjunction and is interpreted as a logically fitting **t-conorm**. As a particular case we can consider $OR \equiv \vee$.

g) The connective "IF-THEN" is a linguistically expressed logical implication.

Fuzzy implications can be constructed with the help of different t-norms [4].

A function $t : [0, 1]^2 \to [0, 1]$ is called a **t-norm** if it is bounded, monotonic, commutative and associative, i.e.,

$t(0, 0) = 0$
$t(a, 1) = t(1, a) = a$
$t(a, b) \leq t(c, d)$ if $a \leq c$ and $b \leq d$
$t(a, b) = t(b, a)$
$t(a, t(b, c)) = t(t(a, b), c)$.

A function $s : [0,1]^2 \to [0,1]$ is called a **t-conorm** iff:

$s(1,1) = 1$

$s(0,a) = s(a,0) = a$

$s(a,b) \leq s(c,d) \; if a \leq c \text{ and } b \leq d$

$s(a,b) = s(b,a)$

$s(a,s(b,c)) = s(s(a,b),c).$

t-norms and t-conorms are related by the following functional equation:

$$s(a,b) = 1 - t(1-a, 1-b).$$

With the help of this relation we can define the t-conorm associated with a given t-norm.

Some usual pairs of dual t-norm and t-conorm are:

$$t_0(x_1, x_2) = min(x_1, x_2); \quad s_0(x_1, x_2) = max(x_1, x_2)$$

$$t_1(x_1, x_2) = x_1 \cdot x_2; \quad s_1(x_1, x_2) = x_1 + x_2 - x_1 \cdot x_2$$

$$t_2(x_1, x_2) = max(0, x_1 + x_2 - 1); \quad s_2(x_1, x_2) = min(1, x_1 + x_2).$$

We note that the union and intersection of fuzzy sets can be expressed with the help of s_0 and t_0.

Some ways of defining fuzzy implication are:

(i) $A \Rightarrow B = t(A, B)$

(ii) $A \Rightarrow B = s(A, B)$

(iii) $A \Rightarrow B = s(\neg A, B)$

(iv) $A \Rightarrow B = s(\neg A, t(A, B))$

(v) $A \Rightarrow B = s(\neg A \neg B, B)$

(vi) $A \Rightarrow B = sup\{c \in [0,1] \mid t(A, c) \leq B\}$

(vii) $A \Rightarrow B = sup\{c \in [0,1] \mid s(A, c) \leq B\}.$

For example, for $t = t_0$ in (i) we get Mandami's implication; for $t = t_1$ in (i) we get Larsen's implication; for $s = s_0$ in (iii) we obtain the implication from classical logic, and for $s = s_2$ in (iii) we get Lukasiewicz's implication, etc.

The approximate reasoning inference called generalized modus ponens is of the following type:

x_1 is A'.

If x_1 is A, then x_2 is B.

x_2 is B'.

where A', A, B are known facts and B' is to be determined (A is not identical to A'). One possibility to find B' is to use the sup-norm rule of composition. If in the sup-norm rule of composition:

$$B' = A' \circ R = \sup_{x_1} t(A'(x_1), R(x_1, x_2))$$

where R is a fuzzy relation (implication), we consider $t = t_0$, then we obtain the well-known composition rule of Zadeh:

$$A'(x_1) \circ R(x_1, x_2) = \bigvee_{x_1 \in X} (A'(x_1) \wedge R(x_1, x_2)).$$

Now let us return to the formal theory of approximate reasoning. As we have discussed, in approximate reasoning, we associate to every syntagm a formula $A(x_1)$ and a fuzzy set $\underline{A}$. In the syntactic version of fuzzy logic we must explicitly specify how a given formula is evaluated, as well as its formal proof. In our case the a_T are degrees evaluating formulas $A_{x_1}(T)$. Therefore, approximate reasoning is a deduction in some fuzzy theory $\mathcal{F}$. We are interested in those a_T that are equal to degrees of provability of the respective formulas $A_{x_1}(T)$ in the given theory $\mathcal{F}$. Hence: $\mathcal{F} \vdash_{a_T} A_{x_1}(T)$.

In the formal theory of approximate reasoning we start with the basis of approximate reasoning at a given moment, i.e., a fuzzy set of special axioms. These axioms are given by a linguistic description. They form a linguistic knowledge base given by:

$$\mathcal{R} = \{\mathcal{R}_1, \mathcal{R}_2, ..., \mathcal{R}_m\}.$$

Here, every $\mathcal{R}_i, i = 1, ..., m$ is a classical logical implication (linguistic if-then rule) of the form:

$$\mathcal{R}_i = \text{IF } x_1 \text{ is } \mathcal{A}_i, \text{ THEN } x_2 \text{ is } \mathcal{B}_i$$

and the clauses x_1 is $\mathcal{A}_i, x_2$ is $\mathcal{B}_i$ are simple syntagms.

This linguistic description may be interpreted as a linguistic conjunction, i.e.,

$$\mathcal{R} = \mathcal{R}_1 \text{ AND } \mathcal{R}_2 \text{ AND...AND } \mathcal{R}_m.$$

Another often used possibility for the $\mathcal{R}_i$ is the following linguistic expression:

$$\mathcal{R}_i = x_1 \text{ is } \mathcal{A}_i \text{ AND } x_2 \text{ is } \mathcal{B}_i.$$

Then, the whole linguistic description is a disjunction:

$$\mathcal{R} = \mathcal{R}_1 \text{ OR } \mathcal{R}_2 \text{ OR...OR } \mathcal{R}_m.$$

It is worth noting that in both situations we do not deal with fuzzy implications. The basic task of approximate reasoning is to find a conclusion from the linguistic knowledge base under consideration, if a premise of the form "x is $\mathcal{A}'$" is given. As we discussed before we can have two situations depending on the form of $\mathcal{R}_i$. Accordingly, we obtain two fuzzy theories. In the first case the fuzzy set of special axioms is given by the following fuzzy set of formulas:

$$\underline{R_i} = \underline{A_i \Rightarrow B_i} = \{[A_{i,x_1}[T] \Rightarrow B_{i,x_2}[S]; c_{TS}] \mid S, T \in \mathcal{T}\}.$$

In the second case it consists of the fuzzy set of formulas:

$$\underline{R_i} = \underline{A_i \bigwedge B_i} = \{[A_{i,x_1}[T] \bigwedge B_{i,x_2}[S]; c_{TS}] \mid S, T \in \mathcal{T}\}.$$

In order to be able to manipulate these formulas we need to have at our disposal some rules of inference. For example, the inference rule of Modus Ponens in fuzzy logic is given by:

$$\underline{A} \wedge \underline{A \Rightarrow B} \Rightarrow \underline{B}.$$

In a fuzzy theory this rule can be translated as:

$$\{[A_{x_1}[T]; a_T] \mid T \in \mathcal{T}\} \bigwedge \{[A_{x_1}[T] \Rightarrow B_{x_2}[S]; c_{TS}] \mid T, S \in \mathcal{T}\} \Rightarrow$$

$$\{[B_{x_2}[S]; \bigvee_{T \in \mathcal{T}} (a_T \odot c_{TS})]\}.$$

If a fuzzy set A_{x_1} represented by the evaluated formulas $[A_{x_1}(T); a_T]$ and the fuzzy set of implication $A_{x_1} \Rightarrow B_{x_2}$ are given, then the fuzzy set B_{x_2} is provided by the following evaluated formulas:

$$[B_{x_2}(S); \bigvee_{T \in \mathcal{T}} (a_T \odot c_{TS})].$$

The Max-Min rule of inference (composition of fuzzy relations) is derived on the basis of fuzzy logic in the case when implication is replaced by conjunction and has the form:

$$\{[A_{k,x_1}[T]; a_{kT}] \mid T \in \mathcal{T}\} \bigwedge \{\bigvee_{i=1}^{m} [A_{i,x_1}[T] \wedge B_{i,x_2}[S]; c_{TS}] \mid T, S \in \mathcal{T}\}$$

$$\Rightarrow \{[B_{k,x_2}[S]; \bigvee_{T \in \mathcal{T}} (a_{kT} \wedge c_{TS})]\}.$$

In approximate reasoning we have to find all the proofs with targets $A_{x_1}[T]$ for all $T \in \mathcal{T}$. Hence, most formulas are only lower estimations of the required provability degree. Therefore, it is important to study the conditions under which these formulas give the highest possible provability degree. The following results give some answers about the strength of the widely used formulas in the theory of approximate reasoning (see [36]).

THEOREM 13.7

[36] Let a theory of fuzzy logic be given by a linguistic description:

$$\mathcal{F} = \{\mathcal{A}_k, \text{ AND}_{j=1}{}^m(\text{IF } \mathcal{A}_j \text{ THEN } \mathcal{B}_j)\}$$

for some k, $1 \leq k \leq m$, where $\mathcal{A}_j, \mathcal{B}_j, j = 1, ..., m$ are syntagms. Let $\mathcal{A}_j, \mathcal{B}_j$ have no common subsyntagms. Let the syntagms be interpreted by fuzzy sets of formulas $\underline{A_k}$, $\bigwedge_{j=1}^{m}(A_j \Rightarrow B_j)$ respectively, i.e., the theory $\mathcal{F}$ is assigned a fuzzy theory $\mathcal{F}t$ given by the following fuzzy set of special axioms:

$$ax = \{[A_{k,x_1}[T]; a_kT\} \cup \{[(\bigwedge_{j=1}^{m}(A_j \Rightarrow B_j))_{x_1,x_2}[T,S]; c_{TS}] \mid T,S \in \mathcal{T}\}.$$

Then there is a conclusion $\mathcal{B}_k$ whose interpretation is

$$\underline{B_k} = \{[B_k[S]; b_S] \mid S \in \mathcal{T}\}$$

where

$$b_S = \bigvee_{T \in \mathcal{T}} (a_kT \odot c_{TS})$$

are the provability degrees $\mathcal{F}t\vdash_{b_S} B_k[S], S \in \mathcal{T}$.

Therefore, if the conditions of the above theorem are fulfilled, then we shall obtain the highest possible truth-degrees when we use the common formulas of approximate reasoning for if-then statements considered as logical implications.

THEOREM 13.8

[36] Let a theory of fuzzy logic be given by a linguistic description:

$$\mathcal{F} = \{\mathcal{A}_k, \text{OR}_{j=1}^{m}(\mathcal{A}_j \text{ AND } \mathcal{B}_j)\}$$

for some k, $1 \leq k \leq m$, where $\mathcal{A}_j, \mathcal{B}_j, j = 1, ..., m$ are syntagms denoting fuzzy numbers interpreted by fuzzy sets of formulas $\underline{A_k}$, $\bigvee^m(A_j \bigwedge B_j)$ respectively, i.e., the theory $\mathcal{F}$ is assigned a fuzzy theory $\mathcal{F}t$ given by the following fuzzy set of special axioms:

$$ax = \{[A_{k,x_1}[T]; a_kT]\} \cup \{[(\bigvee_{j=1}^{m}(A_j \wedge B_j))_{x_1,x_2}[T,S]; c_{TS}] \mid T,S \in \mathcal{T}\}.$$

Furthermore, let $\vdash A_i \wedge A_j \Leftrightarrow 0$ *hold for every* $i \neq j$. *Then there is a conclusion* B_k *whose interpretation is*

$$\underline{B_k} = \{[B_k[S]; b_S] \mid S \in \mathcal{T}\}$$

where

$$b_S = \bigvee_{T \in \mathcal{T}} (a_{kT} \wedge c_{TS})$$

are the provability degrees $\mathcal{F}t \vdash_{b_S} B_k[S], S \in \mathcal{T}$.

This result (see [36]) shows that under the specified conditions the formulas used in approximate reasoning will give the best possible solutions.

Exercises (Part II)

1. For each of the following formal words indicate whether or not it is a formula in predicate logic and, if not, why not. Do not be concerned about parentheses.

   (a) $x \vee P(x)$,   (b) $P(x) \vee P(x)$,

   (c) $P(Q(x), y)$,   (d) $\forall x (\exists P P(x))$,

   (e) $P(x \vee y)$,   (f) $P(x) \Rightarrow Q(y)$.

   The letters P, Q above designate predicate symbols and x, y are individual variables.

2. For each of the following formulas, identify the bound variables and rewrite the formulas so that the bound variables have unique names. Do not change the names of any free variables.

   (a) $\forall x (\exists y (P(x) \vee Q(y))) \Rightarrow R(x, y)$;

   (b) $Q(x, y) \Rightarrow (\exists x P(x) \vee \exists x R(x))$.

3. Let $P(x)$ and $Q(x)$ represent "x is rational number" and "x is a real number", respectively. Symbolize the following sentences:

   (a) Every rational number is a real number.

   (b) Some real numbers are rational numbers.

   (c) Not every real number is a rational number.

4. The order of words in a sentence is important and may change the meaning of sentences. Give the logical formalism for the following sentences:

   (a) Alain likes only Nathalie.

   (b) Alain only likes Nathalie.

   (c) Only Alain likes Nathalie.

5. Consider the abbreviations:

$L(x, y) \equiv$ "x likes y."
$T(x) \equiv$ "x is tall."
$M(x) \equiv$ "x is a man."
$S(x) \equiv$ "x is short."
$j \equiv$ "James"
$m \equiv$ "Michelle"
$W(x) \equiv$ "x is a woman."

Give English translation of the following formulas:

(a) $\forall x \ (M(x) \Rightarrow T(x))$

(b) $\forall x \ (W(x) \wedge S(x) \Rightarrow \forall y \ (L(x, y) \Rightarrow M(y)))$

(c) $\exists x \ (M(x) \wedge T(x) \wedge L(x, m))$

(d) $\exists x \ (W(x) \wedge S(x) \wedge \forall y \ (L(x, y) \Rightarrow T(y) \wedge M(y)))$

(e) $\forall x \ (W(x) \wedge S(x) \Rightarrow L(j, x))$.

6. Let $P(x)$, $L(x)$, $R(x, y, z)$ and $E(x, y)$ represent "x is a point", "x is a line", "z passes through x and y" and "$x = y$", respectively. Translate in symbols the following:
(a) For every two points, there is a line passing through both points.
(b) For every two points, there is only one line passing through both points.

7. Several forms of negation are given for each of the following sentences. Which one is correct?

(a) Some people like computer science.

    i. Some people dislike computer science.

    ii. Everybody dislikes computer science.

    iii. Everybody likes computer science.

(b) Everyone likes fish.

    i. No one likes fish.

    ii. Everyone dislikes fish.

    iii. Someone doesn't like fish.

(c) All people are tall and handsome.

    i. Someone is short and ugly.

    ii. No one is tall and handsome.

    iii. Someone is short or ugly.

8. Consider the following interpretation:

$$\mathcal{D} = \{a, b\}, \ \hat{P}(a, a) = 1, \ \hat{P}(a, b) = 0, \ \hat{P}(b, a) = 0, \ \hat{P}(b, b) = 1.$$

Evaluate the truth value of the following formulas (no free variables) in the above interpretation:

(a) $\forall x \; \exists y \; P(x, y)$

(b) $\forall x \; \forall y \; P(x, y)$

(c) $\exists x \; \forall y \; P(x, y)$

(d) $\exists y \; \neg P(x, y)$

(e) $\forall x \; \forall y \; (P(x, y) \Rightarrow P(y, x))$

(f) $\forall x \; P(x, x)$.

9. Consider the following interpretation:

$$\mathcal{D} = \{1, 2\}, \; \hat{a} = 1, \; \hat{b} = 2, \; \hat{f}(1) = 2, \; \hat{f}(2) = 1,$$

$$\hat{P}(1,1) = 1, \; \hat{P}(1,2) = 1, \; \hat{P}(2,1) = 0, \; \hat{P}(2,2) = 0.$$

Evaluate the truth value of the following formulas (no free variables) in the above interpretation:

(a) $P(a, f(a)) \wedge P(b, f(b))$

(b) $\forall x \; \exists y \; P(y, x)$

(c) $\forall x \; \forall y \; (P(x, y) \Rightarrow P(f(x), f(y)))$.

10. Consider the following interpretation:

$$\mathcal{D} = \{1, 2\}, \; \hat{a} = 1, \; \hat{f}(1) = 2, \; \hat{f}(2) = 1, \; \hat{P}(1) = 0, \; \hat{P}(2) = 1,$$

$$\hat{Q}(1,1) = 1, \; \hat{Q}(1,2) = 1, \; \hat{Q}(2,1) = 0, \; \hat{Q}(2,2) = 1.$$

Evaluate the truth value of the following formula in the above interpretation:

$$\forall x \; (P(x) \Rightarrow Q(f(x), a)).$$

11. Consider the formulas
$F_1 \equiv \forall x \; (P(x) \Rightarrow Q(x))$
$F_2 \equiv \neg Q(a)$.
Prove that $\neg P(a)$ is a logical consequence of F_1 and F_2.

12. Justify that the sequence:

(a) $\forall x P(x)$;

(b) $\forall x Q(x)$;

(c) $\forall x P(x) \Rightarrow P(z)$;

(d) $P(z)$;

(e) $\forall x Q(x) \Rightarrow Q(z)$;

(f) $Q(z)$;.

(g) $P(z) \wedge Q(z)$;

(h) $P(z) \wedge Q(z) \Rightarrow \exists z (P(z) \wedge Q(z))$;

(k) $\exists z (P(z) \wedge Q(z))$

is the syntactic deduction of

$$\exists z \ (P(z) \wedge Q(z)) \text{ from } \{\forall x P(x), \forall x Q(x)\}.$$

13. What is the truth value of each of the following formulas in the interpretation where $\mathcal{D} = \mathbf{Z}$, $O(x) \equiv$ "x is odd", $L(x) \equiv$ "$x \le 10$" and $G(x) \equiv$ "$x > 9$":

   (a) $\exists x \ O(x)$

   (b) $\forall x \ (L(x) \Rightarrow O(x))$

   (c) $\exists x \ (L(x) \wedge G(x))$

   (d) $\forall x \ (L(x) \vee G(x))$.

14. Prove that the following argument is valid: every ambassador speaks only to diplomats, and some ambassador speaks to someone, therefore there exists at least one diplomat.

15. Consider the alphabet $\mathbf{X}_f$ of fuzzy predicate logic in which the literal symbols are chosen such that $x, y, z \in \mathbf{V}$ and $P, Q \in \mathbf{P}$, $\alpha(P) = 2, \alpha(Q) = 1$. If $\mathcal{A} = \{0, \frac{1}{3}, \frac{2}{3}, 1\}$, consider the fuzzy set of axioms $\mathbf{Ax}$ and the following fuzzy set of hypotheses:

$$\mathcal{H}(G) = \begin{cases} \frac{1}{3}, & \text{if } G \text{ contains the variable } x \text{ as free variable} \\ \frac{2}{3}, & \text{otherwise.} \end{cases}$$

Give at least three evaluated deductions from $\mathcal{H}$ of the formula

$$F \equiv (\exists x \exists y P(x, y)) \Rightarrow Q(z)$$

and compute the corresponding values of these deductions.

16. Under the notations and hypotheses from the above exercise, show that

$$\frac{2}{3} \le C^{syn}(\mathcal{H})(P(x, y) \Rightarrow Q(z)) \le C^{syn}(\mathcal{H})(F).$$

17. Explain the resolution method without variables (introduce the notions of atom, literal and clause).

18. For the set of clauses $S = \{p \vee q, \neg q \vee r, \neg p \vee q, \neg r, \neg q\}$ obtain the empty clause by resolution.

19. Prove by using the resolution rule, the following argument:
    If Alain did not meet Raoul the other night, then it is because Raoul is the murderer or Alain is a liar. If Raoul is not the murderer, then Alain did not meet Raoul the other night and the crime happened after midnight. If the crime happened after midnight, then Raoul is the murderer or Alain is not a liar. Therefore, Raoul is the murderer. (see [27], Chapter 1, Section 2, exercises).

20. Show by the resolution method that

$$p \Rightarrow q, q \Rightarrow r \models p \Rightarrow r.$$

21. Use the resolution method to prove the following argument:

    The custom officials searched everyone who entered this country who was not a VIP. Some of the drug pushers entered this country and they were only searched by drug pushers. No drug pusher was a VIP. Therefore, some of the officials were drug pushers.

22. Declare a Prolog knowledge base concerning a family tree (e.g. x is parent of y is expressed as parent(x,y)); Alain and Nathalie are the parents of Michelle and Anthony and Anthony is married to Stephanie and they have a child Richard. Extend the program by adding information on the sex, introduce the offspring relation (rule), add mother, father and grandparent relations.

    Which will be the answer to the query "?- parent(x,y)"?
    Write a query to find all possible solutions for "?- grandparent(x,y)".
    Explain how the query "grandparent(x,richard)" is dealt with.
    Formulate a recursive rule to define the predecessor relation.

23. Consider the query:
    $? - P(x, y), \ Q(x, y).$
    with the following knowledge base:
    $P(1, 1). \ P(1, 2). \ P(3, 2). \ P(4, 4).$
    $Q(1, 2). \ Q(1, 3). \ Q(2, 3). \ Q(3, 2). \ Q(4, 4).$
    What will be the Prolog answers?

24. Consider the following knowledge base:

    noun("Logic Programming").

    noun("Classical Programming").

verb("is").

verb("might be").

adjective("fun").

adjective("friendly").

adjective("natural").

Write a program which will generate all the propositions composed of a noun, a verb and an adjective in this order.

25. Consider the following program:

P(1,one).
P(s(1),two).
P(s(s(1)),three).
P(s(s(s(x))),n) :- P(x,n).

How will Prolog answer the following questions? (Whenever several answers are possible, give at least two.)

(a) ?- P(s(1),y).

(b) ?- P(s(s(1)),two).

(c) ?- P(s(s(s(s(s(s(1)))))),y).

(d) ?- P(x,three).

Appendix A

Boolean Algebras

In the period 1848-1854 the Irish mathematician George Boole trying to find an abstract logical calculus discovered a very appropriate structure for the algebra of binary truth values $\{0, 1\}$. After G. Boole, many mathematicians contributed to the development of this new domain (Jevans (1864), Venn (1881)). In the 20th century a considerable advance in the theory of Boolean algebras was realized by the representation theorem of M. H. Stone (1936) and by the works of Lindenbaum, A. Tarski and others. In contemporary mathematics Boolean algebras are one of the most important chapters of the theory of ordered sets and of the theory of lattices (the reader is invited to consult [11], [39]).

A.1 Boolean Algebras. Definitions. Examples.

In this section we present the first general notions about Boolean algebras and some examples. Two manners of definition of Boolean algebras will be considered, the first one being set-based and the second algebraic. The notion of morphism between Boolean algebras will be also studied.

DEFINITION A.1   *A set B ordered by the relation "$\leq$" is called a* **Boolean algebra** *if:*

1. *For every element $x, y \in B$, there exists an element $x \vee y \in B$ and an element $x \wedge y \in B$ with the following properties:*
   (a') $x \vee y \geq x$ and $x \vee y \geq y$;
   (a") $x \wedge y \leq x$ and $x \wedge y \leq y$;
   (b') $z \geq x$ and $z \geq y$ imply $z \geq x \vee y$;
   (b") $z \leq x$ and $z \leq y$ imply $z \leq x \wedge y$;

2. *Operations "$\vee$" and "$\wedge$" are mutually distributive, i.e.,*

(a) $x \wedge (y \vee z) = (x \wedge y) \vee (x \wedge z)$;
(b) $x \vee (y \wedge z) = (x \vee y) \wedge (x \vee z)$;

3. *B has unit elements, i.e., there are elements $0 \in B$ and $1 \in B$ such that for every $x \in B$ we get*

$$0 \leq x \leq 1;$$

4. *B is complemented, i.e., for all $x \in B$, there exists $\bar{x} \in B$ such that*

(a) $x \vee \bar{x} = 1$;
(b) $x \wedge \bar{x} = 0$.

REMARK A.1   The previous definition represents the set-based modality of definition of a Boolean algebra. An ordered set with properties 1 is called a *lattice*. Therefore, a complemented distributive lattice with unit elements is a Boolean algebra.   ∎

REMARK A.2   From Definition A.1 it results that the elements $x \vee y$ and $x \wedge y$ are uniquely determined for any $x, y \in B$; in the literature they are sometimes denoted by sup{x,y}, respectively by inf{x,y} or $x + y$, respectively $x \cdot y$ (the dot is sometimes dropped).   ∎

REMARK A.3   From properties 1 and 2 (a), results the property 2 (b).   ∎

DEFINITION A.2   *A set B endowed with three algebraic operations, two binary:*

$$(x, y) \mapsto x \vee y \in B$$
$$(x, y) \mapsto x \wedge y \in B$$

*and one unary: $x \mapsto \bar{x} \in B$ is called a **Boolean algebra** if the following properties hold for every $x, y, z \in B$:*

1. **Commutativity**
   (a') $x \vee y = y \vee x$; (a'') $x \wedge y = y \wedge x$;

2. **Associativity**
   (a') $x \vee (y \vee z) = (x \vee y) \vee z$; (a'') $x \wedge (y \wedge z) = (x \wedge y) \wedge z$;

3. **Distributivity**
   (a') $x \vee (y \wedge z) = (x \vee y) \wedge (x \vee z)$; (a'') $x \wedge (y \vee z) = (x \wedge y) \vee (x \wedge z)$;

4. **Unit elements**
   There exist two distinct elements $0 \in B$ and $1 \in B$ such that:
   (a') $x \vee 0 = x$; (a") $x \wedge 1 = x$;
   (b') $x \wedge 0 = 0$; (b") $x \vee 1 = 1$;

5. **Complementarity**
   (a') $x \vee \overline{x} = 1$; (a") $x \wedge \overline{x} = 0$;

6. **Idempotency**
   (a') $x \vee x = x$; (a") $x \wedge x = x$;

7. **Absorption**
   (a') $x \vee (x \wedge y) = x$; (a") $x \wedge (x \vee y) = x$.

REMARK A.4   Definition A.2 represents another way, the algebraic one, for defining the notion of Boolean algebra. ∎

REMARK A.5   One can show that in the previous list of axioms not all are independent. However, we shall continue to use all of the axioms (1-7) in order to keep our considerations short. ∎

PROPOSITION A.1
Definitions A.1 and A.2 are equivalent, i.e., if an ordered structure is given on B with the properties listed in Definition A.1, then one can uniquely define on B three algebraic operations which fulfill the properties from Definition A.2; and conversely.

To prove this, it can be shown that if B is a Boolean algebra in the sense of Definition A.1, then $x \vee y$, $x \wedge y$ and $\overline{x}$ fulfill the conditions 1-7 from Definition A.2. Conversely, if B is a Boolean Algebra in the sense of Definition A.2, then on B one can introduce the relation "$\leq$" in the following way: $x \leq y$ if and only if $x = x \wedge y$. One can verify that "$\leq$" is an ordered relation which satisfies conditions 1-4 from Definition A.1.

PROPOSITION A.2
In a Boolean algebra B for each $x \in B$ the complement $\overline{x}$ is unique. The unit elements are also unique.

PROOF   Let us suppose that for each $x \in B$ there exists two complements $\overline{x} \in B$ and $\overline{y} \in B$. Because both complements satisfy the axiom 5 from Definition A.2 it results that:
$$\overline{x} = \overline{x} \wedge (x \vee \overline{y}) = (\overline{x} \wedge x) \vee (\overline{x} \wedge \overline{y}) = \overline{x} \wedge \overline{y}$$

$$\overline{y} = \overline{y} \wedge 1 = \overline{y} \wedge (x \vee \overline{x}) = (\overline{y} \wedge x) \vee (\overline{y} \wedge \overline{x}) = \overline{y} \wedge \overline{x}$$

thus $\overline{x} = \overline{y}$. The fact that the unit elements are unique is easy to show. ∎

PROPOSITION A.3

If B is a Boolean algebra, then:
(a) the mapping $x \mapsto \overline{x} \colon B \mapsto B$ is an involution, i.e., $x = \overline{\overline{x}}$ for every $x \in B$;
(b) $\overline{x \vee y} = \overline{x} \wedge \overline{y}$ and $\overline{x \wedge y} = \overline{x} \vee \overline{y}$ for every $x, y \in B$.

PROOF   The assertion (a) results from the uniqueness of the complement:

$(x \vee y) \vee (\overline{x} \wedge \overline{y}) = x \vee [y \vee (\overline{x} \wedge \overline{y})] = x \vee [(y \vee \overline{x}) \wedge (y \vee \overline{y})] = x \vee (y \vee \overline{x}) = (x \vee \overline{x}) \vee y = 1$. We also get $(x \vee y) \wedge (\overline{x} \wedge \overline{y}) = (x \wedge \overline{x} \wedge \overline{y}) \vee (y \wedge \overline{x} \wedge \overline{y}) = 0$;
it results that $\overline{x \vee y} = \overline{x} \wedge \overline{y}$.
The other formula can be proved in the same manner.   ∎

REMARK A.6   The relations $\overline{x \vee y} = \overline{x} \wedge \overline{y}$ and $\overline{x \wedge y} = \overline{x} \vee \overline{y}$ are called De Morgan's formulas.   ∎

REMARK A.7   The properties (a) and (b) from Proposition A.3 show that the operations "$\vee$" and "$\wedge$" are connected; therefore, they are not independent. For instance, $x \vee y = \overline{\overline{x} \wedge \overline{y}}$ for every $x, y \in B$.   ∎

DEFINITION A.3   *Let us consider two Boolean algebras B and B'. A function $\varphi \colon B \to B'$ with the properties:*

(a) $\varphi(x \vee y) = \varphi(x) \vee \varphi(y)$, $\forall x, y \in B$;

(b) $\varphi(\overline{x}) = \overline{\varphi(x)}$, $\forall x \in B$

is called a **morphism of Boolean algebras**. *If the morphism φ is a bijection, then φ is called an isomorphism of Boolean algebras.*

REMARK A.8   If $\varphi \colon B \to B'$ is a morphism of Boolean algebras, then $\varphi(x \wedge y) = \varphi(x) \wedge \varphi(y)$, $\forall x, y \in B$.   ∎

PROPOSITION A.4

Let E be an arbitrary set; the set $\mathcal{P}(E)$ of subsets of E, endowed with the operations $\cup, \cap$ and C, is a Boolean algebra.

The proof can be reduced to verifying the axioms of Boolean algebras.

REMARK A.9 In what follows the Boolean algebra from Proposition A.4 will be denoted simply by $\mathcal{P}(E)$. ∎

DEFINITION A.4 *The Boolean algebras $\mathcal{P}(E)$ are called standard Boolean algebras.*

Example A.1

Let E be an arbitrary infinite set; we shall denote by $\mathcal{P}_f(E)$ the set of finite subsets of E, and by $\mathcal{P}_{cof}(E)$ the set of subsets with finite complement in E. It is easy to notice that the set $\mathcal{P}_f(E) \cup \mathcal{P}_{cof}(E)$ is a Boolean algebra with the operations $\cup, \cap$ and C. ⬚

Example A.2

Let E be an arbitrary infinite uncountable set; we shall denote by $\mathcal{P}_{\aleph_0}(E)$ the set of all at most countable subsets of E, and by $\mathcal{P}_{co\aleph_0}(E)$ the set of all subsets with at most countable complement in E. The set $\mathcal{P}_{\aleph_0}(E) \cup \mathcal{P}_{co\aleph_0}(E)$ endowed with the operations $\cup, \cap$ and C is a Boolean algebra. ⬚

Example A.3

Let X be a topological space and $\mathcal{D}$ the set of open sets in X. If we denote by $\mathcal{H}$ the set of all subsets $M \in X$ which are simultaneously open and closed, i.e., $\mathcal{H} = \{M \in X \mid M \in \mathcal{D} \text{ and } CM \in \mathcal{D}\}$, then $\mathcal{H}$ is a Boolean algebra with respect to the operations $\cup, \cap$ and C. ⬚

REMARK A.10 The algebras from Examples 1, 2 are not standard Boolean algebras. ∎

A.2 Atoms, Atomic and Finite Boolean Algebras

The aim of this section is to analyze the structure of a remarkable class of Boolean algebras called *atomic Boolean algebras*.

DEFINITION A.5 *Let B be a Boolean algebra; an element $a \in B, a \neq 0$ is called an* **atom** *if $x \leq a$ implies $x = 0$ or $x = a$.*

PROPOSITION A.5

Let B be a Boolean algebra; an element $a \in B, a \neq 0$ is an atom if and only if for every $y \in B$ we get $a \wedge y = 0$ or $a \wedge y = a$.

PROOF   Let us suppose that the element $a \in B$ is an atom and let us consider an arbitrary element $y \in B$. Let us denote by $x = a \wedge y$; obviously, $x \leq a$. From Definition A.2 it results that $x = 0$ or $x = a$, i.e., $a \wedge y = 0$ or $y \wedge a = a$. Thus, for every $y \in B$ we get $a \wedge y = 0$ or $y \wedge a = a$ (thus $y \geq a$). Conversely, let us consider $x \in B$ with $x \leq a$, i.e., $x = a \wedge x$. By assuming the condition is fulfilled we get $a \leq x$ or $a \wedge x = 0$, i.e., $a = x$ or $x = a \wedge x = 0$, thus a is an atom. ∎

Example A.4

In the standard Boolean algebra $\mathcal{P}(E)$ where E is a nonempty set, the subsets composed by a single element ($\{e\}, e \in E$) are the atoms and other atoms do not exist.   ☐

DEFINITION A.6   *A Boolean algebra B is called **atomic** if for all $x \in B, x \neq 0$, there exists an atom $a \in B$ with $a \leq x$.*

REMARK A.11   Every standard Boolean algebra $\mathcal{P}(E)$, with $E \neq \emptyset$ is atomic. Also the Boolean algebra $\mathcal{P}_f(E) \cup \mathcal{P}_{cof}(E)$ is atomic, for all infinite sets E. ∎

DEFINITION A.7   *Let be B a Boolean algebra and $S \subset B$ an arbitrary subset of B. If for S there exists an element $x^* \in B$ having the properties*

(a) $x^* \geq s$, for all $s \in S$;

(b) if $y \geq s$, for all $s \in S$, then $y \geq x^*$,

then the element x^* is called the lowest upper bound (l.u.b.) of S and is denoted by $\sup S$ or by $\bigvee_{s \in S} s$. The same definition holds for an arbitrary family $(x_i)_{i \in I}$ of elements in B; in this case, the lowest upper bound of the family (if it exists) is denoted by $\sup_{i \in I} x_i$ or by $\bigvee_{i \in I} x_i$.

Analogously, we can define (if it exists) the greatest lower bound of a set S or of a family $(x_i)_{i \in I}$ and this greatest lower bound (g.l.b.) will be denoted by $\inf S = \bigwedge_{s \in S} s$ and by $\inf_{i \in I} x_i = \bigwedge_{i \in I} x_i$.

The elements $\sup S$, $\inf S$ are unique, if they exist.

DEFINITION A.8   *A Boolean algebra B having the property that for each subset $S \subset B$ there exist* sup S *and* inf S *is said to be a* **complete** *Boolean algebra.*

REMARK A.12   From the previous definition it follows that every standard Boolean algebra $\mathcal{P}(E)$ is complete. It is easy to show that if E is an infinite set, then the Boolean algebras $\mathcal{P}_f(E) \ \cup \ \mathcal{P}_{cof}(E)$ and $\mathcal{P}_{\aleph_0}(E) \ \cup \ \mathcal{P}_{co\aleph_0}(E)$ are not complete.   ∎

PROPOSITION A.6
Every finite Boolean algebra B is complete and atomic.

PROOF   It is obvious that for every finite subset S in an arbitrary Boolean algebra B, there exist

$$\bigvee_{s \in S} s \in B$$

and

$$\bigwedge_{s \in S} s \in B.$$

If the Boolean algebra B is itself finite, then every subset S of B is still finite. Therefore, there exist

$$\bigvee_{s \in S} s \in B$$

and

$$\bigwedge_{s \in S} s \in B.$$

It follows that a finite Boolean algebra is complete. Let us show that a finite Boolean algebra is also atomic. Let $x \in B, x \neq 0$; let us prove that there exists an atom $a \in B$ such that $a \leq x$. Let us number the nonempty elements from B such that

$$B = \{0, x_1, x_2, ..., x_n\}$$

where

(1) $x_1 = x.$

Now we define the elements $y_1, y_2, ..., y_n \in B$ (distinct or not) by the following relations:

(2) $$y_1 = x_1$$

(3) $$y_k = \begin{cases} y_{k-1} \text{ if } x_k \not\le y_{k-1} \\ x_k \text{ if } x_k \le y_{k-1} \end{cases}$$

$(k = 2, 3, ..., n)$. One can see that

(4) $$y_1 \ge y_2 \ge ... \ge y_n$$

and therefore,

(5) $$y_n \le y_1 = x.$$

Let us show that y_n is an atom. Let $y \in B$ such that

(6) $$y \le y_n.$$

If $y \ne 0$, then $y = x_k$ $(1 \le k \le n)$. In the case $k = 1$ we get $y = x_1 = y_1 \ge y_n$ and from (6) we also get $y \le y_n$. Therefore, $y = y_n$. In the case $k > 1$ we get $x_k = y \le y_n \le ... \le y_{k-1}$. Thus, $x_k \le y_{k-1}$ and therefore, $y_k = x_k$. It results that $y = x_k = y_k \ge y_n$ and from (6) we get $y \le y_n$; thus $y = y_n$. To conclude, for all $y \le y_n$ we get $y = 0$ or $y = y_n$, which means that y_n is an atom. Taking into account that $y_n \le x$, it results that B is an atomic Boolean algebra. ∎

PROPOSITION A.7

If B is an atomic Boolean algebra and A is the set of all atoms from B, then for all $x \in B, x \ne 0$ we get

$$x = sup \ A_x$$

where

$$A_x = \{a \in A \mid a \le x\}.$$

PROOF   It is obvious that x is a majorant of the set A_x; let us show that if y is an arbitrary majorant of A_x, then $y \ge x$, i.e., that $x = sup \ A_x$. Let $y \in B$ such that

(1) $$y \ge a, \quad \text{for all} \ a \in A_x.$$

Taking into account that $x = x \wedge (y \vee \bar{y}) = (x \wedge y) \vee (x \wedge \bar{y})$ it results that if $x \wedge \bar{y}$, then $x = x \wedge y \le y$. Therefore, assuming by contradiction that

$y \not\geq x$ it results that $x \wedge \overline{y} \neq 0$. Because Boolean algebra B is atomic, there exists an atom $b \in A$ such that:

(2) $$b \leq x \wedge \overline{y}.$$

Thus, $b \leq x$, i.e., $b \in A_x$; but from (1) it also results:

(3) $$b \leq y.$$

Therefore, from (2) we obtain:

$$b = b \wedge y \leq (x \wedge \overline{y}) \wedge y = x \wedge (\overline{y} \wedge y) = 0,$$

i.e., $b = 0$ which is in contradiction with $b \in A$. This contradiction shows that $y \geq x$ and therefore, $x = \sup A_x$. ∎

PROPOSITION A.8

If the conditions of Proposition A.7 are fulfilled, then the function:

$$\varphi : B \to \mathcal{P}(A), \text{ defined as:}$$
$$\varphi(x) = A_x \text{ if } x \neq 0 \text{ and } \varphi(0) = \emptyset$$

is an injective morphism from the Boolean algebra B to the standard Boolean algebra $\mathcal{P}(A)$.

PROOF   Let us show that for every $x \neq 0$

(1) $$A_{\overline{x}} = CA_x.$$

Let $a \in A_{\overline{x}}$, i.e., $a \leq \overline{x}$; if we suppose $a \in A_x$, then $a \leq x$ and it results that $a \leq x \wedge \overline{x} = 0$, thus $a = 0$ which is impossible because $a \in A$. Therefore,

(2) $$a \in A_{\overline{x}} \Rightarrow a \in CA_x.$$

Conversely, if $a \in CA_x$, then $a \not\leq x$ and therefore, $a \wedge \overline{x} \neq 0$; indeed, if $a \wedge \overline{x} = 0$, then

$$a = a \wedge (x \vee \overline{x}) = (a \wedge x) \vee (a \wedge \overline{x}) = a \wedge x$$

thus, $a = a \wedge x \leq x$, which is in contradiction with the hypothesis that $a \in CA_x$. Thus, if $a \in CA_x$, then $a \wedge \overline{x} \neq 0$ and there exists an atom $b \in A$ such that $b \leq a \wedge \overline{x}$. Therefore, $b \leq a$ and $b \leq \overline{x}$. But a and b being atoms it results that $b = a$; thus, $a \leq \overline{x}$. This means that $a \in A_{\overline{x}}$. In other words,

(3) $$a \in CA_x \Rightarrow a \in A_{\overline{x}}.$$

(1) results from (2) and (3). Let us now show that

(4) $$A_{x \wedge y} = A_x \cap A_y.$$

If $a \in A_{x \wedge y}$, then $a \le x \wedge y$, thus $a \le x$ and $a \le y$, i.e., $a \in A_x \cap A_y$. Conversely, if $a \in A_x \cap A_y$, then $a \in A_x$ and $a \in A_y$, i.e., $a \le x$ and $a \le y$, thus $a \le x \wedge y$ or $a \in A_{x \wedge y}$. Therefore, we get $A_{x \wedge y} \subseteq A_x \cap A_y$ and $A_x \cup A_y \subseteq A_{x \wedge y}$ which is the relation (4). The relations (1) and (4) may be rewritten as:

(5) $\varphi(\overline{x}) = C\varphi(x), \quad \varphi(x \wedge y) = \varphi(x) \cap \varphi(y).$

From the relations (5), using De Morgan's laws we deduce:

$$\varphi(x \vee y) = \varphi(\overline{\overline{x} \wedge \overline{y}}) = C(\varphi(\overline{x}) \cap \varphi(\overline{y})) = C(C\varphi(x) \cap C\varphi(y)) = \varphi(x) \cup \varphi(y),$$

which together with the relations (5) and with the equalities $\varphi(0) = \emptyset, \varphi(1) = A$ show that φ is a morphism of Boolean algebras. The morphism φ is also injective. Indeed, if for $x, y \in B$ we get $\varphi(x) = \varphi(y)$, i.e., $A_x = A_y$, then from Proposition A.7 we deduce that $x = \sup A_x = \sup A_y = y$, which completes the proof. ∎

PROPOSITION A.9

Let B be a Boolean algebra and $x, y, z \in B$; the following assertions are true:

(a)   if $x \le y$, then $x \wedge z \le y \wedge z$;
(b)   if $x \le y$, then $x \vee z \le y \vee z$;
(c)   if $x \le y$, then $\overline{y} \le \overline{x}$.

PROOF   (a) Assuming $x \le y$, it results that $x = x \wedge y$ and therefore, $x \wedge z = (x \wedge y) \wedge z = x \wedge (y \wedge z)$; thus

$$x \wedge z \le y \wedge z.$$

(b) Assuming that $x \le y$, we get as above that $x = x \wedge y$ and therefore, $x \vee z = (x \wedge y) \vee z = (x \vee z) \wedge (y \vee z)$; thus

$$x \vee z \le y \vee z.$$

(c) Assuming that $x \le y$, it results that $x \wedge \overline{y} \le y \wedge \overline{y} = 0$, thus $x \wedge \overline{y} = 0$. On the other hand, $\overline{y} = \overline{y} \wedge (x \vee \overline{x}) = (\overline{y} \wedge x) \vee (\overline{y} \wedge \overline{x}) = \overline{y} \wedge \overline{x}$, thus $\overline{y} \le \overline{x}$. ∎

PROPOSITION A.10

If B is a complete Boolean algebra, then for any family $(x_i)_{i \in I}$ of elements from B the following equalities hold:

(a)
$$x \vee \left(\bigvee_{i \in I} x_i \right) = \bigvee_{i \in I} (x \vee x_i);$$

(b)
$$x \wedge \left(\bigwedge_{i \in I} x_i \right) = \bigwedge_{i \in I} (x \wedge x_i);$$

(c)
$$x \wedge \left(\bigvee_{i \in I} x_i \right) = \bigvee_{i \in I} (x \wedge x_i);$$

(d)
$$x \vee \left(\bigwedge_{i \in I} x_i \right) = \bigwedge_{i \in I} (x \vee x_i).$$

PROOF   Obviously we get:

$$x_j \leq x \vee x_j \leq \bigvee_{i \in I} (x \vee x_i), \quad \text{for all } j \in I$$

thus

(1)
$$\bigvee_{i \in I} x_i \leq \bigvee_{i \in I} (x \vee x_i).$$

Also

(2)
$$x \leq x \vee x_j \leq \bigvee_{i \in I} (x \vee x_i).$$

From (1) and (2) we deduce:

(3)
$$x \vee \left(\bigvee_{i \in I} x_i \right) \leq \bigvee_{i \in I} (x \vee x_i).$$

On the other hand,

$$\bigvee_{i \in I} x_i \geq x_j, \quad \text{for all } j \in I,$$

hence:

$$x \vee \left(\bigvee_{i \in I} x_i \right) \geq x \vee x_j, \quad \text{for all } j \in I,$$

hence

(4)
$$x \vee \left(\bigvee_{i \in I} x_i \right) \geq \bigvee_{i \in I} (x \vee x_i).$$

From (3) and (4) the equality (a) results.

(b) Obviously we get

$$x_j \geq \bigwedge_{i \in I} x_i, \quad \text{for all } j \in I,$$

hence

$$x \wedge x_j \geq x \wedge \left(\bigwedge_{i \in I} x_i \right), \quad \text{for all } j \in I,$$

and therefore,

(5)
$$\bigwedge_{j \in I} (x \wedge x_j) \geq x \wedge \left(\bigwedge_{i \in I} x_i \right).$$

On the other hand,

$$\bigwedge_{j \in I} (x \wedge x_j) \leq x \wedge x_i, \quad \text{for all } i \in I,$$

hence:

(6)
$$\bigwedge_{j \in I} (x \wedge x_j) \leq x$$

and

$$\bigwedge_{j \in I} (x \wedge x_j) \leq x_i, \quad \text{for all } i \in I,$$

therefore

(7)
$$\bigwedge_{j \in I} (x \wedge x_j) \leq \bigwedge_{i \in I} x_i.$$

From (6) and (7) we deduce

(8)
$$\bigwedge_{j \in I} (x \wedge x_j) \leq x \wedge \left(\bigwedge_{i \in I} x_i \right).$$

From (5) and (8) we obtain the equality (b).

(c) Taking into account that $x_j \leq \bigvee_{i \in I} x_i,$ for all $j \in I$, it results that

$x \wedge x_j \leq \bigvee_{i \in I} x_i$, hence

(9)
$$\bigvee_{j \in I} (x \wedge x_j) \leq x \wedge \left(\bigvee_{i \in I} x_i \right).$$

On the other hand,

$$x \wedge x_j \leq \bigvee_{i \in I} (x \wedge x_i), \quad \text{for all } j \in I;$$

but $x_j = (x \vee \overline{x}) \wedge x_j = (x \wedge x_j) \vee (\overline{x} \wedge x_j)$, hence

(10)
$$x_j \leq \bigvee_{i \in I} ((x \wedge x_i) \vee (\overline{x} \wedge x_i)), \quad \text{for all } j \in I.$$

Taking into account the relations

$$(x \wedge x_j) \vee (\overline{x} \wedge x_j) = [(x \wedge x_j) \vee \overline{x}] \wedge [(x \wedge x_j) \vee x_j]$$

$$\leq \bigvee_{i \in I} ((x \wedge x_i) \vee \overline{x}) \wedge \bigvee_{i \in I} ((x \wedge x_i) \vee x_i)$$

it results that:

$$\bigvee_{i \in I} ((x \wedge x_i) \vee (\overline{x} \wedge x_i)) \leq \bigvee_{i \in I} ((x \wedge x_i) \vee \overline{x})$$

hence from (10) and (a) we obtain

$$x_j \leq \left[\bigvee_{i \in I} (x \wedge x_i) \right] \vee \overline{x}, \quad \text{for all } j \in I$$

therefore

$$\bigvee_{i \in I} x_i \leq \left(\bigvee_{i \in I} (x \wedge x_i) \right) \vee \overline{x}.$$

It results that:

$$x \wedge (\bigvee_{i \in I} x_i) \leq x \wedge [(\bigvee_{i \in I} (x \wedge x_i)) \vee \overline{x}] = x \wedge \bigvee_{i \in I} (x \wedge x_i),$$

hence:

(11)
$$x \wedge (\bigvee_{i \in I} x_i) \leq \bigvee_{i \in I} (x \wedge x_i).$$

From (9) and (11) equality (c) results. Equality (d) can be proved in a similar manner. ∎

PROPOSITION A.11

If B is a complete Boolean algebra, then for any family $(x_i)_{i \in I}$ of elements from B, the De Morgan's equalities hold:

(a)
$$\overline{\bigvee_{i \in I} x_i} = \bigwedge_{i \in I} \overline{x_i};$$

(b)
$$\overline{\bigwedge_{i \in I} x_i} = \bigvee_{i \in I} \overline{x_i}.$$

PROOF   (a) Because

$$x_j \leq \bigvee_{i \in I} x_i, \quad \text{for all } j \in I$$

it results that $\overline{x_j} \geq \overline{\bigvee_{i \in I} x_i}$,   for all $j \in I$, hence

(1)
$$\bigwedge_{i \in I} \overline{x_i} \geq \overline{\bigvee_{i \in I} x_i}.$$

On the other hand,

$$\bigwedge_{i \in I} \overline{x_i} \leq \overline{x_j}, \quad \text{for all } j \in I,$$

hence

$$\overline{\bigwedge_{i \in I} \overline{x_i}} \geq \overline{\overline{x_j}} = x_j, \quad \text{for all } j \in I,$$

from which results

$$\overline{\bigwedge_{i \in I} \overline{x_i}} \geq \bigvee_{i \in I} x_i.$$

By passing to the complements we obtain:

$$(2) \qquad \bigwedge_{i \in I} \overline{x}_i \leq \overline{\bigvee_{i \in I} x_i}.$$

From (1) and (2) we get equality (a). Equality (b) can be proved in a similar manner. ∎

PROPOSITION A.12
Stone's Theorem

Every complete and atomic Boolean algebra is isomorphic to the standard Boolean algebra $\mathcal{P}(A)$, where A is the set of atoms of B. In particular, every finite non-degenerate algebra is isomorphic to a standard Boolean algebra.

PROOF From Proposition A.8 it results that the function $\varphi : B \to \mathcal{P}(A)$ is an injective morphism of Boolean algebras. It is sufficient to show that if B is complete, then the morphism φ is surjective, hence a bijection. Therefore, if the hypotheses of the theorem are true, let us consider an arbitrary subset $S \in \mathcal{P}(A)$ and let us denote:

$$z = \bigvee_{s \in S} s.$$

Denoting by $A_z = \{a \in A \mid a \leq z\}$ obviously we get:

$$S \subseteq A_z.$$

Let us prove the converse implication. For this let us assume by contradiction that there exists $a \in A_z$ such that $a \notin S$. In this situation, obviously $a \neq b$, for all $b \in S$ and hence

$$a \wedge z = a \wedge (\bigvee_{b \in S} b) = \bigvee_{b \in S} (a \wedge b) = 0$$

and therefore, $a = a \wedge (z \vee \overline{z}) = (a \wedge z) \vee (a \wedge \overline{z}) = a \wedge \overline{z} \leq \overline{z}$. It results that $a \leq z$ (because $a \in A_z$) and $a \leq \overline{z}$ hence $a \leq z \wedge \overline{z} = 0$ or $a = 0$, which is impossible because $a \in A$. The contradiction shows that

$$(2) \qquad A_z \subset S.$$

From (1) and (2) it results that $S = A_z = \varphi(z)$, which means that the morphism φ is surjective. ∎

A.3   Boolean Algebra of Boolean Functions

In bivalent mathematical logic and in other domains we frequently use the Boolean algebra with just two elements $\Gamma = \{0, 1\}$ and the algebra $B(N; \Gamma)$ of Boolean functions with N Boolean arguments. There exists a small Boolean algebra which can be reduced to a single element and which is characterized by the fact that the unit elements coincide; this algebra is called a degenerate Boolean algebra or null Boolean algebra. A Boolean algebra which contains two or more elements is called proper. We can note easily that every proper Boolean algebra with the unit elements 0 and 1 contains the Boolean subalgebra $\{0, 1\}$. In this regard, the algebra $\Gamma = \{0, 1\}$ is the smallest proper Boolean algebra.

In the following we shall briefly present the algebras Γ and $B(N; \Gamma)$. Concerning the algebra $B(N; \Gamma)$ we shall show that its elements have a simple but remarkable structure; we shall prove that every Boolean function $f \in B(N; \Gamma)$ is a Boolean polynomial.

In the set $\Gamma = \{0, 1\}$ we introduce an order relation "$\leq$" by the conventions:

$$0 \leq 0, \ 0 \leq 1, \ 1 \leq 1.$$

The set Γ together with the relation "$\leq$" is an ordered, reflexive set. For every pair $x, y \in \Gamma$ we denote $x + y = \max\{x, y\}, xy = \min\{x, y\}$ and

$$\overline{x} = \begin{cases} 0, \text{ if } x = 1 \\ 1, \text{ if } x = 0. \end{cases}$$

The following assertion is established by direct verification.

PROPOSITION A.13
The set Γ together with the operations $x + y, xy$ and $\overline{x}$ is a Boolean algebra.

Let us consider $N \geq 1$ a natural number; we denote the cross product of the set Γ N-times by Γ^N; hence $x \in \Gamma^N$ if and only if

$$x = (x_1, x_2, ..., x_N), \ x_i \in \Gamma \ (1 \leq i \leq N).$$

Any function

$$f : \Gamma^N \to \Gamma : (x_1, x_2, ..., x_N) \mapsto f(x_1, x_2, ..., x_N)$$

is called a **Boolean function of N Boolean variables**. The set of all Boolean functions of N variables will be denoted by $B(N; \Gamma)$, i.e.,

$$B(N; \Gamma) = \{f : \Gamma^N \to \Gamma\}.$$

We endow the set $B(N;\Gamma)$ with three internal algebraic operations in the following manner: if $f, g \in B(N;\Gamma)$, then the elements $f + g, fg, \overline{f} \in B(N;\Gamma)$ are defined in a natural way as follow:

$$(f + g)(x_1, x_2, ..., x_N) = f(x_1, x_2, ..., x_N) + g(x_1, x_2, ..., x_N),$$
$$(fg)(x_1, x_2, ..., x_N) = f(x_1, x_2, ..., x_N) \, g(x_1, x_2, ..., x_N),$$
$$\overline{f}(x_1, x_2, ..., x_N) = \overline{f(x_1, x_2, ..., x_N)}.$$

By direct verification one can establish the following assertion:

PROPOSITION A.14
The set $B(N;\Gamma)$ with the operations $f + g, fg, \overline{f}$ is a Boolean algebra.

Let us consider two elements $x, \sigma \in \Gamma$ and let us denote

$$x^\sigma = \begin{cases} x & \text{if } \sigma = 1 \\ \overline{x} & \text{if } \sigma = 0. \end{cases}$$

It is easy to see that $\overline{x}^\sigma = x^{\overline{\sigma}}$ and

$$x^\sigma = \begin{cases} 1 & \text{if } \sigma = x \\ 0 & \text{if } \sigma \neq x. \end{cases}$$

For each two systems $x_1, x_2, ..., x_N \in \Gamma$ and $\sigma_1, \sigma_2, ..., \sigma_N \in \Gamma$, the product $a x_1^{\sigma_1} x_2^{\sigma_2} ... x_N^{\sigma_N}$, where $a \in \Gamma$ is called a Boolean monomial (conjunctive); a is called the monomial coefficient.

If $(a_{\sigma_1 \sigma_2 ... \sigma_N})_{\sigma_1 \sigma_2 ... \sigma_N \in \Gamma}$ is a set of coefficients, then the sum

$$P(x_1, x_2, ..., x_N) = \sum_{\sigma_1 \sigma_2 ... \sigma_N \in \Gamma} a_{\sigma_1 \sigma_2 ... \sigma_N} x_1^{\sigma_1} x_2^{\sigma_2} ... x_N^{\sigma_N}$$

is called a **disjunctive Boolean polynomial**. Analogously (by duality) one can define **the conjunctive Boolean polynomial**.

PROPOSITION A.15
Representation theorem for Boolean functions
Every Boolean function $f \in B(N;\Gamma)$ can be written as a disjunctive Boolean polynomial and if $f \neq 0$, then the function can be represented by:

$$f(x_1, ..., x_N) = \sum_{f(\sigma_1, ..., \sigma_N)=1} x_1^{\sigma_1} x_2^{\sigma_2} ... x_N^{\sigma_N}, \quad \text{for all } x_1, x_2, ..., x_N \in \Gamma.$$

PROOF   Let us consider $x = (x_1, x_2, ..., x_N) \in \Gamma$ and an arbitrary element $\sigma = (\sigma_1, \sigma_2, ..., \sigma_N) \in \Gamma$. Let us denote $f(\sigma) = a_\sigma$, for all $\sigma \in \Gamma^N$.

We intend to compute the sum:

$$\sum = \sum_{\sigma \in \Gamma^N} a_\sigma x_1^{\sigma_1} x_2^{\sigma_2} ... x_N^{\sigma_N}.$$

Let us express the sum $\sum$ as:

(1) $$\sum = a_x x_1^{x_1} x_2^{x_2} ... x_N^{x_N} + \sum{}',$$

where

(2) $$\sum{}' = \sum_{\sigma \in \Gamma^N, \sigma \neq x} a_\sigma x_1^{\sigma_1} x_2^{\sigma_2} ... x_N^{\sigma_N}.$$

For the computation of monomial $a_x x_1^{x_1} x_2^{x_2} ... x_N^{x_N}$ we note that $a_x = f(x)$ and $x_i^{x_i} = 1$ $(1 \leq i \leq N)$ hence

(3) $$a_x x_1^{x_1} x_2^{x_2} ... x_N^{x_N} = f(x) \, 1 \, 1 ... \, 1 = f(x).$$

For the computation of the sum $\sum{}'$ let us remark that every monomial $a_\sigma x_1^{\sigma_1} x_2^{\sigma_2} ... x_N^{\sigma_N}$ with $\sigma \neq x$ contains at least a factor $x_i^{\sigma_i}$ with $\sigma_i \neq x_i$; it results that $a_\sigma x_1^{\sigma_1} x_2^{\sigma_2} ... x_N^{\sigma_N} = 0$, for all $\sigma \in \Gamma^N$, $\sigma \neq x$. Consequently,

(4) $$\sum{}' = 0.$$

From (3) and (4) it results that $\sum = f(x)$. We also note that, if $f \neq 0$, then in $\sum$, the monomials which have null coefficients can be omitted, hence $\sum$ is composed of the monomials which have $a_\sigma = 1$, i.e., $f(x) = 1$. Finally we obtain:

$$\sum_{f(\sigma)=1} x_1^{\sigma_1} x_2^{\sigma_2} ... x_N^{\sigma_N} = f(x_1, x_2, ..., x_N),$$

which proves the assertion. ∎

REMARK A.13   An analogous result (dual) can be obtained for the representation of Boolean functions as conjunctive polynomials. ∎

Appendix B

MV-Algebras

We shall present below some basic elements about MV-algebras: definitions of MV-algebras, MV-subalgebras, homomorphisms of MV-algebras, examples, the MV-algebras $\mathcal{A}_M = \{0, \frac{1}{M}, \frac{2}{M}, \ldots, \frac{M}{M-1}, 1\}$ (with $M \geq 2$), etc. The topics we have selected are mainly the ones used in the fundamental papers of J. Pavelka, V. Novák and others and also in our approach to fuzzy logic. The present treatment follows closely and summarizes the extensive paper of C.C. Chang [6]. Important contributions in the field are due to D. Mundici, A. Di Nola and others.

B.1 Definitions of MV-algebras, Basic Properties; Examples

In this section we shall introduce the notion of MV-algebra; we shall point out some fundamental properties of this algebraic structure and we shall give some important examples of MV-algebras.

DEFINITION B.1 *An* **MV-algebra** *is a system* $(A, \oplus, \odot, ^-, 0, 1)$, *where* A *is a nonempty set, 0 and 1 are distinct (fixed) elements of* A, $\oplus$ *and* $\odot$ *are binary operations on* A, *and* $^-$ *is a unary operation on* A, *which obeys the following eleven axioms:*

    Ax.1. (a) $x \oplus y = y \oplus x$; *(b)* $x \odot y = y \odot x$;

    Ax.2. (a) $x \oplus (y \oplus z) = (x \oplus y) \oplus z$; *(b)* $x \odot (y \odot z) = (x \odot y) \odot z$;

    Ax.3. (a) $x \oplus \bar{x} = 1$; *(b)* $x \odot \bar{x} = 0$;

    Ax.4. (a) $x \oplus 1 = 1$; *(b)* $x \odot 0 = 0$;

    Ax.5. (a) $x \oplus 0 = x$; *(b)* $x \odot 1 = x$;

Ax.6. (a) $\overline{x \oplus y} = \bar{x} \odot \bar{y}$ *; (b)* $\overline{x \odot y} = \bar{x} \oplus \bar{y}$;

Ax.7. $\bar{\bar{x}} = x$;

Ax.8. $\bar{0} = 1$.

For the needs of further axioms, we introduce the following definitions of two new (derived) binary operations, $\vee$ and $\wedge$, on A:
$$x \vee y = (x \odot \bar{y}) \oplus y;$$
$$x \wedge y = (x \oplus \bar{y}) \odot y.$$

Ax. 9. (a) $x \vee y = y \vee x$ *; (b)* $x \wedge y = y \wedge x$;

Ax.10. (a) $x \vee (y \vee z) = (x \vee y) \vee z$ *; (b)* $x \wedge (y \wedge z) = (x \wedge y) \wedge z$;

Ax.11. (a) $x \oplus (y \wedge z) = (x \oplus y) \wedge (x \oplus z)$ *; (b)* $x \odot (y \vee z) = (x \odot y) \vee (x \odot z)$.

REMARK B.1   Not all of the axioms in Definition B.1 are independent; they are given in the above form for their intuitive content and for more symmetry.   ■

In the following we shall present some basic properties of MV-algebras. We shall also introduce a canonical order relation between the elements of an MV-algebra, as we did in the case of Boolean algebras in APPENDIX A.

Let $(A, \oplus, \odot, ^-, 0, 1)$ be an arbitrary MV-algebra.

PROPOSITION B.1
For each $x, y \in A$ we have:

(1) $x \vee 0 = x = x \wedge 1$ *, $x \wedge 0 = 0$ and $x \vee 1 = 1$;*

(2) $x \vee x = x = x \wedge x$;

(3) $\overline{x \vee y} = \bar{x} \wedge \bar{y}$ *and $\overline{x \wedge y} = \bar{x} \vee \bar{y}$;*

(4) $x \wedge (x \vee y) = x = x \vee (x \wedge y)$;

(5) If $x \oplus y = 0$, then $x = y = 0$;

(6) If $x \odot y = 1$, then $x = y = 1$;

(7) If $x \vee y = 0$, then $x = y = 0$;

(8) If $x \wedge y = 1$, then $x = y = 1$.

PROOF   (1)-(3) are obvious from Definition B.1.

(4) $x \wedge (x \vee y) = x \wedge (y \vee x) = (y \vee x) \wedge x = ((y \odot \bar{x}) \oplus x) \wedge x =$
$((y \odot \bar{x}) \oplus x \oplus \bar{x}) \odot x = (y \odot \bar{x} \oplus 1) \odot x = 1 \odot x = x.$
The other equality of (4) follows similarly.

(5) $0 = x \wedge 0 = x \wedge (x \oplus y) = (x \oplus 0) \wedge (x \oplus y) = x \oplus (0 \wedge y) = x \oplus 0 = x.$
The fact that $y = 0$ results in a similar way.

(6) follows from (5).

(7) If $x \vee y = 0$, then $(x \odot \bar{y}) \oplus y = 0$, hence by (5) $y = 0$ and also $x = 0$.

(8) follows in the same way as (7).

∎

In an MV-algebra A we can introduce a natural order relation "$\leq$".

DEFINITION B.2   *If $x, y \in A$, we say that $x \leq y$ if $x \vee y = y$. By $x < y$ we shall mean $x \leq y$ and $x \neq y$.*

The following properties of the order relation are easy consequences of Definitions B.1, B.2 and of Proposition B.1.

PROPOSITION B.2
For any x, y, z in A the following relations hold:

(1) $0 \leq x \leq 1$;

(2) $x \leq x$;

(3) If $x \leq y$ and $y \leq z$, then $x \leq z$;

(4) If $x \leq y$ and $y \leq x$, then $x = y$;

(5) $x \leq y$ if and only if $x \wedge y = x$;

(6) $x \leq y$ if and only if $\bar{y} \leq \bar{x}$.

REMARK B.2   The assertions (2), (3) and (4) above show that "$\leq$" is a genuine order relation, i.e., it is reflexive, transitive and antisymmetric.
∎

Next, we give a list of propositions which can be proved in succession without too much difficulty.

PROPOSITION B.3
If $x \leq y$, then $x \vee z \leq y \vee z$ and $x \wedge z \leq y \wedge z$.

PROPOSITION B.4
$x \wedge y \leq x \leq x \vee y$.

PROPOSITION B.5
If $x \leq y$ and $x' \leq y'$, then $x \vee x' \leq y \vee y'$ and $x \wedge x' \leq y \wedge y'$.

PROPOSITION B.6
If $x \leq y$, then $x \oplus z \leq y \oplus z$ and $x \odot z \leq y \odot z$.

PROPOSITION B.7
$x \odot y \leq x \leq x \oplus y$.

PROPOSITION B.8
If $x \leq y$ and $x' \leq y'$, then $x \oplus x' \leq y \oplus y'$ and $x \odot x' \leq y \odot y'$.

PROPOSITION B.9
The elements $x \vee y$ and $x \wedge y$ are, respectively, the l.u.b. and the g.l.b. of the elements x and y with respect to the ordering "$\leq$".

PROPOSITION B.10
$x \odot y \leq x \wedge y \leq x \leq x \vee y \leq x \oplus y$.

PROPOSITION B.11
The following conditions are equivalent:

$$(1) \quad x \leq y; \quad (2) \quad y \oplus \bar{x} = 1; \quad (3) \quad x \odot \bar{y} = 0.$$

PROPOSITION B.12
If $x \oplus z = y \oplus z$, $x \leq \bar{z}$ and $y \leq \bar{z}$, then $x = y$.

PROPOSITION B.13
The following conditions are equivalent:

(1) $x \oplus y = y$;

(2) $x \odot y = x$;

(3) $y \vee \bar{x} = 1$;

(4) $x \wedge \bar{y} = 0$.

The proposition below follows immediately from Proposition B.13 and gives an interesting characterization of the elements of an MV-algebra which are idempotent with respect to the operations "$\oplus$" or "$\odot$": they are exactly those which satisfy the law of the excluded middle with respect to the operations "$\vee$" or "$\wedge$".

PROPOSITION B.14
The following conditions are equivalent:

(1) $x \oplus x = x$;

(2) $x \odot x = x$;

(3) $\bar{x} \oplus \bar{x} = \bar{x}$;

(4) $\bar{x} \odot \bar{x} = \bar{x}$;

(5) $x \vee \bar{x} = 1$;

(6) $x \wedge \bar{x} = 0$.

Examples of MV-algebras

(a) Let **C** be the axiomatic system of Lukasiewicz and Tarski, presented in Section 5.3. We have introduced there an equivalence relation, between the formulas belonging to the language of this system, which decomposes the set of all formulas in equivalence classes. The set $\mathcal{L}/\sim$ of these equivalence classes becomes an MV-algebra by defining the operations and the unit elements as follows:

$$\xi_F \oplus \xi_G = \xi_{\neg F \Rightarrow G};$$
$$\xi_F \odot \xi_G = \xi_{\neg(F \Rightarrow \neg G)};$$
$$\overline{\xi_F} = \xi_{\neg F};$$
$$\xi_F = 1 \text{ if and only if } \vdash_\mathbf{C} F;$$
$$\xi_F = 0 \text{ if and only if } \vdash_\mathbf{C} \neg F.$$

(b) Let us consider a set $\mathcal{A}$ of real numbers situated between 0 and 1, which satisfies the conditions:

(1) $0 \in \mathcal{A}$ and $1 \in \mathcal{A}$;

(2) If $x, y \in \mathcal{A}$, then $\min(1, x + y) \in \mathcal{A}$;

(3) If $x, y \in \mathcal{A}$, then $\max(0, x + y - 1) \in \mathcal{A}$;

(4) If $x \in \mathcal{A}$, then $1 - x \in \mathcal{A}$.

If we define now for $x, y \in \mathcal{A}$, the operations

$x \oplus y = \min(1, x + y)$,

$x \odot y = \max(0, x + y - 1)$,

$\bar{x} = 1 - x$,

then the system $(\mathcal{A}, \oplus, \odot, ^{-}, 0, 1)$ is an MV-algebra. One can check easily that the elements $x \vee y$ and $x \wedge y$ in $\mathcal{A}$ are, respectively, the *l.u.b.* and the *g.l.b.* of elements x, y from $\mathcal{A}$. In other words, we have

$$x \vee y = \max(x, y); \ x \wedge y = \min(x, y).$$

The order relation "$\leq$" coincides in this case with the natural ordering of real numbers. Particular cases of such sets $\mathcal{A}$ are the following: (α) $\mathcal{A} = \{0, 1\} = \Gamma$; ($\beta$) $\mathcal{A} = [0, 1]$ (the set of all real numbers x with $0 \leq x \leq 1$); (γ) $\mathcal{A} = [0, 1] \cap \mathbf{Q}$ where $\mathbf{Q}$ is the set of all rational real numbers.

(c) Very important examples from a theoretical and a practical point of view are the MV-algebras $\mathcal{A}_M$ presented below, where $M \geq 2$ is a given natural number. We shall see that many important MV-algebras are "isomorphic" to $\mathcal{A}_M$. On the other hand, the MV-algebras $\mathcal{A}_M$, in our approach to fuzzy logic, represent the sets of "truth values", i.e., the algebras $\mathcal{A}_M$ play an analogous role in fuzzy logic to that played by the Boolean algebra $\Gamma = \{0, 1\}$ in classical (crisp) logic.

The subset $\mathcal{A}_M = \{0, \frac{1}{M}, \frac{2}{M}, \ldots, \frac{M-1}{M}, \frac{M}{M} = 1\}$ of $[0, 1]$ obviously satisfies conditions (1)-(4). Together with the unit elements 0 and 1 and defining the internal operations:

$$\frac{p}{M} \oplus \frac{q}{M} = \min\left(1, \frac{p+q}{M}\right)$$

$$\frac{p}{M} \odot \frac{q}{M} = \max\left(0, \frac{p+q-M}{M}\right)$$

$$\overline{\frac{p}{M}} = \frac{M-p}{M},$$

where $p, q \in \{0, 1, \ldots, M\}$ are arbitrary, the set $\mathcal{A}_M$ becomes an MV-algebra. Axioms (1)-(11) are obviously verified.

B.2   MV-Subalgebras and Homomorphisms of MV-Algebras

In this section other theoretical notions concerning MV-algebras are introduced; some immediate properties involving these notions are mentioned. A remarkable result which reveals a deep and interesting relation between MV-algebras and Boolean algebras (presented in APPENDIX A) will be also presented.

DEFINITION B.3

(a) *Given an MV-algebra* $(A, \oplus, \odot, ^-, 0, 1)$, *we say that B is an* **MV-subalgebra** *of A if* $B \subseteq A$, $0, 1 \in B$ *and B are closed under the operations* $\oplus$, $\odot$ *and* $^-$;

(b) *A system* $(B, \oplus, \odot, ^-, 0, 1)$ *is a* **homomorphic image** *of A (or A is homomorphic to B) if there is a mapping f of A onto B such that* $f(0) = 0$, $f(1) = 1$ *and f preserves the operations* $\oplus$, $\odot$ *and* $^-$; *if the function f is bijective, then f is called an isomorphism of A onto B. In this case we say that the systems* $(A, \oplus, \odot, ^-, 0, 1)$ *and* $(B, \oplus, \odot, ^-, 0, 1)$ *are* **isomorphic**;

(c) *Given a family of MV-algebras* $(A_i)_{i \in I}$ *(where I is an arbitrary nonempty set), we denote by* $\prod_{i \in I} A_i$ *the Cartesian (or direct) product of the family of MV-algebras* $(A_i)_{i \in I}$; *this Cartesian product can be, in a natural manner, organized as an MV-algebra if the element 0 is the function* f_0 *such that* $f_0(i) = 0$ *for each* $i \in I$, *the element 1 is the function* f_1 *such that* $f_1(i) = 1$ *for each* $i \in I$, *the addition of two functions f, g is the function h such that* $h(i) = f(i) \oplus g(i)$ *for each* $i \in I$, *the product and the complement of functions being similarly defined.*

PROPOSITION B.15

A subalgebra of an MV-algebra is an MV-algebra; a homomorphic image of an MV-algebra is an MV-algebra; the direct product of MV-algebras is an MV-algebra.

PROPOSITION B.16

Given an MV-algebra $(A, \oplus, \odot, ^-, 0, 1)$, *let B be the set of elements of A which are idempotent with respect to the operation* $\oplus$ *(or equivalently which are idempotent with respect to operation* $\odot$). *Then B is closed under the operations* $\oplus$, $\odot$ *and* $^-$; *moreover, for each* $x, y \in B$, *we have*

$$x \oplus y = x \vee y \qquad x \odot y = x \wedge y.$$

Furthermore, the system $(B, \oplus, \odot, ^-, 0, 1)$ *is not only an MV-subalgebra of A, but is also the largest MV-subalgebra of A which is at the same time a Boolean algebra with respect to the operations* $\oplus, \odot$ *and* $^-$.

PROOF   Let $x, y \in A$ be two idempotent elements of A with respect to the additive operation in A. Then, by Proposition B.14 one has also

$x \odot x = x$;

$\bar{x} \oplus \bar{x} = \bar{x}$;

$\bar{x} \odot \bar{x} = \bar{x}$;

$x \vee \bar{x} = 1$;

$x \wedge \bar{x} = 0$;

and the corresponding equalities for y. Using the above equalities and also the axioms of MV-algebras one gets

$(x \oplus y) \oplus (x \oplus y) = (x \oplus x) \oplus (y \oplus y) = x \oplus y$;

$(x \odot y) \odot (x \odot y) = (x \odot x) \odot (y \odot y) = x \odot y$, hence, by Proposition B.14,

$(x \odot y) \oplus (x \odot y) = x \odot y$;

$\bar{x} \oplus \bar{x} = \bar{x}$.

We obtained that $x \oplus y$, $x \odot y$ and $\bar{x}$ are idempotent elements of A, too, hence B is closed under the operations $\oplus, \odot$ and $^-$. It is easy to check that B is also closed under the operations $\vee$ and $\wedge$. Since $x \leq x \vee y \in B$ and $y \leq x \vee y \in B$, one obtains by Proposition B.8 that $x \oplus y \leq (x \vee y) \oplus (x \vee y) = x \vee y$, by Proposition B.10 one has $x \vee y \leq x \oplus y$ and by Proposition B.2, (4) one gets $x \vee y = x \oplus y$. In the same way we can obtain $x \odot y = x \wedge y$.

The axioms of a Boolean algebra for B are now easy to verify.

If C is an MV-subalgebra of A which is also a Boolean algebra with respect to the operations $\oplus, \odot$ and $^-$, then every element x of C must satisfy the equality $x \oplus x = x$. Hence $C \subseteq B$. It follows that B is the largest MV-subalgebra of A which is at the same time a Boolean algebra with respect to the operations $\oplus, \odot$ and $^-$.   ∎

B.3   Finite MV-Algebras

In this section, some technical results (containing other properties of MV-algebras) and some definitions will be presented. These results will allow us to prove the main result of this section, concerning the characterization of those MV-algebras which are isomorphic to the MV-algebras $\mathcal{A}_M$.

We begin with some technical results. Let x, y, z be arbitrary elements of a given MV-algebra A.

PROPOSITION B.17

$y \odot (x \oplus z) \leq x \oplus y \odot z$.

PROPOSITION B.18

If $x \wedge \bar{y} = 0$, then $x \oplus y \odot z = y \odot (x \oplus z)$.

PROPOSITION B.19

$(x \oplus \bar{y}) \vee (y \oplus \bar{x}) = 1$.

PROPOSITION B.20

If $x \vee y = 1$, then $(x \odot x) \vee (y \odot y) = 1$.

We shall now define the "natural multiples" and the "natural powers" of an arbitrary element $x \in A$.

DEFINITION B.4 (a) $0 \cdot x = 0$ and $(n+1) \cdot x \overset{def}{=} n \cdot x \oplus x$;

(b) $x^0 = 1$ and $x^{n+1} \overset{def}{=} (x^n) \odot x$.

DEFINITION B.5 The order of the element x, noted ord(x), is the least integer m such that $m \cdot x = 1$ and we shall denote ord(x)= m; if no such integer exists, then by definition, ord(x)= ∞.

PROPOSITION B.21

(a) $\overline{n \cdot x} = \bar{x}^n$, $\overline{x^n} = n \cdot \bar{x}$;

(b) $m \cdot (n \cdot x) = (m \cdot n) \cdot x$, $x^{m+n} = (x^m) \odot (x^n)$, $x^{(m \cdot n)} = (x^m)^n$.

PROPOSITION B.22

If ord($x \odot y$)$< \infty$, then $x \oplus y = 1$.

PROPOSITION B.23

If ord(x)> 2, then ord($x \odot x$)$= \infty$.

DEFINITION B.6 We say that the MV-algebra is **totally (linearly) ordered** if for every $x, y \in A$, either $x \leq y$ or $y \leq x$.

PROPOSITION B.24

If A is totally ordered, then $x \oplus z = y \oplus z$ and $x \oplus z \neq 1$ implies $x = y$.

We now introduce the function

$$d : A \times A \to A$$

by defining $d(x, y) = \bar{x} \odot y \oplus \bar{y} \odot x$.

PROPOSITION B.25
For each $x, y, u, v \in A$ we have:

(1) $d(x, x) = 0$, $d(x, y) = d(y, x)$, $d(x, y) = d(\bar{x}, \bar{y})$, $d(x, 0) = x$ and $d(x, 1) = \bar{x}$;

(2) *If $d(x, y) = 0$, then $x = y$;*

(3) $d(x, z) \leq d(x, y) \oplus d(y, z)$;

(4) $d(x \oplus u, y \oplus v) \leq d(x, y) \oplus d(u, v)$;

(5) $d(x \odot u, y \odot v) \leq d(x, y) \oplus d(u, v)$.

PROPOSITION B.26
If $x \leq y$, then $d(x, y) = \bar{x} \odot y$ and $x \oplus d(x, y) = y$.

PROPOSITION B.27

(1) *If $n \cdot y \leq x \leq (n + 1) \cdot y$, then $d(x, n \cdot y) \leq y$ and $d(x, (n + 1) \cdot y) \leq y$;*

(2) *If $n \cdot y \leq x < (n + 1) \cdot y$, then $d(x, n \cdot y) < y$;*

(3) *If $n \cdot y < x \leq (n + 1) \cdot y$, then $d(x, (n + 1) \cdot y) < y$.*

PROPOSITION B.28
If A is totally ordered and ord$(y) = m < \infty$, then $d(\overline{n \cdot y}, (m - n) \cdot y) < y$ for each $n \leq m$.

DEFINITION B.7   *An element x of an MV-algebra A is an **atom** if $x \neq 0$ and if $0 \leq y \leq x$ implies $y = 0$ or $y = x$.*

Now we are ready to prove the main result of this section.

PROPOSITION B.29
If A is a totally ordered MV-algebra which contains an atom of order M, then A is isomorphic with A_M.

PROOF   Let y be an atom of A of order M; we get by Proposition B.24 that:
$$0 < y < 2 \cdot y < \ldots < (M - 1) \cdot y < M \cdot y = 1.$$
Indeed, for $i \in \{0, 1 \ldots, M - 1\}$ it is obvious that $i \cdot y \leq (i + 1) \cdot y$ and that it is not possible to have $i \cdot y = (i + 1) \cdot y$ (this would imply $y = 0$, by

Proposition B.24). Since A is totally ordered, it follows that any element $x \neq 1$ is such that $n \cdot y \leq x \leq (n+1) \cdot y$ for some $0 \leq n \leq M - 1$. From Proposition B.27, (2) we have that $d(x, n \cdot y) < y$ and, since y is an atom, $d(x, n \cdot y) = 0$. This result, by using Proposition B.25, (2), implies $x = n \cdot y$. Thus we have proved that any element of A must be a multiple of y. By Proposition B.28 we get $d(\overline{n \cdot y}, (M-n) \cdot y) < y$ and, again since y is an atom, $\overline{n \cdot y} = (M-n) \cdot y$. Clearly, $n \cdot y \oplus l \cdot y = (n+l) \cdot y$. It is now obvious that the function f defined by $f(n \cdot y) = \dfrac{n}{M}$ will map A isomorphically onto $\mathcal{A}_M$ and the proposition is proved. ∎

REMARK B.3   Taking into account that the algebra $\mathcal{A}_M$ is totally ordered and that the element $\frac{1}{M}$ is an atom of order M, Proposition B.29 shows that MV-algebras $\mathcal{A}_M$ $(M \geq 2)$ are the canonical representatives for the totally ordered MV-algebras containing an atom of finite order M. ∎

Appendix C

General Considerations about Fuzzy Sets

In this section, we will look at one theory that may be used to deal with the imperfection of knowledge based on fuzzy logic. Fuzzy logic is an extension of classical logic and uses degrees of membership, rather than the strict membership classifications of "yes/no". The essential purpose of fuzzy logic is to create a formal investigative system of concepts and techniques that allows approximate reasoning. Fuzzy logic is based on the theory of fuzzy sets, introduced by Professor L. Zadeh. The main reason for the introduction of fuzzy set theory was the fact that this theory can pick up on the vagueness of knowledge such as it is expressed through natural language. This theory has developed at an impressive rate, due to its usefulness in fields such as control theory, medical diagnosis, economy, pattern recognition, operations research, computational linguistics, psychology, etc.

C.1 Introduction to Fuzzy Set Theory

We know that mathematics is the most logical and the most formal of all the sciences. In past years, however, a disconcerting question has been troubling mathematicians: could it be that precision, in some contexts, is in fact a vice and that vagueness is a virtue? For a number of logicians, the answer is "yes". The explanation for this phenomenon can be found in the complicated process of concept development. The human brain is singularly flexible, has great capacities for filtering and parallelism and is able to interpret information that is vague and ambiguous. The flow of information gathered by the brain through visual, audio, tactile and other senses is subjected to rational transformation (analysis, synthesis, generalization, abstraction), after which only essential characteristics of this information are retained, and concepts are developed. Our senses consider reality only through vague models of reality. What goes on in the human brain is an exchange between vague, global considerations and precise, logical considerations. Between the analytic power of the left lobe and the imagination of the right lobe of the human brain, a dynamic equilibrium evolves, which leads to thought. Human thought is vague, yet we wish it to be as crisp (logical) as possible. This is true, for instance, for the purposes of negotiating, arguing, meeting, orientating, demanding, resolving doubts, delving into philosophical problems, etc. Natural human logic, however, carries many more nuances than computer logic. Human logic continually associates logic and semantics and also pragmatics: logic tends to reduce entropy (disorder), while semantics tends to augment it. The law of the excluded middle or contradiction is a law of Boolean logic which is too restrictive. In other words, while a computer instruction must contain no entropy, this is not true for the communication between humans, where fuzziness intervenes. This fuzziness is the very essence of semantics. One word does not correspond to a single object of thought or of perceived reality. The meaning of this word depends on the individual, the place, the time, the environment, etc. Thanks to these variable aspects, the imagination is able to considerably enrich the perceived message and the individual's personality can play its part.

Subjectivity and imprecision must be taken into consideration. The majority of today's computers are precise, objective and certain, but they have no *imagination*. *"In order to invent, one must be subjective and vague."* (A. Kaufmann)

The brain's tolerance for imprecision leads to the formation of fuzzy labels that are attached to pieces of information in order to approach the primary data. For human beings, the ability to store information manifests itself through the use of natural languages. In all natural languages, the meaning of words is often fuzzy. This fuzziness may be of an intrinsic nature (tall man) or of an informative (subjective) nature (solid client). The transition from membership to non-membership is rather gradual in a number of concepts such as intelligence, competence, truth, democracy, love, hope, etc. There is a divergence between the possibilities afforded by our thoughts and the weak power of the language being used. For instance, every word x in a natural language L may represent a summary description of a vague statement F in the universe of discourse U, where $F(x)$ represents the meaning of x. Hence, language represents a means by which we assign vague atomic and composite labels of U to primary data, with the help of words, expressions and phrases. The notion of color, for example, corresponds to a fuzzy variable that may take the value of red, blue, green, yellow, etc. When we speak of a color red, it is then a fuzzy term, since the difference between color red and color brick red is clear. Many nuances of color red and color brick red exist, however, and we appreciate the color red according to our internal scale. The same color red is interpreted differently by different persons, according to their human subjectivity. Of course, it is important to distinguish between subjectivity and will. In order to get an appropriate picture of reality, we must get the opinion of a group of experts. The ability to handle fuzzy information is one of the main qualities of human intelligence, while computers generally work only with precise information.

We see this ability to handle fuzzy information in our daily behavior. When establishing a diagnostic, parking the car, crossing the road or buying a house, we use a part of the subconscious, we make attempts based on our past experiences, we take heuristic steps, we grope our way around or we proceed to things without quite knowing how we do them. The decision to buy a house, for example, depends on a series of fuzzy factors, such as its distance from our place of work, the existence of essential services, schools, property taxes, etc. This is, therefore, a problem-solving process which entails a number of fuzzy criteria.

The idea of set is primary in the classical theory of sets. We can know a set by listing its elements or when we know a property common to all its elements. Consequently, it is always possible to ascertain whether or not an element belongs to the set. Logically speaking, belonging or non-belonging is designated by the values of true (1) or false (0) in 2-valued logic. As we have seen, however, the reality is that we generally think and reason through multivalent logic, containing various degrees of belonging (membership) which fall somewhere between total belonging (1) and non-belonging (0). For example: "the class of tall men", "the class of long streets" of a city, "the class of beautiful women" of a city, "the class of intelligent men" ,"the class of competent men", "the class of real numbers much larger than 1", assertions such as "x is near x_0", "x is approximately equal to y", or "many students are in first year", "many first year students live close to campus", "which is the proportion of students who live near campus?"

Complex problems and/or ill-defined problems generate imperfect knowledge. Imperfect knowledge makes it difficult to make decisions due to missing and inappropriate information. Which brings us to the following classification:

Complex and ill-defined problems →imperfect knowledge
Imperfect knowledge   →uncertain and/or imprecise knowledge
Imprecise knowledge   →inexact and/or vague knowledge
Vagueness   →fuzziness and/or ambiguity

A fragment of knowledge is said to be imprecise if the value of certain parameters is only partially or roughly specified. The inability to ascertain the validity of a piece of knowledge comes from our inability to definitely establish its truth or falsehood within a given context. In natural language, uncertainty is expressed through words such as probable, possible, necessary, plausible, credible; while imprecision is conveyed by words such as vague, fuzzy, general, ambiguous, confused, without boundaries, etc.

REMARK C.1   There exists a number of classical techniques designed to represent the imperfection of knowledge: error calculation, theory of probabilities, Shannon theory, Dempster-Shafer theory, Zadeh's fuzzy sets theory, etc. The main types of error linked to imperfect information are: ambiguity, incompleteness, incorrectness (human error, equipment failure), error of measurement (lack of precision, lack of exactitude), random aspect, erroneous reasoning (inductive or deductive error), etc. The calculus of errors does not allow for nuances and is used exclusively for numerical parameters. The theory of probabilities is not flexible, since it is based on the fundamental axiom regarding the probability of mutually exclusive events. The Shannon theory is used mainly in the field of telecommunications. The Dempster-Shafer theory of belief functions, used for imperfect reasoning, has a sound theoretical basis but also has its limits. In a series of articles, Professor L. A. Zadeh, of the University of California at Berkeley, outlines the basis of fuzzy set theory. This theory does not apply to all situations. In order to treat different aspects of the same problems, we must therefore apply various theories related to the imprecision of knowledge.
∎

C.2   Basic Definitions and Examples

Fuzzy set theory is rigorously mathematical. This theory is adapted to the treatment of that which is subjective and/or imprecise. It is not a revolutionary theory, but it is adapted to our times.

In order to study imprecise knowledge (not inexact knowledge), we can introduce certain membership functions. Imprecise knowledge is to be distinguished from inexact knowledge; inexact knowledge misses the existing precise value by a small amount; in the case of imprecise knowledge there does not exist a precise corresponding knowledge. The values of these functions, contained within the interval (0,1) will indicate the degree of membership of an element in a fuzzy set. To better understand the notion of fuzzy sets, we consider the space X of generic element x. We know that the classical (crisp) subset family of X, $P(X) = \{A|A \subseteq X\}$, with its reunion, intersection and complementation operations $\{P(X), \cup, \cap, \neg\}$, has a structure of Boolean algebra. We can easily establish the bijection that $A \rightarrow \Psi_A(x) = 1$, if $x \in A$ and 0 if $x \notin A$. $(\{0,1\}^X, \vee, \wedge, \neg)$, where $\vee =$ max,

$\wedge$ =min, $\neg\Psi_A = 1 - \Psi_A$ has also a Boolean algebra structure. Since this bijection preserves the operations, the two sets are isomorphic, as Boolean algebras. We can, therefore, identify a crisp set A with its characteristic function.

The class of fuzzy subsets of X is identified with the richest class of all functions $F : X \rightarrow [0, 1]$. Taking into account that each such function F is uniquely determined by its graph $G_F = \{(x, F(x)|x \in X\}$ which is a (crisp) subset of the Cartesian product $[0, 1]^X$, we can formulate the following two equivalent definitions:

(a) Each function $F : X \rightarrow [0, 1]$, valued in $[0, 1]$ and defined on X is called **a fuzzy subset of** X;

(b) A subset $F \subset [0, 1]^X$ is a fuzzy subset of X if and only if F has the following two properties:

   (i) $pr_X(F) = X$;

   (ii) $(x, y_1) \in F$ and $(x, y_2) \in F$ implies $y_1 = y_2$, where $pr_X(F)$ is the projection along X of the set $F \subset [0, 1]^X$.

On both cases (a) and (b), the fuzzy set F is denoted often by $F = (x, F(x))$, where $F(x)$ is called the membership degree of x in F.

Remarks.

(1) The theory of fuzzy subsets of X is a generalization of the theory of CRISP (classical) subsets of X: a fuzzy subset F of X is no longer a characteristic function (which has at most two values, 0 and 1), while the fuzzy subset F generally takes values in the whole interval $[0, 1]$.

(2) The notion of membership function is crucial to this whole theory. In practice, the construction of a fuzzy set F, i.e., of the function $F(x)$, is subjective (for instance, through taking a survey of expert opinions) and reflects the context within which the concrete problem is being studied.■

We can associate four important crisp sets to a fuzzy set F:

- the support $S(F) = \{x \in X|F(x) > 0\}$,

- the transom $T(F) = \{x \in X|F(x) \in (0, 1)\}$,

- the height $H(F) = \{x \in X|F(x) = 1\}$, and

- the zero part $Z(F) = \{x \in X|F(x) = 0\}$.

Of course, we have $S(F) = T(F) \cup H(F)$. These sets play an important role in the introduction of certain measures (for example, the measure of transom) on the fuzzy sets which characterize the degree of fuzziness of these sets. It is obvious that it is sufficient to indicate the value $F(x)$ only for $x \in S(F)$.

Examples.

(1) The set of real numbers much larger than 1 can be given by the fuzzy subset A of **R** with the membership function $A(x) = \{x \in \mathbf{R} | x >> 1\}$, defined as follows: $A(x) = 0$, if $x \le 1$ and $A(x) = \frac{x-1}{x}$, if $x > 1$. Using the discrete description on **N**, we get:

$$A = \{(1,0), (2,0.5), (3,0.6), (4,0.7), \ldots\}.$$

(2) The membership function of the proposition $A =$ "John is between 20 and 25 years old" can be given, for instance, by $A(x) = 1$, if $x \in [20, 25]$ and $A(x) = 0$, if $x \notin [20, 25]$.

(3) The imprecise proposition $F =$ "the class of young person" can be given by: $F(x) = 1$, if $x \le 25$ and $F(x) = (1 + (x - 25)^2)^{-1}$, if $x > 25$.

In the following, the space X is identified with the function 1, defined by $1(x) = 1$, $(\forall)x \in X$, while the empty set is identified with the function 0, defined by $0(x) = 0$, $(\forall)x \in X$.

If $\{Fi\}_{i \in I}$ is a class of fuzzy sets from X, then $(\forall)i, j \in I$ and $(\forall)x \in X$ and we can define the following operations:

(a) $F_i = F_j \Leftrightarrow F_i(x) = F_j(x)$;

(b) $F_i \le F_j \Leftrightarrow F_i(x) \le F_j(x)$;

(c) $A = \bigcup_{i \in I} F_i \Leftrightarrow A(x) = \bigvee_{i \in I} F_i(x) = \sup_{i \in I} F_i(x)$;

(d) $B = \bigcap_{i \in I} F_i \Leftrightarrow B(x) = \bigwedge_{i \in I} F_i(x) = \inf_{i \in I} F_i(x)$;

(e) $F_i' = \neg F_i \Leftrightarrow F_i'(x) = 1 - F_i(x)$.

REMARK C.2   The principle of excluded middle is no longer valid in this context. For example, $F(x) = 2/3$, $x \in [0, 2]$, $F'(x) = 1/3$ and $F \cup F' = 2/3 \ne 1$.   ∎

Examples

(1) Let be $X =$ {Alain, John, Jacques, Richard, Eugene, Raoul} and let us consider the fuzzy subsets P and Q of X,

$(x, P(x))$, $P(x) \equiv$ "x is handsome"

$(x, Q(x))$, $Q(x) \equiv$ "x is rich"

given respectively by

P : {(Alain,0.7), (John,0,2), (Jacques,0.8), (Richard,1), (Eugene,0.1), (Raoul,0.5)}

Q : {(Alain,0.8), (John,0,7), (Jacques,0.3), (Richard,0.5), (Eugene,1), (Raoul,0.5)}.

Then we get

(a) $P \cap Q$: {(Alain,0.7), (John,0,2), (Jacques,0.3), (Richard,0.5), (Eugene,0.1), (Raoul,0.5)};

(b) "John is not handsome"$=P'(\text{John})=1\text{-}P(\text{John})=1\text{-}0.2=0.8$;

(c) "Jacques is handsome but not rich"$=(P \cap Q')(\text{Jacques})=$ $P(\text{Jacques}) \wedge Q'(\text{Jacques})=(0.8) \wedge (1\text{-}0.3)=0.7$.

(2) Let $X = \{1, 2, 3, 5, 6, 7\}$ and consider the fuzzy subsets A and B of X:

$$A = \{(3, 0.8), (5, 1), (6, 0.6)\} \text{ and } B = \{(3, 0.7), (4, 1), (6, 0.5)\} .$$

Then we obtain:

$$A \cup B = \{(3, 0.8), (4, 1), (5, 1), (6, 0.6)\} ;$$
$$A \cap B = \{(3, 0.7), (6, 0.5)\} ;$$
$$A' = \{(1, 1), (2, 1), (3, 0.2), (4, 1), (6, 0.4), (7, 1)\} .$$

References

[1] Bender, A. E., *Mathematical Methods in Artificial Intelligence*, IEE Computer Society Press, Los Alamitos, California, 1996.

[2] Boicescu, V., Filipoiu, A., Georgescu, G., Rudeanu, S., *Lukasiewicz-Moisil Algebras*, North-Holland, 1991.

[3] Bourbaki, N., *Éléments de Mathématique; Théorie des Ensembles*, Hermann, Paris, 1970.

[4] Butnariu, D., Klement, E. P., *Triangular Norm-Based Measures and Games with Fuzzy Coalitions*, Kluwer, Dordrecht, 1993.

[5] Carnap, R., *The Logical Syntax of Language*, Routledge & Kegan Paul, London, 1937.

[6] Chang, C. C., Algebraic analysis of many-valued logics, *Transactions of the Am. Math. Society*, vol. 88, 467, 1958.

[7] Clocksin, W. F., Mellish, C. S., *Programmer en PROLOG*, Éditions Eyrolles, Paris, 1985.

[8] Di Nola, A., MV-algebras in the treatment of uncertainty, *Proceedings of the International IFSA Congress*, Bruxelles, 1991, Löwen, P., Roubens, E., Eds., Kluwer, Dordrecht, 1993, 123.

[9] Dubois, D., Prade, H., An introduction to possibilistic and fuzzy logics, in: *Non-Standard Logics for Automated Reasoning*, Academic Press, New York, 1988, 287.

[10] Dumitriu, A., *Logica Polivalentă*, Ed. Enc. Rom., Bucarest, 1971 (in Romanian).

[11] Faure, R., Heurgon, E., *Structures Ordonées et Algèbres de Boole*, Gauthier-Villars, Paris, 1971.

[12] Gersting, J. L., *Mathematical Structures for Computer Science*, Computer Science Press, New York, 1993.

[13] Goguen, J. A., The logic of inexact concepts, *Synthese*, 19, 325, 1968-1969.

[14] Goodstein, R. L., *Mathematical Logic*, Leicester, 1957.

[15] Gottwald, S., *Mehrwertige Logik. Eine Einführung in Theorie und Anwendungen*, Akademie-Verlag, Berlin, 1989.

[16] Gottwald, S., *Fuzzy Sets and Fuzzy Logic*, Vieweg Verlag, Braunschweig, and Tecnea, Toulouse, 1993.

[17] Hájek, P., Fuzzy logic and arithmetical hierarchy, *Fuzzy Sets and Systems*, 73, 359, 1995.

[18] Hájek, P., Godo, L., Deductive systems of fuzzy logic, *Tatra Mountains Math. Publ.*, 73, 359, 1995.

[19] Hilbert, D., Ackermann, W., *Grundzüge der Theoretischen Logik*, 2nd. ed., Springer, Berlin, 1938.

[20] Höhle, U., Monoidal logic, *Systems in Computer Science*, Vieweg, 33, 1994.

[21] Kleene, S. C., *Introduction to Metamathematics*, D. Van Nostrand, New York, 1952.

[22] Kleene, S. C., *Mathematical Logic*, John Wiley & Sons, New York, London, Sidney, 1967.

[23] Lloyd, J. W., *Foundations of Logic Programming*, Springer-Verlag, Berlin, 1993.

[24] Łukasiewicz, J., Tarski, A., Untersuchungen über den Aussangenkalkül, *Comptes Rendus des Séances de la Société des Sciences et des Lettres de Varsovie*, Classe III, Vol. 23, 30, 1930.

[25] Lyndon, R. C., *Notes on Logic*, D. Van Nostrand, Princeton, New Jersey, 1966.

[26] Maliţa, Mir., Maliţa, Mih., *Bazele Inteligenţei Artificiale*, 1, Ed. Tehnică, Bucarest, 1987 (in Romanian).

[27] Mendelson, E., *Introduction to Mathematical Logic*, D. Van Nostrand, Princeton, Toronto, New York, London, 1964.

[28] Moisil, G., Recherches sur les logiques non-chrysippiennes, *Ann. Sci. Univ. Jassy*, 26, 431, 1940.

[29] Moisil, G., *Essais sur les logiques non-chrysippiennes*, Éditions de l'Académie Roumaine, Bucharest, 1972.

[30] Mundici, D., Satisfiability in many-valued sentential logic in NP-complete, *Theoretical Computer Science*, 52, 145, 1987.

[31] Negoiţă, C. V., Ralescu, D. A., *Applications of Fuzzy Sets to System Analysis*, Birkhauser, Stuttgart, 1975.

[32] Nerode, A., Shore A. R., *Logic for Applications*, Springer-Verlag, New-York, 1993.

[33] Novák, V., Pedrycz, W., Fuzzy sets and t-norms in the light of fuzzy logic, *Int. J. Man.-Mach. Stud.*, 29, 113, 1988.

[34] Novák, V., On the syntactico-semantical completeness of first order fuzzy logic, Part I-Syntactical aspects; Part II-Main results, *Kybernetika*, 26, 47, 134, 1990.

[35] Novák, V., A new proof of completeness of fuzzy logic and some conclusions for approximate reasoning, in *Proc. of Int. Conference IEFS '95*, Yokohama, 1995.

[36] Novák, V., Paradigm, formal properties and limits of fuzzy logic, *Int. J. General Systems*, vol. 24(4), 377, 1996.

[37] Pavelka, J., On fuzzy logic I, II, III, *Zeit. Math. Logic Grundl. Math.*, 25, 45, 119, 447, 1979.

[38] Post, E. L., Introduction to a general theory of elementary propositions, *Am. J. Math.*, vol. 43, 163, 1921.

[39] Rasiowa, H., Sikorski, K., *The Mathematics of Metamathematics*, PWN-Polish Scientific Publishers, Warszawa, 1970.

[40] Reghiş, M., *Elemente de Teoria Mulţimilor şi de Logică Matematică*, Ed. Facla, Timişoara, 1981 (in Romanian).

[41] Reghiş, M., Roventa, E., Boolean algebra of "sup-inf" operations on sets of characteristic functions and some applications to iterated quantifications in predicate logic, *An. Univ. din Timişoara*, Fasc. 2, 1994.

[42] Reghiş, M., Roventa, E., Order relations for iterated quantifiers and some deducibility properties, *An. Univ. din Timişoara*, 1995.

[43] Reghiş, M., Roventa, E., Caşu, I., About an algorithmic treatment of the semantic interpretations in first-order logic, *An. Univ. din Timişoara* (to appear).

[44] Rose, A., Rosser, J. B., Fragments of many-valued statement calculi, *Transactions of the Am. Math. Society*, vol. 87, 1, 1958.

[45] Rubin, J. E., *Mathematical Logic: Applications and Theory*, Saunders College Publishing, Philadelphia, 1990.

[46] Schöning, U., *Logic for Computer Scientists*, Birkhauser, Boston, 1989.

[47] Shoenfield, J. R., *Mathematical Logic*, Addison-Wesley, Reading, MA, 1967.

[48] Thayse, A. and co-authors, *Approche Logique de l'Intelligence Artificielle 1*, Dunod, Paris, 1990.

[49] Yager, R. R., Ovchinnikov, S., Tong, R. M., Nguyen, H. T., *Fuzzy Sets and Applications: Selected Papers by L. A. Zadeh*, John Wiley & Sons, New York, 1987.

[50] Zadeh, L. A., Fuzzy sets, *Information and Control*, 8, 338, 1965.

[51] Zadeh, L. A., The concept of a linguistic variable and its application to approximate reasoning, I, II, III, *Inf. Sci.*, 8, 199, 301; 9, 43, 1975.

[52] Zimmermann, H. I., *Fuzzy Set Theory and its Applications*, 2nd. ed., Klumer, Nijhof, 1990.

Index